Collaborative Governance of Forests
Towards Sustainable Forest Resource Utilization

Motomu Tanaka and Makoto Inoue, Editors

UNIVERSITY OF TOKYO PRESS

First Edition is published February 2015 by the University of Tokyo Press.
© 2015 Motomu Tanaka and Makoto Inoue

All rights reserved. This book, or parts thereof,
may not be reproduced in any form or by any means,
electronic or mechanical, including photocopying,
recording or any information storage and
retrieval system now known or to be invented,
without written permission from the University of Tokyo Press.

University of Tokyo Press
4-5-29 Komaba, Meguro-ku
Tokyo, 153-0041, Japan
Website: http://www.utp.or.jp

Printed in Japan
ISBN 978-4-13-077011-8

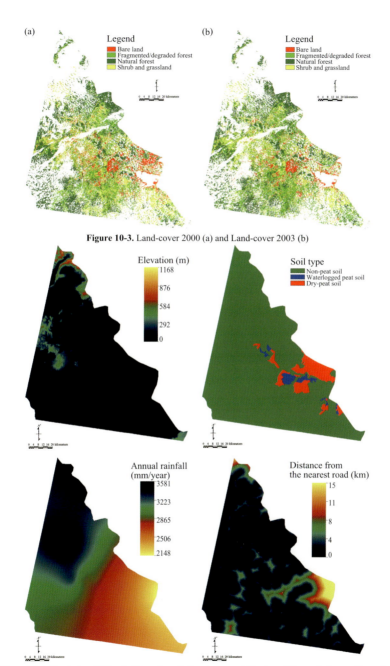

Figure 10-3. Land-cover 2000 (a) and Land-cover 2003 (b)

Figure 10-4. Physical and Climatic Properties of a Landscape and Proxies of Human Influence Properties

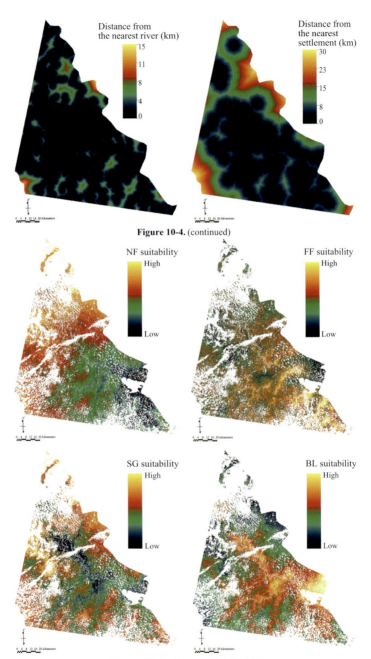

Figure 10-4. (continued)

Figure 10-6. Land-cover Suitability Map

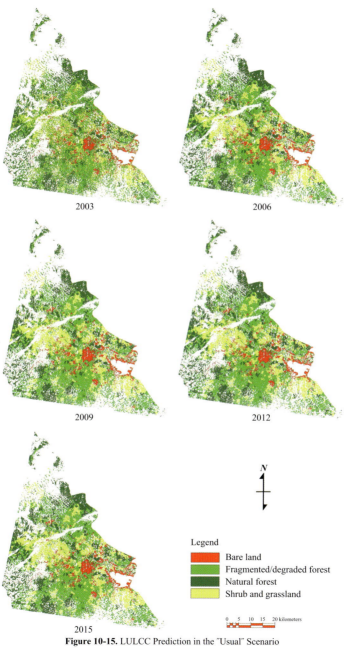

Figure 10-15. LULCC Prediction in the "Usual" Scenario

Contents

List of Contributors   vi

Acknowledgement   vii

Introduction   3

Part I   Policies, Institutions, and Rights to Share
Prerequisites for Collaborative Governance

Chapter 1   Historical Typology of Collaborative Governance
Modern Forest Policy in Japan / 15
1. Introduction: Policy Change in Forest Governance in Japan / 15
2. Problems of Forest Governance Research in Japan / 16
3. Japan's Forest Administration System and Policies / 19
4. Considering Japanese Forest Policy Problems and Forest and Forestry Revitalization Plans / 26
5. Conclusion: Policy Challenge to Reform in Forest Governance / 34

Chapter 2   Endogenous Development and Collaborative Governance in Japanese Mountain Villages / 39
1. Challenges for the Promotion of Mountain Villages / 39
2. Endogenous Development Theory and Collaborative Governance Theory / 40
3. The Network for Supporting Lives in Sawauchi-mura / 42
4. Network for Growing *Quercus serrata* Trees in Tsukimoushi-machi / 48
5. The Network for Conserving the Environment of the Forest Commons / 54
6. The Network for Preserving and Nurturing a Beautiful Townscape in Kaneyama-machi / 57
7. Endogenous Development and Collaborative Governance / 66

Chapter 3   Collaborative Forest Governance in Mass Private Tree Plantation Management
Company–Community Forestry Partnership System in Java, Indonesia (PHBM) / 71

1. Introduction / 71
2. An Outline of PHBM in the Madiun Forest District / 76
3. Local People's Response to PHBM in the Madiun Forest District / 82
4. Strategy Analysis from Collaborative Governance Perspective / 89

Chapter 4   Legitimacy for "Great Happiness"
Communal Resource Utilization in Biche Village, Marovo Lagoon in the Solomon Islands / 99

1. Collaborative Governance and Dynamics of Local Societies / 99
2. Rights, Ownership, and Utilization in Biche / 104
3. The Concept of *Noro* Underpins the Legitimacy of the Communal Utilization of Resources / 106
4. Christianity and Commercial Logging and Conservation / 111
5. Is Collaborative Governance Able to Produce "Great Happiness?" / 129

Part II   Sharing Interests, Roles and Risks
The Process of Collaborative Governance

Chapter 5   Task-sharing, to the Degree Possible
Collaboration between Out-migrants and Remaining Residents of a Mountain Community Experiencing Rural Depopulation / 135

1. Rural Depopulation and Problem of Underuse / 135
2. Research Challenges and Focus / 137
3. Overview and Migration History of Mogura / 141
4. Out-migrants and Community Association / 146
5. Changes in Festivals due to the Migration of Young People / 149
6. Discussion and Conclusion / 156

Chapter 6   Collaborative Governance for Planted Forest Resources
   Japanese Experiences / 163
   1. Forest Ownership and the Characteristics of Planted Forest Resources in Japan / 163
   2. An Outline of the Case Study Areas and their Location / 167
   3. Governance of Large-Scale Non-reforestation Lands Owned by Non-Resident Owners: The Example of KUVM in Kumamoto Prefecture / 169
   4. Participation in State-Owned Profit-Sharing Forests: Former Kitago Town Municipality (KITM), Miyazaki Prefecture / 182
   5. Conditions for the Collaborative Governance of Forests from Two Examples in Japan / 194

Chapter 7   Forest Resources and Actor Relationships
   A Study of Changes Caused by Plantations in Lao PDR / 199
   1. Introduction / 199
   2. Merits and Demerits of Plantation Projects / 202
   3. The Rubber Plantation Boom in Laos / 208
   4. Contract Plantations for Eucalyptus / 212
   5. Collaborative Governance under Plantation Programs / 217

Chapter 8   Whom to Share With?
   Dynamics of the Food Sharing System of the Shipibo in the Peruvian Amazon / 223
   1. Various Stakeholders Surround Tropical Forest in Amazon / 223
   2. How did the Sharing System Change? / 225
   3. Changes in the Sharing System / 228
   4. Historical Changes in the Sharing System / 229
   5. The Current Sharing System / 233
   6. Toward Dynamics and Complexity in Sharing System / 240

Part III Sharing Information
Extending Collaborative Governance

Chapter 9 Providing Regional Information for Collaborative Governance
Case Study Regarding Green Tourism at Kaneyama Town, Yamagata Prefecture / 249

1. Introduction / 249
2. Current State of Green Tourism in Japan / 250
3. Method for Providing Regional Information / 257
4. Results and Discussion / 262
5. Future Challenges / 272

Chapter 10 Simulating Future Land-cover Change
A Probabilistic Cellular Automata Model Approach / 273

1. Introduction / 273
2. Materials and Methods / 275
3. Results and Discussions / 280
4. Conclusion / 288

Chapter 11 Potential of the Effective Utilization of New Woody Biomass Resources in the Melak City area of West Kutai Regency in the Province of East Kalimantan / 291

1. Introduction / 291
2. Types and Volumes of Forest (Tropical Timber) Resources and New Woody Biomass Resources in Melak City, East Kalimantan Province, West Kutai Regency / 292
3. Present State of New Woody Biomass Resources from Rubber Trees and Oil Palms in Melak City / 300
4. Utilization of New Woody Biomass Resources (Rubber Trees and Oil palms) in Melak City / 303
5. Conclusion / 309

Final Chapter Multifaceted Significance of Collaborative Governance and
Its Future Challenges / 311
  1. Definition and Design Guidelines of Collaborative Governance / 311
  2. Role of Government in Collaborative Governance / 312
  3. Collaborative Governance as Resource Management Theory / Ownership Theory / 314
  4. Collaborative Governance as Community Theory / Civil Society Theory / 316
  5. Collaborative Governance as Publicness Theory / 318

Appendix Prototype Design Guidelines for "Collaborative Governance" of
Natural Resources / 323

Index / 335

List of Contributors

Darmawan Arief, Lecturer, Lampung University; Technical Assistant, REDD-plus Agency of Republic of Indonesia

Fujiwara Takahiro, Assistant Professor, Institute of Decision Science for a Sustainable Society, Kyushu University

Harada Kazuhiro, Graduate School of Bioagricultural Sciences, Nagoya University

Hyakumura Kimihiko, Associate Professor, Institute of Tropical Agriculture, Kyushu University

Inoue Makoto, Professor, Graduate School of Agricultural and Life Sciences, The University of Tokyo

Nguyen Vinh Quang, Program Analyst, Forest Carbon and Finance, Forest Trends

Kakizawa Hiroaki, Professor, Research Faculty of Agriculture, Hokkaido University

Ohashi Mariko, PhD Student, Graduate School of Agricultural and Life Sciences, The University of Tokyo

Oktalina Silvi Nur, Vocational School, Gadjah Mada University

Okubo Mika, Curator, Lake Biwa Museum

Okuda Hironori, Researcher, Kansai Research Center, Forestry and Forest Products Research Institute

Rohman, Faculty of Forestry, Gadjah Mada University

Sato Masatoshi, Professor, Graduate School of Agricultural and Life Sciences, The University of Tokyo

Sato Noriko, Professor, Faculty of Agriculture, Kyushu University

Tanaka Motomu, Associate Professor, Institute of Decision Science for a Sustainable Society, Kyushu University

Tanaka Nobuhiko, Professor, School of Tourism, Tokai University

Tsuyuki Satoshi, Associate Professor, Graduate School of Agricultural and Life Sciences, The University of Tokyo

Yokota Yasuhiro, Researcher, Kyushu Research Center, Forestry and Forest Products Research Institute

Wiyono, Vocational School, Gadjah Mada University

Acknowledgement

The publication of this book was facilitated by a Grant-in-Aid for Scientific Research from the Japan Society for the Promotion of Science (JSPS) from FY2007 to FY2011, and titled *Conditions for Collaborative Forest Governance in accordance with the Locality* (Grant Number:19208014, Principal Investigator : Makoto Inoue). The translation and publication of the book was also facilitated by a JSPS Grant-in-Aid for Publication of Scientific Research Results (Grant Number: 256006, Principal Investigator: Makoto Inoue). We extend our heartfelt appreciation to Mr. Kensuke Goto and Mr. Itaru Saito at University of Tokyo Press and Ms. Yuko Furuya at Crimson Interactive Japan for their editorial contribution to the publication of this book.

January 2015
Motomu Tanaka and Makoto Inoue

Collaborative Governance of Forests

Introduction

Motomu Tanaka

Throughout the modern era, the local people have moved away from the forests of Asia and there has been an exploitation of resources and excessive devastation of the natural forest environment. During the latter half of the 20th century, due to serious environmental problems and growth limitations in developed countries, the "commons" approach to forests emerged; therefore since the end of the 20th century efforts have been made on a trial basis to formulate detailed measures relating to that concept. The heart of the problem is how best to manage coordination between the parties that participate in the planning of forest management and the stakeholders that may occasionally be in conflict with them to manage the forests as an inclusive resource for all.

How does local society connect with outsiders?
This book is a result of research conducted with the aim of seeking the most appropriate forms of collaborative governance for forests formulated when local society has connected with outsiders. The involvement of people with the forests has included a wide range of aspects, such as ascertaining policies and management, local society and culture, geographical information and scenery, and untouched resources. This has involved the collaborative effort of many researchers involved in fields such as forest policy and business administration, meteorology, forest products, sociology, and anthropology.

In tropical countries, though forest management has been conducted by the governmental agencies, since the 1990s, the move to include local residents in forest management has been emphasized, consequently promoting activities such as afforestation. Japan has many privately owned forests that have been managed over many years. On the other hand, there have been problems such as stagnation in the price of domestic lumber and the lax attitude toward forest management. This book considers examples of the new networks and concepts of local society that are evolving among the diverse actors involved in forest management as well as the associated problems and possibilities. This connecting of the local society with outsiders has been absent from existing commons research. The trials and tribulations experienced in the collaborative governance of forests in Japan contain many lessons that will help tropical countries, such as Indonesia and Laos, overcome some of the long-term problems they are facing and the discord that arises between local society and a diverse range of actors.

On the other hand, Japanese mountain villages are already losing aspects such as cultural resources that have been formed out of the joint utilization of resources and fear and respect for nature. Using examples from places such as the Solomon Islands and Peru, where such things still play a central role in society, this book re-examines how a local society based on nature should be and provides suggestions regarding how Japan's mountain villages can re-formulate their functions as a society. It is hoped that this book will provide food for thought on discussions regarding how to create a society where resource management is organized through networks of people based on their involvement with nature and that it will be an important first step in solving the problems of resource management and the rebuilding of society.

The Future Envisaged by Collaborative Governance
Society is formed through the relationships that people have with each other, such as with family, kin, friends, colleagues, and other residents. The question arises: why is it necessary for people to form societies?

People use plants and animals, whether for their livelihood or sustenance, to sustain and raise their children; thus, the natural environment of the area where they live constitutes an important foundation for their lives.

Using nature—and, at times, being beaten by nature—in their lives, one of peoples' survival strategies is to build relationships and help each other without fighting over natural resources. Thus, nature fascilited these relationships among people who utilize natural resources. Local society has been formed by people who jointly use their area's natural environment, such as the forests, sea, and rivers, as the foundation for their lives and support each other.

However, local societies are currently undergoing a tidal wave of changes. Outsiders from various backgrounds whose livelihoods are not based on such societies' natural environment and who have no direct relationship with the local areas have become involved through initiatives such as development, conservation of nature, resource management, and policies and subsidies. Moreover, these outsiders have actually induced transformation of these societies. In fact, some societies no longer manage their own natural environment; instead, the norm is to entrust its management to the government or outsiders. Networks of people transcending local boundaries are now acknowledging the risks associated with global environmental problems, such as deforestation, rising sea levels, and climate change. There are movements afoot to solve these problems.

This book examines the impact of such movements on individuals and local society, considering examples of trial and error in resource-related collaborative governance, specifically of forests.

Collaborative governance is, "the mechanism of resource management through the collaboration of various agents (stakeholders), such as central governments, local governments, residents, enterprises, NGOs/NPOs, and global residents" (Inoue, 2004: 140). In collaborative governance, the prime focus of discussions is how best to collaboratively manage natural resources, which could only be achieved because collaborative governance is based on commons theory.

Until now, discussions in commons theory have mainly focused on the joint management of natural resources. Discussions regarding the tragedy of the commons, which arose due to open access to resources, commenced with how to restrict ownership in order to be able to continue using natural resources without depleting them. Such discussions became linked with the legitimization of state or private ownership of resources that had been subject to customary joint use by local residents. In contrast, there have been demands to reevaluate customary joint use of resources, for example, in areas where a diverse range of natural resources have been jointly used while engaging in long-term traditional shifting cultivation or in areas of communal forest (*Iriai*) in Japan.

This is also linked to efforts aimed at solving problems such as those related to the restriction or elimination of use in large-scale developments, such as commercial logging, and nature conservation measures, creating national parks, or the fact that the customary use of resources by local residents was not considered seriously. This has also become the driving force responsible for collaborative governance theory.

Compared with governments and enterprises, local residents have little power in aspects such as legal rights, influence regarding resources, and information and funding. Moreover, in many cases, it is difficult to fight usage restrictions for resources that form the basis of residents' lives. Interested persons, such as outsiders, including researchers and NGOs/NPOs, who support such local people and make efforts to determine the best form of collaborative governance are an example of measures that have been taken to bring together the stakeholders related to the resources in question around the discussion table. This is one of the aspects of the new commons theory that describes the changes in the state of the commons through the involvement of a diverse range of actors.

Cultural Resources and the Commons

Efforts are also being made to include not only natural resources but also cultural resources, such as festivals and dances, in discussions regarding commons theory[1]. Much research has been carried out, and there are many case studies regarding the commons as natural resources. On the other hand, due to the increasing pace of rural depopulation, there is increasing concern about the declining cultural resources, such as festivals, because of the lack of people available to continue these practices.

If we consider this from the local residents' viewpoint, then maybe such things can be taken for granted. For example, for residents of mountain villages, while they recognize the forests, grasslands, and rivers that have been passed down to them by their ancestors, and are important foundations for their lives, as assets, their culture, such as festivals, that has been developed by being in awe of nature, enjoying the harvest, inviting the rains, and seeking to avoid insects and wild animals, are also important assets.

It is ironic that, in some cases, festivals that were born out of this involvement with nature have now become nothing more than noisy entertainment. Furthermore, a pseudo-scientific emphasis on the sentimental or emotional conservation of certain wild animals has, to a great extent, twisted and overshadowed the original diversity of local society, the culture of which, by contrast, emphasized the non-wasteful use of wild animals. Thus, there appears to be an aspect of conflict between the ecological and genetic values and the social and cultural values of natural resources[2]. This book describes the involvement and transformation of local and outside societies in relation to cultural resources, dealing with scenery and interpersonal relationships in Chapter 2, legitimacy in Chapter 4, festivals in Chapter 5, and sharing systems in Chapter 8. These dynamics result in problems such as the difficulty experienced by outsiders in fully understanding the logic, values, and culture of local society and the resulting friction between local and outside societies.

---

1) Natural resources are "resources that are reliant on nature for many of their formative foundations, with people having some involvement." Cultural resources are "resources that are reliant on people and the relationships among people for many of their formative foundations, with nature having some involvement." In some cases, natural resources are used and processed by people and, in doing so, derive social and cultural value and assume the nature of cultural resources. There are also cultural resources that are formed through expression and information, such as movies, comics and music.

2) To a certain extent, festivals with social or cultural values have a closed nature, and the fact that they have been conducted within the constraints of geographical or blood ties implies the possibility that the number of people who acknowledge their value is limited. In contrast, ecological and genetic values have the potential to be acknowledged by a larger group of people with a common awareness because though there may be differences in ideology and emphasis, there are objective scientific criteria. Therefore, restrictions on the use of natural resources are probably one of the factors that threaten cultural resources.

Globalization and the Commons

Such problems are thought to have emerged in the midst of globalization as the range of actors saying, doing, paying for, or seeking to be involved in some way with certain resources has become increasingly diversified. Collaborative governance is one of the means being utilized to seek solutions for such problems through trial and error. However, a major problem in this potential solution is the fact that the diverse range of actors involved has led to a diversity of aims. There will be people wanting to use these resources, people wanting to stop others from using them, people wanting to consume and exhaust them, and others only wanting to be mere onlookers or circumferential users. Chapter 9 investigates ways of digitizing cultural resources, so that they can be utilized by diverse actors, including outsiders and for tourism. Chapter 10 presents materials that enable diverse actors to develop relationships through a common awareness of problems. This includes the analysis of satellite images to visualize natural resources for providing a new perspective on risk avoidance in local society.

It is probably not too complicated a problem if the most effective way of managing such natural resources is sought to simply and ecologically sustain the resource. All that is left to be determined is how to isolate the resource. However, what is difficult is to determine how to sustain the social and cultural values of a resource. Cultural resources derive their value from being utilized and have an aspect of being inherited. Ecological and genetic values of natural resources are threatened by overuse, and thus, controlled use of natural resources does not greatly decrease their value[3]. However, as cultural resources are being lost through underutilization due to reduced opportunities for festivals, etc., it is important to consider how their use can be sustained[4]. In Chapter 5, the turmoil among local residents is described, as they seek to find ways of passing on their festivals and try to decide whether and what they should partially renounce.

In the midst of globalization, various stakeholders are becoming increasingly involved with natural resources and cultural assets. Moreover, discussing management merely in

---

3) The necessity of adaptive management with moderate utilization and disturbance, such as diversification of the understory through measures such as the [selective] felling of tall trees and diversification of other plants and animals through capturing and killing predators, is also being debated in the context of the management of natural resources.

4) It is possible to preserve cultural resources through exhibitions in museums and by documenting them both in words and on video. However, only when cultural resources are used in the context of interpersonal relationships do they take on value as living culture and have value as one of the foundations of society. Cultural resources that have simply been preserved as a record are "dead culture" and could be considered to have largely lost their value. On the other hand, cultural resources that have been preserved as a record or for information purposes have the advantage of being accessible to many people.

terms of sustainable or controlled use is proving to be inadequate. With a diverse range of resources and an even more diverse range of values among the people involved, discovering how stakeholders should collaborate will be a major challenge.

The principle of involvement is being emphasized within the overall concept of collaborative governance as a means of solving this problem. This could also be termed "localism," which seeks to bring local residents to the forefront as they are the ones who have been involved with such natural resources. The commitment principle, in particular, which in the decision-making process provides the right to make decisions according to the depth of involvement with the resources in question, is an important principle of collaborative governance (Inoue, 2009: 14–15). However, elucidating the state of collaborative governance, together with its possibilities and problems, needs to be further examined in detail.

The Possibility of "Sharing"

Herein, this challenge is approached from the perspective of sharing. Collaborative governance is not merely an attempt to discover how resource management should be implemented or how it could be successful. Rather, collaborative governance is a new vision of society, one in which all stakeholders can interact in a collaborative manner to govern certain resources. Societies are formed through the collaboration of individuals. It is vitally important, therefore, to consider how diverse stakeholders with diverse values seeking to be involved with diverse resources can be accepted or excluded by local societies and how they can collaborate and share roles, benefits, and risks—in fact, share the future.

"Sharing", in this case, refers to (1) sharing a portion of profits, (2) sharing work and cost burdens, or (3) sharing of roles and resources as well as collaborative utilization. This includes sharing the reasons for involvement, goals, roles, and responsibilities and the resulting gains. The key to everything—including the prerequisite, which is systems regarding resources; the purpose of collaborative governance, which is the entrance; the role that actors play in collaborative governance, which is the process; and what they obtain, which is the conclusion and exit—is sharing.

In terms of sharing, this book consists of three parts. First, the sharing of rights and systems regarding the resources in question as a prerequisite to collaborative governance is discussed. Second, the sharing of roles and the benefits and risks in the process of establishing collaborative governance is discussed. Finally, the sharing of information (including the latest in geography, groundcover, and technology) that could lead to an expansion of the range of stakeholders involved in collaborative governance, simultaneously maximizing benefits and reducing risks, is discussed.

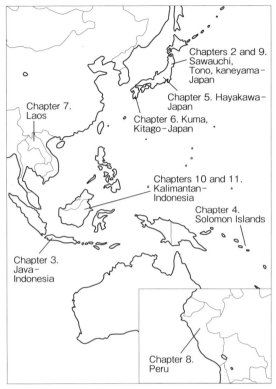

Figure 0–1. Research Field of this Book

In the design guidelines for collaborative governance, Inoue (2009) highlighted the relevance of the commitment principle that grants decision-making rights to actors involved in collaborative governance according to a scale of graduated membership based on the degree of involvement. These design guidelines for collaborative governance have been suggested by numerous researchers, including those involved in commons research (Ostrom 1990; McKean 1999; Stern et al. 2002; Ostrom 2005). The first and second parts of this book describe the possibilities and difficulties that concern involving outsiders with local residents, who are supposed to be important members in terms of their involvement in local forests but have left the area, resulting in the decline of local communities. The final chapter presents the roles of government and residents together with associated problems and discusses collaborative governance in light of resource management theory, community, and the public nature of the commons.

Where will collaborative governance lead?

As a concept for society, collaborative governance may be a balancing act. Why should outsiders with different values be included? How can it be fair for everyone? For how long will outsiders be involved? What are the long-term benefits of involving outsiders with different values? These are some of the questions that must be answered before collaborative governance can be advanced.

On the one hand is the image of local residents, sometimes seen as the underdog—oppressed and excluded by the state and corporations; therein lies the ease of understanding the problems. However, collaborative governance will have to be realized through trial and error, where in the whole world must join together through the market economy, finance, and information, tackling issues such as global environmental problems within the overall flow of globalization. This could lead to the prosperous, sustainable society envisioned—or it could lead to chaos. In the midst of all this, it is quite probable that questions will be asked regarding the overall flow of globalization itself.

For hundreds and thousands of years, human beings have formed societies, developed cultures, and based their livelihoods on the natural environment. Over the recent decades, the sweeping tide of globalization has demanded a collaboration that goes beyond anything from the past. The need arises to forge as yet untried ties with people all over the world. Will the natural environments and cultures of these diverse areas be engulfed in this wave, or will they transform our current state of globalization and the collaborative governance it has spawned? Will this lead to a new era of harmony, diversity, and prosperity? This book will open your eyes and challenge your preconceptions, compelling you to reflect upon the issues of globalization and collaborative governance.

References

Inoue, M. 2009. "Shizen Shigen '*kyouchi*' no Sekkei Shishin: Rokaru kara Gurobaru he" [The Design Guidelines of Natural Resources 'Collaborative Governance': Connecting the Local and the Global]. In: Murota, T. (ed.) *Gurobaru Jidai-no Rokaru Komonzu* [*The Local Commons in the Global Era*]. Kyoto: Minerva Shobo, pp. 3–25.

McKean, M. A. 1999. "Common Property: What Is It, What Is It Good for, and What Makes It Work?" In: Gibson, C., M. A. McKean, and E. Ostrom (eds.) *Forest Resources and Institutions*. Rome: FAO, pp. 27–55.

Ostrom, E. 1990. *Governing the Commons: The Evolution of Institutions for Collective Action*. New York, NY: Cambridge University Press.

Ostrom, E. 2005. *Understanding Institutional Diversity*. Princeton, NJ: Princeton University Press.

Stern, P. C., T. Dietz, N. Dolšak, E. Ostrom, and S. Stonich 2002. "Knowledge and Questions after 15 Years of Research". In: E. Ostrom, T. Dietz, N. Dolsak, P. C. Stern, S. Stonich, and E. U. Weber (eds.) *The Drama of the Commons*. Washington, D.C.: National Academy Press, pp. 445–489.

Part I

Policies, Institutions and Rights to Share
Prerequisites for Collaborative Governance

Chapter 1

Historical Typology of Collaborative Governance
Modern Forest Policy in Japan

Hiroaki Kakizawa

1. Introduction: Policy Change in Forest Governance in Japan

Centralized administrative policy measures have dominated and have strengthened the foundation of forest resources in Japan. This system is being split in to two fronts. First, because resources, especially planted forests, have reached their utilization front, there has been a demand for transitioning to a collaborative system involving government organizations, policies, and measures that are appropriate for this stage of sustainable forest management. Reforms have been progressing on the basis of the Forest and Forestry Revitalization Plan since the Democratic Party of Japan assumed power in 2009. Second, as forests are being recognized as local public goods, there is a demand for policies that involve the participation of various stakeholders. Concerned organizations are becoming increasingly diversified in forest-related activities.

One of the purposes for this joint research project is to suggest new systems for supporting sustainable forest management, in other words, moving away from rigid and centralized systems to a flexible and de-centralized system. Forest administrative institutions and policies in Japan are subject to change, and the restructuring of these systems to achieve sustainable forest management is an urgent task and is important for laying the foundation for forest governance development.

This chapter will first review the related research on forest governance in Japan and then determine why it has failed to place appropriate focus on government. Next the problems confronting Japan's forest policies will be revealed. The concluding section will then consider how these issues may be tackled in light of policy changes under the Forest and Forestry Revitalization Plan and how these issues impact forest governance development. The following sections will primarily discuss the manner in which governance should be established for forest policies and management as a whole in Japan, with focus on private forests[1].

---
1) National forest and other public forest will not be discussed in this chapter.

2. Problems of Forest Governance Research in Japan

2.1 Why Reseach Forest Governance?

Almost all the studies on forest governance in Japan have a common tendency—to move toward a collaborative system of forest management and policy with citizens as its primary constituents. Behind this tendeney lies the primary intention of reconsidering the ties between the forest and people. Forest policy and management has been authorized with its implementation by government officials, forest owners, and forestry specialists. However, as forests are being recogmzed as local public goods, there is an increasing shared awareness that forest beneficiaries should be involved in policymaking. Furthermore, amida tough economic situation for influencing forestry, its revitalization is reconsiclooed almost impossible without public support of initiatives. To ensure public participation in forest management and in turn revitalize forestry, the collaboration of a diverse range of stakeholders is essential. Research has revealed the establishment of governance as an important goal (Kakizawa, 2007).

Forest governance research aims to create improved relationships between the forest and society and establish a fundamental social change related to forests. Therefore, the revision of existing systems and policies has naturally become the central theme of such research.

Corsidering the developments in such research, discussion about governance generally begins with the limitations of the government and, progresses to reconsiderations of the government's role the importance of creating collaborative relationships (Kakizawa, 2002). However, despite such theoretical background and the necessity for discussion of forest governance, substantial fact-finding research on forest governance in Japan has focused only on peripheral issues such as civil movements, and has not yielded any conclusions regarding the "limitations of the government."

2.2 The Challenges of Research on Forest Governance

Research about forest governance that has been conducted can be broadly divided into the fields of forest governance theory, ideology and concepts, and empirical research regarding new efforts aimed at establishing of forest governance.

The former has already been discussed and will not be repeated here. This section emphasizes that most research verified the necessity of restructuring the relationship between the forest and society and reforming policies to establish forest governance. Empirical social science research on forests in Japan approaches forest governance from two different perspectives: public participation and the commons (Kakizawa, Yamamoto, and Saito, 2006). On examining forests following one of these approaches, the use of keywords such as "forest volunteer activities", "Satoyama[2] conservation", "wildlife management", "upstream and downstream collaboration for watershed conservation", and "utilization of local timber in residential construction" is noted. Such research is conducted with a common focus on efforts by forest-related movements and partnerships that are proceeding with the aim of establishing the direction of forest governance and aims to examine the state of collaborative forest management and the conditions related to the structures of such collaborations.

The most actively researched topics are those regarding the activities of forest volunteers, public participation and social movements, and the conservation of Satoyama[3]. In addition to Yorimitsu (1984, 1999), who was one of the first researchers to focus on citizen action as a means of causing change in forest policies and management, there has been abundant research inoorporating factual surveys regarding the development of forest volunteer activities with regard to the management of plantations, Satoyama, and suburban forests[4]. Furthermore, forest volunteers have conducted research on the basis of their experiences and framed their own policy proposals (Uchiyama, 2001). Through research, such as those outlined above, many case studies have been conducted on public activities and the development of collaboration among various groups. The most important results of such studies have been the elucidation of the present state of the public as new participants and the progress related to forest management made by social movements.

The influence on changes in forest policies is undeniably minor. Of course, there might be cases where social movements have majorly influenced the formation of forest conservation policy in suburban local governments; however, it has remained a peripheral movement within the Japan's forest policy and management development process as a whole. Moreover, efforts to utilize local timber have not been able to shake their reputation as a niche industry. Research on forest governance, despite aiming to change forest policy, has focused on peripheral movements and efforts, without focusing on research concerning policy formation and implementation processes.

On the other hand, research on forest governance has been oriented changing the balance of power between the forest, society, and policy, and it focused on movements that were believed to have the potential of causing changes that would shape future reforms.

---

2) Satoyama is an area of secondary forests (plantation forests surrounding human settlements) that is located between more natural, deep mountainous areas and urban areas of intensive human activities. Satoyama has been formed through various human interventions such as forestry activities over a long period of time, where people have practiced land use in cyclic resource utilization.
3) For major research literatures, refer to the list specifieel by Kakizawa, Yamamoto, and Saito (2006).
4) Refer to, for example, Yamamoto (2003) who poovides summary of these results.

However, these efforts are still uncommon in Japan and as such have remained peripheral movements. Thus, corresponding research has also remained peripheral in nature.

Furthermore, the examination was conducted using sociological techniques, that focused on social movements aimed at new forms of forest management; this was typically research involving forest volunteers. However, few reports clearly state the research techniques implemented, and almost none discuss appropriate types of research techniques. Case studies and analyses progressed without adequate methodological consideration.

This type of research tendency is not confined to researchers affiliated with the Japanese Forest Economic Society. For example, Nishio et al. (2008) used the keyword "coexistence" in terms of public administration to connect these issues with the governance discussion. For forest policy formation, they proposed moving away from state dependence toward collaboration among neighboring local governments, suggesting that these local municipalities pursue better performance on common arena. However, with regard to future directions and alternatives, they have only discussed "local production for local consumption" forestry. Thus, while developing the theory of governance that aims for fundamental change, their empirical research ultimately chose the easy way by citing the unbalanced research pathway of niche industries.

Collaborative forest governance research is not necessarily unimportant. In particular, it clarifies social movements such as that of the forest volunteers —— one of the stakeholders in forest management. Discussions regarding their significance and limitations can also be considered a major achievement. However, the fact remains that forest-related institutions and policies as well as administrative structures and organizations have yet to be comprehensively discussed from the perspective of governance.

Research on forest governance in Japan probably progressed because of an affinity with the "governance without government" theory. Gaining popularity during the latter half of the 1990s, the subjects of this research comprised the pioneering activities that dealt with those problems that the government was unable to solve. These activities forged a future for forest management through collaboration among various groups formed by social movements rather than the government. These activities were orchestrated by the leaders who played key roles in such movements and the groups that collaborated with them. Thus, although it is possible to clarify through research the conditions that lead to the success of these movements, extending these ideas to other arenas. The management and utilization of forests by such social movements in their respective regions has been the exception to the rule, and developments cannot be expected in other regions. In such areas, the government played a major role in the establishment and operation of local forest management systems. Further research is required to evaluate the role played by the government[5] as well as the current issues to be addressed.

The government certainly plays a major role in the development of actual forest policies; however, research on forest governance related to this topic is clearly lacking. This indicated that there was no concrete basis regarding how to achieve the transition from "the government to governance". Thus, concrete strategies for formulating forest governance as the basis for such movements could not be created. Because of the government's vital role in developing forest policies, as long as it is inconceivable to develop governance without the government, it is necessary to research the government itself to achieve governance.

3. Japan's Forest Administration System and Policies

3.1 Introduction: The Forest and Forestry Revitalization Plan after Change of Power

This section clarifies the challenges that forest administration systems and policies have faced in Japan and deliberate what type of governance structure is required.

In 2009, the Democratic Party of Japan rose to power and policy changes followed. On December 25th of the same year, the Ministry of Agriculture, Forestry and Fisheries formulated the Forest and Forestry Revitalization Plan with its slogan "From a Concrete Society to a Forest Society"[6]. The Forest and Forestry Revitalization Plan was formulated as part of emergency employment measures. It deals with the economic revitalization of forestry as one of its main challenges, as observed from the timber self-sufficiency ratio set at a 10-year goal of 50% (which was 28% in 2009). However, in addition to revitalizing the forestry and forest industry, this plan incorporates aspects aimed at the creation of a sustainable society. This is achieved through the sustainable and appropriate management of forests, which forms the basis of forest and forestry revitalization. This revitalization will, in turn, contribute to the realization of a low-carbon society through sustainable realization of the multifaceted functions of forestry and the utilization of timber and energy.

A Forest and Forestry Revitalization Plan task force was established to formulate and implement this plan, and detailed studies were conducted from February 2010. Based on the final report that was released the following December, the Forest Act was amended in April 2011. Reforms are now well underway, with activities such as programs for human resource development already in progress[7].

---

5) Throughout the rest of this chapter, the word "government" is used as a generic term to refer to public institutions, systems, and policies.

6) For materials related to the Forest And Forestry Revitalization Plan, see the following website of the Forestry Agency: http://www.rinya.maff.go.jp/j/kikaku/saisei/index.html

The forest policy entered a period of great change with the transition in government from the Liberal Democratic Party to the Democratic Party of Japan. However, this chapter emphasizes that first, regardless of which party is in power, the existing forest policy is entering a time of reform. Second, the abovementioned reform should not be treated as an extension of the existing systems and policies; rather, they should be viewed as requiring a fundamental change, including in areas such as the legal system and organizational structures pertaining to the framework of forest laws.

The reason for the need of fundamental reforms is that the system of forest policies, with its emphasis on forest establishment, is absolutely inappropriate for the times. This section will briefly discuss the postwar state of Japan's forests and the related policy trends. It will also discuss why the existing forest policy needs to be reformed according to the times and the direction of such required reforms.

3.2 The Development Process of Japan's Postwar Forest Policy

After World War II, an urgent challenge was the restoration of forests, which were devastated by excessive logging during wartime, for the purpose of soil and water conservation. Cousequently, activities such as afforestation were developed to revitalize the forests. In the Forest Act, amended in 1951 in an attempt to conserve forest resources, regulations for logging of forests of regular final age were changed to require permission under the forest planning system. Furthermore, the designation of protection forests[8] and logging permissons of protection forest were assimilated into forest planning system.

The development of postwar forest policies that initially aimed at rehabilitating this type of devastation changed from the mid-1950s to support the period of high economic growth. Although the demand for timber rapidly increased with this economic growth, there was no growth in the domestic supply, leading to a tight supply and sharp increases in prices —— a threatening factor hindering economic growth. At that point, domestic production and imports of timber from overseas are to be expanded to support growth. In terms of imports, it could be said that there was an improvement in the state of foreign exchange reserves. In 1959, the import of timber was liberalized and port facilities were improved to facilitate the smooth import of foreign timber.

The government increased logging in national forests and introduced a policy to encourage increased domestic production in private forests. In national forests, where natural forest resources existed in abundance and it was possible to directly reflect the will

---

7) The Democratic Party of Japan suffered a major defeat in the general election held in December 2012. Though the Liberal Democratic Party assumed power, efforts are continuing unhindered in the basic direction in terms of forest and forestry revitalization.

8) Protection forest system was established in 1897 by Forest Act for ensuring public welfare.

of the State, policy development for quantitative increase production volumes was prioritized. Plans such as "the Plan for Increase of the Productive Capacity of National Forests" verified these developments. While increasing the logging of natural forests, such plans aimed to expand and develop areas of highly productive conifer plantations through aggressive afforestation.

The amendment to the Forest Act in 1962 eliminated the system of logging permissions, and all unrestricted forest logging (indicating forests other than protection forests) was changed to a notification system. This amendment separated the designation of protection forests and logging approvals from the forest planning system, and the regulatory aspects of the forest planning system were disregrded. Furthermore, regarding the designation of protection forests, which was practically the sole policy for regulating forestry practices under the Forest Act system, objectives that covered broad areas without imposing severe restrictions were adopted. In addition, expansive afforestation, including clear cutting of broad-leaf secondary forests that were used as fuelwood forests, and the establishment of highly productive conifer plantations for timber production, was actively promoted. Measures such as afforestation subsidies and profit-sharing afforestation were also developed. The system of regulatory forest policies for resource conservation, which was formed immediately after the war, was turned largely into a system of policies fostering growth in support of forestry production and strengthening the foundation of forest resources.

Amid all these reforms, the development of forestry industries and endeavors to improve the socioeconomic standing of forestry owners and workers were recognized as important policy challenges that were addressed by the Forestry Basic Law enacted in 1964. Until that time, the Forest Act formed the foundation for forest policies in Japan. Forest plans regarding planting, ensuring public welfare, and the development of infrastructure such as forestry roads were established under the Forest Act. The enactment of the Forestry Basic Law placed the formulation of timber supply plans under the law, and branched forest plans were placed underneath. Resource plans were classfieel under the realm of economic plans, and forest plans strongly resembled economic plans. Although the initial policy objective under the Basic Law endeavored to expand the scale of individual forest owners to establish autonomous forestry management, the effectiveness of the policies did not increase and no fundamental change occurred in the ownership or management structure of small-scale and micro-ownership entities. Further, the policy mainly emphasized the fostering and strengthening of the Forest Owners Association.

As a result of policy developments aimed at strengthening the foundation of forest resources, the plantation ratio grew from approximately 20% in 1950 to approximately 43% by 1990. Currently, there are more than 10 million ha of plantations in Japan.

These huge plantations, which were established within a short period of time, require appropriate management such as tending and thinning. However, adequate progress was achieued with regard to efforts such as the development of road network, realization of productivity improvement forestry operations, and consolidation of forest practices. Thus, compared with foreign timber, the competitiveness of domestic timber continued to decline, because of which forest owners increasingly lost motivation to manage the forests, and the number of plantations where appropriate thinning was not being performed continued to increase. The greatest challenge was development and management of these plantations. Since the 1980s, policy development, such as the formulation of plans to promote thinning and the development of subsidy schemes, was actively executed. Furthermore, under the Kyoto Protocol, which placed importance on thinning in the context of achieving forest carbon dioxide sink goals, further generous subsidies for thinning were incorporated.

The importance of the diverse functions of the forest is now widely known in the society, and forest policies are facing pressure to incorporate these functions. Since the 1970s, nature conservation movements opposing the logging of state-owned old-growth forests have become active, and policies have been drafted that consider the environment in the forest practices adopted for national forests. Furthermore, urban development and the conversion of forests into golf courses and farm land progressed at a rapid pace during the 1970s. Owing to growing concerns regarding the wanton development of the forests, the Forest Act was partially amended in 1974 to create a permit system for the conversion of forest land. Permits were required for the conversion of areas of forest land that were 1 ha or greater in size. in addition, the focus on forestry production in the laws and policies under the Forestry Basic Law was criticized. As stated earlier, the conditions surrounding forest management deteriorated and it became difficult for forest management to meet peoples' needs. Increasing numbers of forests that were not being properly managed indicated a need for fundamental changes in the policy. Further, internationally, there was an increasing emphasis on the conservation of forest ecosystems and sustainable forest management. Therefore, in 2001, the Forestry Basic Law was amended as the Forest and Forestry Basic Act to create an awareness of the multiple functions of forests and pursue sustainable and healthy development of forestry.

3.3 Considering Future Forest Policy in Japan

As mentioned above, forest policy has been facilitative in nature, and the forestry resources that are the basis of forestry production have developed through policies to expand plantations and promote thinning. The challenges confronting Japanese forest policies at present are related to how the plantation resources that have been developed can be sustainably managed and utilized.

Though the standard cutting age[9] determined in Japanese forest plans is a short rotation of approximately 40–50 years, no real model has been agreed upon in terms of the target cutting age or diameter class for forest practices. Pertaining to the management of plantations that are reaching the regeneration cutting age, some maintain the opinion that a long rotation should be adopted; however, long rotation practices have not yet been established. Furthermore, since plantations of this age class have been rare till date, timber-processing facilities have been developed for handling small- and medium-diameter trees. A limited number of facilities can actually process large-diameter timber is limited. And also, demand of building material has been shifted from lumber to engineered wood, such as laminated lumber, demand for large-diameter timber has decreased. Therefore, the price per cubic meter of large-diameter, unsawn timber is approximately the same as that of small- or medium-diameter timber. It is uncertain whether an appropriate price or adequate demand can be expected if long rotation forestry is adopted in the future. In light of such facts, it is possible that, for the time being, short rotation forestry will continue as the norm and short rotation clear cutting forestry will become established as the standard form of forestry practices in Japan. In Kyushu and Hokkaido, clear cutting is already being actively performed around the standard cutting age.

On the other hand, timber prices continue to decline, and motivation among owners to manage the forests is at an all-time low. This has given rise to the increasing problem of people abandoning the land after clear cutting without implementing any measures of regeneration. In southern Kyushu alone, there have been instances where areas of more than 100 ha have been abandoned after clear cutting.

Another problem is that the age class structure is extremely irregular. Figure 1–1 illustrates the age class structure of Japanese plantations. The rapid postwar artificial reforestation mentioned above can be clearly as being concentrated in age class 9. In this situation, if a forest is logged when it reaches the standard cutting age, then it is feared that plantation resources will be consumed all at once. That is, plantations will disappear as a single-cycle resource.

As seen in the Figure 1–1, though the Japanese forest policy has till date focused on the expansion of tendering and thinning plantation, it can now focus on the best methods

---

9) The standard cutting age is determined as an index regarding the minimum cutting age (regeneration cutting age) of the standard standing trees in an area. It is used in forest management plan certification standards and in criteria for determining the suitability of logging of protection forests. It is not used for promoting the logging of forests once they reach the standard cutting age.

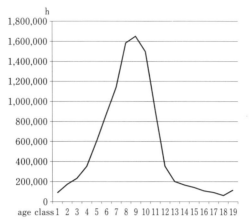

Figure 1–1. Age Class Structure of Plantations in Japan
Source: Forestry Agency of Japan.

of harnessing and utilizing those resources. Therefore, this section examines the forest policies implemented until now as resource creation stage policies and policies to be implemented as sustainable management as well as utilization stage policies and discuss what type of policy change is required.

Table 1–1 shows a comparison of the content of resource creation stage policies and the requirements for future sustainable resource management and utilization stage policies. From this, it is observed that a major paradigm shift is required.

At the forest resource creation stage, the expansion, tendering, and thinning of plantations were the main areas of focus, and it was relatively easy to composee manuals containing such techniques. Furthermore, it was possible to set forest policy goals as simple indices related to planting, maintenance, and thinning areas. There was a growing awareness that the creation of these resources was contributing to environmental conservation. At that stage, a high level of expertise about forest management was not required, and as long as subsidies were provided, it was generally possible to achieve goals. Rather than skilled foresters, forest management and administration required an administrator at its forefront with an expert understanding of how to access subsidies.

Incidentally, at the sustainable forest management and utilization stage, the above-mentioned type of policy structure no longer functions, as exemplified by the problem of clear cutting large areas and their subsequent abandonment after logging. The problems here are (1) how to convey the vision of future sustainable forest management and (2) how to control the impact of forest practices on the environment and disaster prevention. In recent years, attention to the environment has been particularly emphasized, such as biodiversity conservation, in addition to global warming countermeasures.

Table 1–1. Past and Future Forest Policies in Japan

	Past policy; Resource creation stage	Future policy; Sustainable forest management and utilization stages
Technique required	Manual	Expert knowledge
Implementation system	Subsidy	Governance
Human resources required	Administrator	Forester
Policy-making process	Top-down	Bottom-up
Policy goal setting	Simple	Complex
Environmental concern	Easy	Difficult

Thus, at this point, control over forest practices is being sought through combining various methods such as regulations and guidance to set forest management goals and create a vision for forest management that considers the local natural environment and the state of the society and economy. To implement these controls, collaboration among foresters[10] with fair knowledge of local forest and the surrounding socioeconomical conditions, those involved in the local forest and forestry, and nonprofit organizations (NPOs) and individuals interested in the forest will be essential.

At the resource creation stage, access to subsidy schemes was all that was required. However, presently, there is need for an ability to promote forestry revitalization, and simultaneously control logging activities to achieve environmental conservation.

Thus, at the sustainable forest management and utilization stage, the following three points need to be considered:

(1) Master Plan Formulation: The primary direction of forest management in consideration of the nature of a specific region—the establishment of a vision.
(2) Establishment of Rules for Forest Practices: Ensuring not only the sustainability of the resource but also a minimum level of sustainability, including the conservation of ecosystems, prevention of disasters, and conservation of water quality.
(3) The development of human resources and administrative organizations that can not only achieve objectives (1) and (2) mentioned above but also collaborate with a broad range of stakeholders.

Concerning these three points, the existing problems pertaining to forest policy will

---

10) This chapter will use the word "foresters" to refer to specialized staff engaged in local forest administration and management as administrators.

be discussed first. Then, how these problems were addressed during the Forest and Forestry Revitalization Plan deliberation process and the types of policies that were subsequently adopted will be examined. Numbers (1) and (2) are problems related to the forest planning and protection forest systems, which will be discussed along with the manner in which forest practices should be controlled.

4. Considering Japanese Forest Policy Problems and Forest and Forestry Revitalization Plans

After addressing the forest policy problems prior to the reforms of 2011, this section discusses how these problems were addressed under Forest and Forestry Revitalization Plans and what type of new forest policies were drafted.

4.1 Problems associated with the Control of Forest Practices
4.1.1 Present Problems
The system that controls Japanese forest practices includes the forest planning system and the protection forest system. The former deals with the forest as a whole, whereas the latter deals with specific forests that are designated as protection forests. This section addresses the state of each system and its associated problems prior to the reform of the Forest and Forestry Revitalization Plan in April 2011.

The Forest Planning System
The forest planning system comprises of four different levels: national, prefectural, municipal, and forest owners. In the Forest and Forestry Basic Act, which defines the direction of basic measures for forests and forestry in Japan, the government has formulated the Basic Plan for Forests and Forestry that determines the basic policy for forests and supply goals for forest products. Furthermore, the forest plan system is defined in the Forest Act, which is the law that forms the basis for forest administration in Japan. First, at the national level, the National Forest Plan is formulated in line with the Basic Plan for Forests and Forestry to clarify the national guidelines regarding forest management and conservation over the long term and for a wide area. Then, at its local level, prefectural governors formulate the Regional Forest Plan in line with the National Forest Plan for each of the 158 watersheds in the country, setting guidelines for forest management in each watershed. The head of each municipality then establishes a Municipal Forest Plan in line with the Regional Forest Plan[11]. The Municipal Forest Plan provides long-term goals for the forest and establishes rnles and guidelines for implementing these goals. Forest owners are required to provide notification to the head of the municipality during regeneration cutting and regeneration, and when a notification does not comply with the Municipal Forest Plan, the head of the

municipality may demand compliance. Though formulation of the abovementioned plans is mandatory, forest owners may also set a long-term policy for 40 years or more, formulate Forest Management Plans detailing the forest practices to be implemented for the next five years, and acquire approval for these from the head of the municipality. Owners who have formulated Forest Management Plans may receive preferential treatment such as access to additional subsidies or tax reductions.

This system has three functions: (1) setting the direction of forest management; (2) establishing regulatory tools for forest practices, and (3) determining business plans for forest practices. Furthermore, ordinances related to the Forest Act require detailed notes about the planning matters to be included in plans at each of these stages, and such plans have a detailed systematic structure to ensure consistency throughout the stages.

However, major problems appear in terms of the effectiveness of this planning system.

First, there are inadequate areas concerning the setting and implementation of rules for promoting sustainable forest management, including those related to environmental conservation. Under the forest planning system in Japan, rules to be applied on-site are set under Municipal Forest Plans. Based on these plans, logging notifications are considered and recommendations regarding forest practices are made. However, it is unusual for a municipality to competently implement the forest policy on its own initiative, which will be discussed later. Therefore, most municipalities are unable to establish systematic regulations regarding ecosystems and environmental conservation or set clear standards for the regulations pertaining to forest practices. They are also unable to adequately handle the logging notifications of owners within the region or provide appropriate guidance in response to such logging notifications. On the other hand, municipalities that are able to competently implement forest policies include their own rules in the plans. However, in Japan, where private property rights are strong, when the regulation enforcement is disputed in court, there owises the problem of whether regulations based on municipal plans—which are not legally backed by the state—will be recognized[12]. The system is also absolutely nonfunctional in terms of management recommendations because municipalities do not have specialist staff, and in most cases, recommendations are not made even in case of events where recommendations should be given[13].

Second, Municipal Forest Plans have not become master plans for forests. Municipal Forest Plans should indicate the vision and goals of forest management and present their vision of an ideal forest to local forest stakeholders; however, it is unusual for municipali-

---

11) The Municipal Forest Plan was introduced in the amendment to the Forest Act in 1998 in line with the government decentralization policy. Until then, prefectures were at the forefront of the forest planning system and were responsible for the logging notification system.

ties to create plans with this type of content. Therefore, within the framework provided by the state, these plans became formalized and were mere forms created by the prefectures into which target values were later incorporated. As for zoning, though the state requested that all forests be classified into three zones—soil and water conservation forests, nature conservation and public health protection forests (symbiotic forests), and production forests—it was practically impossible at the municipal level, to classify all forests throughout the country under these three categories. Thus, it was never really functional.

Third, regarding the effectiveness of this planning system, it did not function from the standpoint of business plans. Under the National Forest Plan and Regional Forest Plan, target (planned) values concerning forest management and conservation, such as those pertaining to thinning and forest road development, were set; however, in many cases, they were set as ideal values with no adequate budgetary or policy support. This indicates that they were unable to fulfill their role as plans or targets because they had lost touch with reality[14]. The same is true of Forest Management Plans, which were formulated as means of acquiring additional subsidies in many cases, because forest owners mostly do not conduct maintenance in accordance with these plans and it was normal for maintenance such as thinning and regeneration cutting to be conducted according to the state of the timber market. Furthermore, because these plans did not consider geographically large forests into account in many cases, it was difficult to implement measures for reducing costs, such as the planned maintenance of road networks and conducting forest maintenance in a concentrated manner.

The Protection Forest System

This segment reviews some characteristics of protection forests. Under the Protection Forest System in Japan, specific forests are designated and in addition to prohibiting conversion as a rule, maintenance is regulated through a logging permit system to achieve certain goals in public interest, such as watershed protection, prevention of damage from landslides, and realization of areas for health and recreation.

---

12) Forest owners whose logging notification were rejected or who were requested to make changes in accordance with rules determined independently under Municipal Forest Plans should possibly have their claims accepted in the court when suing municipalities for violation of property rights.

13) Under the Forest Act, when it is acknowledged that proper forest management is not being conducted in accordance with Municipal Forest Plans the head of the municipality may order the forest owner and others to conduct appropriate management in cases where it is necessary to comply with Municipal Forest Plans. Furthermore, in forests where thinning is not being properly performed in accordance with Municipal Forest Plans and where the forests have been neglected, if it is feared that this will significantly disrupt the various functions of the forest and cause wind and snow damage, or leaching of topsoil. In such cases, the head of the municipality may notify regarding the method and timing of such thinning and list recommendations for maintenance.

14) Because there is no connection between establishing these plan values and budgetary measures, values that are completely different from those that are actually achievable are set.

First, though they cover a designated area, the content of the regulations is relatively moderate. Presently, approximately 20% of privately owned forests and 87% of national forests are designated as protection forests; in total, nearly half of the forests in Japan were designated as protection forests. The content of regulations pertaining to forest practices is generally moderate. Considering watershed protection forests, which account for approximately 70% of all protection forests, under the current system, clear cutting is possible up to an area of 20 ha, and whether designated goals can be achieved remains questionable. Further, comparing regulations pertaining to forest practices for all forests in the European countries with those for protection forests in Japan, it is observed that in many cases, the level of such regulations is higher in European countries than in Japan.

Second, though the number of different types of protection forests is increasing, the current system is not able to cope with modern challenges. Since protection forests were first established under the Forest Act in 1897, the number of different types of forests has continued to increase. Presently, there exist headwater conservation forests, soil conservation forests, erosion control forests, shifting sand protection forests, wind protection forests, flood control forests, tide damage protection forests, drought prevention forests, snowbreak forests, mist protection forests, avalanche prevention forests, falling stone protection forests, firebreak forests, protection forests for fish gathering, navigation target forests, public health protection forests, and esthetic forests (17 different varieties in total). However, the present system is unable to cope with the challenges faced today; for example, there is no mention of riparian areas or any network of systematic protection regarding biodiversity.

Therefore, the Protection Forest System is unable to fulfill its role in defining zoning in areas as required in accordance with regulations.

4.1.2 The Result of Deliberations and the Direction of Reform

This section examines how the problems abovementioneel discussed during the Forest and Forestry Revitalization Plan deliberation process and how new policies were consequently drafted.

There are major problems with the Forest Planning System and Protection Forest System, and there was a need to consider the fundamental reform of both systems from the viewpoint of how to reformulate the means of controlling forest practices. Because these systems had been added to and partially amended from time to time over the long term to suit various situations without changing the basic framework, the systems themselves had become very complex. Furthermore, because these systems are closely related to other systems such as subsidies and the tax system, when changing these systems, it is inevitably necessary to consider how to change other systems as well.

Though a detailed review of the Forest and Forestry Revitalization Plan commenced in February 2010, the plan was aimed to be implemented from FY2011. To implement the legal amendments and budgetary measures required before starting, i.e., before the summer of 2010, it was not only necessary to indicate the basic direction of the reforms but also necessary for these deliberations to be conducted under an extremely tight schedule. Therefore, there was insufficient time to discuss the plan and clarify the course of the fundamental changes, which means that the scope of such deliberations had to be limited. Thus, the issue of protection forests was not on the agenda, and the revision of the Forest Planning System became the subject of discussion.

Furthermore, because private property rights are strongly defended in Japan, it was not possible to establish rules with a legal force with regard to forestry practices. Unless it is possible to scientifically explain the maior influence of these regulations on public life and property, it will be difficult to include even upper limits on clear cutting in laws and regulations because they will not pass the hurdle of whether they infringe private property rights in the Cabinet Legislation Bureau. This means that the crucial problem of how to establish even minimum regulations regarding forestry practices was left untouched. The main issue here is not merely an internal problem of forest administration, but the nature of land use policies and private property rights in Japan.

Examining the content of the final reforms as well as the 2011 amendment to the Forest Act, the following points can be observed. Basically, the Forest Planning System has adopted this approach to make it easy to understand and achievable. The formulation periods for the Basic Plan for Forest and Forestry and the National Forest Plan were aligned and formulated in an integrated manner, clearly summarizing the state's vision. Furthermore, Municipal Forest Plans have been positioned as "master plans" for local forest management. The three-function zoning that had been uniform throughout the country was abolished and municipalities were able to freely perform zoning according to the state of the local area. A comprehensive plan for the development of the road networks that were required to implement intensive and effective forest practices was also positioned in the Municipal Forest Plan. However, with regard to regulations concerning forest practices, which are important to achieve zoning goals, no legal support could be provided in the Forest Law from the perspective of the protection of private property rights.

The content of the Forest Management Plan was also fundamentally revised. As mentioned previously, forest management plans were not implemented effectively. The new system stressed on the consolidation of forestry practices among small forest owners, and paid attention to geographic considerations for each forest compartment. Furthermore, attention was also paid to forest managers other than forest owners who can formulate plans, such as forest owners' associations and the logging contractor, who are entrusted by forest owners and are motivated to conduct forest management responsibly. Through this, it is hoped that motivated forest managers will supervise construction of road networks and efficiently manage the forest, thus resulting in the achievement of Forest Management Plans. Furthermore, forest owners will no longer receive subsidies if they are not a part of the Forest Management Plan; this is a powerful means of encouraging forest owners to participate in the Forest Management Plan.

Because the development of road networks is a pressing issue and, the identity of forest owners is ambiguous in some cases, procedures have been introduced that enable establishment of roads in cases where the landowner's identity is unknown. Measures to accurately ascertain the identity of forest owners have also been introduced, such as requiring new forest landowners to submit notification.

Thus, from the standpoint of increasing discretion on the municipal level, restricting unregulated forest practices, developing conditions to promote the consolidation of forest practices, and accurately ascertaining the identity of forest owners, it appears that some steps have been taken in the right direction; However, no major changes have been made to the manner in which rules are formulated.

To promote appropriate forest practices with regard to making recommendations to forest owners, Though there was a system for providing such recommendations, it was essentially dysfunctional. This was mainly because municipalities had no specialized administrative staff for forests and their forest administration implementation systems were weak. It was also difficult for them to implement recommendations as they were not prepared to handle the friction caused by dealing with forest owners. Furthermore, though zoning could be performed by local entities, Municipal Forest Plans had until then been mere forms and lacked substance because of the inadequate competence of the municipalities.

Considering the given facts, the amendments to the Forest Planning System represented a step forward in terms of the overall system. For these amendments to be truly achievable, it is important to strengthen not only municipalities but also local forest management systems. Without this step, these amendments to the system will be merely specified. What really poses a problem at this stage, however, is the second problem—how to restructure Japan's forest administration system.

4.2 Problems on Institutions, Human Resource Development

4.2.1 Forest Administration Institutions and Foresters

The most important aspect pertaining to developing a vision for forest management and for controlling forest practices is the development of supportive forest administration institutions.

The relationship between the central government and local government in terms of Japanese administrative institutions vastly differs from that in many European and North

American countries. In many of these countries, depending on the field of administration, there are numerous instances of local authorities and central government dividing responsibilities. Forest administration is one such area where the central government (or the state government) assumes responsibility and municipalities are generally not involved. Thus, forest administration can be handled by a specialized organization. In Japan, however, the central government, prefectural governments, and municipalities all perform these activities under a centralized structure. In forest administration, the chain of command begins with the Forestry Agency and leads to various prefectures and municipalities, with municipalities being positioned on the front line of forest policy implementation. The Forestry Agency and prefectures employ teams of staff specializing in forests (foresters) to develop specialized measures. However, because most municipalities are relatively small in size, they lack the capacity to employ foresters for forest administration.

In January 2011 a survey of municipalities in Hokkaido (Kakizawa and Kawanishi, 2011) revealed that, on average, 1.2 staff members were responsible for forest administration in each municipality. Whereas 48.5% of staff members had no more than 2 years of experience in forest administration, barely 7.5% of staff members had received any special education regarding forests or forestry. It is normal for general administrative staff to occasionally be involved in forest administration during the course of their regular assignment to different sections. Moreover, most municipalities faced major problems in implementing systems for the administration of forest management—73% of municipalities struggled to cope with specialized forest administration problems and 91% of municipalities indiceteel problems in dealing with forest planning matters, such as logging notification. Most municipalities are unable to take an initiative to promote forest management administration or support local forest management.

Therefore, the stark reality is that forest administration by municipalities is possible only if prefectures provide strong guidance. For example, the Municipal Forest Plans have only become a reality because prefectures created the templates into which the figures of various municipalities are then incorporated. Consequently, these plans have become mere forms with no substance and do not reflect local policies. In addition, logging notification and forest practice recommendation systems have simply fallen into a state of dysfunction.

At the forest resource creation stage, it would be possible to handle forest administration with these types of institutions and personnel. However, in the resource sustainable management and utilization stage, these institutions and personnel are completely incapable of handling such tasks. The development of administrative systems for sustainable forest management is a major challenge, which requires the placement of staff with the required degree of specialization in each local area.

4.2.2 The Result of Deliberations and the Direction of Reform

The method of dealing with the weak state of municipal forest administration was an important point of discussion concerning the development of the Forest and Forestry Revitalization Plan. Because the structure and fiscal scale of municipalities is generally small, they were aware that it was difficult to reformulate systems of forest administration by employing forest specialists and it was suggested that these responsibilities be returned to the prefectures. However, decentralization is the most important policy challenge in Japan, and the government's basic policy is to decentralize power to local authorities. Thus, it was difficult to form policies that opposed the flow of decentralization and passed the power to formulate plans and control logging and replanting activities into the hands of the prefectures. Furthermore, prefectures were losing staff under administrative reforms. In addition, many prefectural forest management bureaus maintained that even if such powers were returned to them, they would be unable to cope. Therefore, it became necessary to consider strengthening the existing forest administration systems of municipalities as a prerequisite to decentralization.

Thus, to ensure a degree of specialization in the existing system as a prerequisite, the fostering of personnel became an important point for consideration. As explained in detail below, because it was difficult to place specialized staff in municipalities, a system was created wherein prefectural extension staff members who had been providing guidance to owners of privately owned forests were retained to support municipalities. This is known as "the Japanese Forester System", and extension staff members with a certain level of experience were certified as Japanese Foresters after passing an examination. These Japanese Foresters supported municipal forest administration by formulating and managing Municipal Forest Plans for municipalities and certifying Forest Management Plans, etc.

To encourage small-scale and micro-size forest owners to participate in Forest Management Plans of a certain geographical size, calls were made for proposing efficient and structured forest management plans to such owners and obtain their consent. For this purpose, the fostering of Forestry Management Planners is being promoted.

In the forest policy till date, human resource development has not been a central tenet of policies; rather, policy development has focused on "hard" policies such as the provision of subsidies. It is believed that with these major changes in the forest policy, placing an emphasis on "soft" policies, human resource development that can form the foundation of the governance structure is now essential.

On the other hand, there are various problems surrounding the Japanese Forester System. Although Japanese Foresters support the formulation, implementation, and monitoring of Municipal Forest Plans implemented by municipalities and the certification of Forest Management Plans, they do not possess any real power because the authority for such actions belong to the authority. Therefore, both the power and position of Japanese Foresters remains unclear. What type of relationship should be formed between these trained foresters and the municipalities and how to support them to best use their expertise are questions that remain to be elucidated in the future.

Further, the fostering of personnel is not possible through training alone and will not produce much result in the short term. What is required of foresters is that they possess a broad range of knowledge and abilities—not only with respect to techniques concerning forest management but also the ability to communicate with owners, residents, and other stakeholders to formulate Municipal Forest Plans in related matters, such as zoning that is appropriate to the local area. Foresters are unable to learn everything regarding knowledge and techniques through training alone, and forming relationships with local people, including forest owners, requires time.

Therefore, it is necessary to consider the process of human resource development in itself as one of the processes involved in the formulation of governance. Currently, it is difficult for Japanese Foresters to cover all technical areas, and it will be necessary for them to form cooperative relationships with other specialized staff in prefectures and training and research institutions to perform their duties. Furthermore, to actualize support for municipalities, it is necessary to build a trusting relationship with municipal staff and collaborate with other specialized staff, as mentioned above, as well as promote the systematic development of forest administration at the municipal level. Regarding support for the certification of Forest Management Plans, it is necessary to consider their duties, including the proper implementation of plans and the improvement of plan content as well as their support for and collaboration with forest managers and owners concerning matters pertaining to plans. It is extremely important to foster specialists with expert knowledge regarding forest administration and forest management, who also possess the necessary leadership skills, and to build collaborative relationships with a diverse range of entities through such specialists.

To foster such specialists, in the short term, the intention is to retrain the extension staff of the prefecture. However, in the long term, it is important to consider the overall framework of the forest-related human resource development system, including university and technical college.

5. Conclusion: Policy Challenge to Reform in Forest Governance

Presently, forest resources in Japan are entering the sustainable management and utilization stage, and fundamental reforms are required in forest policies. As discussed above, reforms are proceeding under the Forest and Forestry Revitalization Plan, and the promotion of achievable reforms is a challenge that will persist for some time. However, there are limitations to the reforms under the Forest and Forestry Revitalization Plan. These reforms are not being implemented on the basis of a review of the overall system of forest policy in Japan, which includes protection forests, and do not present any clear goals regarding human resource development, which is the future forest administration structure. In this respect, the following points should be considered as long-term challenges that are part of a more fundamental approach to reform.

First, research that will form the basis of reform should be promoted. Scarce research exists on the formation of systems and policies regarding the forest, particularly concerning forest plans and protection forests that are the backbone of forest policy. Reviewing the development of policy formation and clarifying the causes of barriers to policy formation are essential for promoting future policy reform. Furthermore, national forests are not the subject of this discussion, and to endeavor to forge links with national forests, it is necessary to promote such research.

Second, cooperating with similar fields to formulate better policy development will be necessary. As a common belief, problems related to land use restrictions and the ability of municipalities to formulate and implement the policies are shared by other natural resource management fields such as wildlife management. It is necessary to define such shared problems and the direction of reforms and to cooperate to increase the influence and advocate policy change.

Third, Japanese Foresters and forest management planners should be fostered. The success or failure of not only the Forest and Forestry Revitalization Plan but also of the establishment of sustainable forest management itself depends on systems and policies as well as on how well we can foster specialists with roots in the community, specialized knowledge, and passion can be fostered.

There are foresters at both the prefectural and the municipal level who are aware of the serious problems regarding the current situation and the fact that real forest management is impossible when bound by a centralized system. These foresters are engaged in various efforts aimed at realizing better forest management. They engage in networking and make efforts to present detailed concepts regarding the next steps that should be taken. Supporting these people to foster and promote future reform is important. The aim should be to create positions and systems that would enable specialized staff members to use their expert knowledge, actively engage in their work with a sense of satisfaction, and contribute to sustainable forest management and the revitalization of forestry.

Fundamental reform with firm roots in the workplace will be possible only after these foresters have gained a deeper awareness of the problems inherent in the state of the current system through their training and work. As for forest policies, the degree of power that foresters have on site should determine the direction of future reform.

This chapter will summarize the relationship between the direction of the abovementioned reform and its relationship with forest governance. As has been reiterated, the major problem in the development of forest policy and management in Japan is the weakness of the municipal forest administration structure and uncommonness of specialized staff members. Furthermore, though the focus of this discussion has been limited to forest administration, another problem, which this chapter has been unable to cover, is the inadequate initiation of business entities such as the forest owners' associations engaged in organizing forest owners and conducting forest practices. Thus, the fundamental problem facing the formulation of governance is the revitalization of specialists and institutions engaged in the forest administration and management.

Though existing policies regarding human resource development are being developed, it is important for such personnel to form networks and help and encourage each other to become forest administrators and technicians who will play a central role in local forest management. It is necessary to form a foundation for local-based forest management through collaboration and by allowing people to excel in their areas of expertise, These foresters can cover a vast range of activities such as the formulation of plans and regulations regarding forest practices, detailed matters related to the maintenance and utilization of forests such as the development of road networks and the planning and implementation of activities, and promotion and guidance. It can be said that the formulation of collaborative relationships with foresters, who are active on the front lines—everywhere from prefectures and municipalities to the business entities—is the first step toward the formulation of governance.

To present a vision for the creation of local forests related to Municipal Forest Plan formulation, the challenge will be threefold—first, to achieve collaboration among local residents and people related to such forests in order to create a master plan as a consensus of the local community; second, to achieve collaboration with forest owners to promote collective forest management; and third, to build relationships with business entities and front-line technicians for promoting the development of appropriate forest management. Foresters are expected to coordinate the formation of these relationships, and this relationship building process will probably be the process used for building forest governance in Japan. However, the goals of forest management and the socioeconomic conditions surrounding such forests are diverse, and there is no formula for the division of roles. Thus, it is a case of accumulating detailed case studies and determining such things as foresters learn from each other.

As stated before, research conducted on forest governance in Japan has been centered on the analysis of outstanding examples, with a focus on social movements. In most areas, movements for the formulation of governance, as seen in such examples, cannot be found, making it difficult to apply such related analysis. Thus, fostering foresters and revitalizing local forest management and forestry centered on such foresters can be considered a new process in governance formulation. Research is expected to create a clear vision for governance formulation in cooperation with foresters while learning from the efforts of each region.

It is also important to highlight the relationship of forest administration reform and governance with the property rights of forest owners. As mentioned previously, because private property rights in Japan are strongly defended, at present, it is nearly impossible at present to establish regulations for environmental conservation such as those restricting clear cutting areas through laws and ordinances. In such situations, building a consensus among the stakeholders in the concerned area, such as forest owners and forest owners' associations, incorporating rules into Municipal Forest Plans, providing education, and ensuring that the revalant rules are followed is essential. In situations where noneconomic functions are not adequately secured through regulatory methods, it is necessary for communities as a whole to create and implement rules. In this respect, promoting forest governance is essential for promoting sustainable forest management and foresters should play a central role in governance formulation. While clarifying the essential factors in promoting local sustainable forest management from a specialist's viewpoint, foresters will also be expected to formulate forest management systems in collaboration with forest owners, business entities, and the local community. Through such efforts, forests will be recognized as local public goods, which will in turn form the foundation for the promoting forest governance formulation.

References

Inoue, M. 2004. *Komonzu no Shiso wo Motomete* [*In Search of Commons*]. Tokyo: Iwanami Shoten, Publisher.

Kakizawa, H. 2000. *Ekoshisutemu Manejimento* [*Ecosystem Management*]. Tokyo: Tsukiji Shokan Publishing.

Kakizawa, H. 2002. "Chiiki Kankyo Seisaku Keisei no tame ni Motomerareru Mono: Chiiki Kankyo Gabanansu no Shiten kara [Towards Community Bases Ecosystem Management: From the Viewpoint of Regional Environmental Governance]". *The Toshi Mondai* [*Municipal Problems*], 93(10): 15–28.

Kakizawa, H., K. Yamamoto, and K. Saito 2006. "History of Study on Research of Nature Conservation and Public Participation". The Japanese Forest Economic Society (ed.) *History of the Japanese Forest Economic Society*. Tokyo: J-FIC.

Kakizawa, H. 2007. "Shinrin Gabanansu no Kochiku ni Mukete: Kyodo no Chikara de Yori Yoi Shinrin Dzukuri wo" [Towards Development of Forest Governance: Seeking for Better Forest Management through Collaporation]. *Sanrin*, 1478: 2–9.

Kakizawa, H., and Kawanishi, H. 2011. "Current State of Municipal Forestry Administration: Result of Questionnaire Survey on Municipalities in Hokkaido", *Forest Economy* 64(9): 1–14.

Kitao, K. 2009. *Shinrin Shakai Dezain-gaku Josetsu* [*Designing Forest and Society*]. Tokyo: J-FIC.

Nishio, T. (ed.) 2008. *Bunken Kyosei Shakai no Shinrin Gabanansu: Chisan Chisho no Susume* [*Forest Governance under Decentralized and Harmonious Regime*]. Tokyo: Fuko-sha Publishing.

Ushiyama, T. (2001) *Forest Policy Reecommendations from Forest Volunteers, Commons*.

Yamamoto, S. (ed.) 2003. *Shinrin Boranthiaron* [*Forest Volunteers*]. Tokyo: J-FIC.

Yorimitsu, R. 1984. *Nihon no Shinrin / Midori Shigen* [*Forest and Green Resources of Japan*]. Tokyo: Toyo Keizai Publishing.

Yorimitsu, R. 1999. *Mori to Kankyo no Seiki: Jumin Sanka Sisutemu wo Kangaeru* [*Forest and Environmental Management based on Public Participation System*], Nihon Keizai Hyouronsha.

Chapter 2
Endogenous Development and Collaborative Governance in Japanese Mountain Villages

Hironori Okuda and Makoto Inoue

1. Challenges for the Promotion of Mountain Villages

Promotion of mountain villages encounters several problems such as fragility of social organizations and lack of public facilities for daily subsistence and industrial infrastructure. Although various measures, whether effective or not, have been implemented to address this problem, there are no specific measures that address the "fragility of social organizations". To strengthen and support this fragility of social organizations, it is necessary to promote social change to form new identities in areas that have, thus far, held little attraction for people to live in.

In Sawauchi-mura, Iwate Prefecture in North Japan, local residents and networks of volunteer groups are effecting social change and leading an endogenous development of the local community by revitalizing support activities that protect "the lives of the elderly and the physically challenged" and by promoting activities for exchange, such as *Furusato Takkyubin* [delivery services from Sawauchi-mura], between the mountain village and urban areas (Okuda et al., 2001).

In Tsukimoushi-machi in Tono City, Iwate Prefecture, shiitake mushrooms comprises an important industry. To secure the bed logs required for their production, residents of Tsukimoushi-machi approached the manager of national forests, the prefecture, and the National Forest Wood Production Cooperative (hereafter, *Kokuseikyo*), and with their assistance and guidance, started growing *Quercus serrata trees* in the national forests, thus leading to social change and endogenous development in the local society (Okuda et al., 1999).

Similarly, based on suggestions from new residents from large cities, the residents of Tsukimoushi-machi approached the manager of national forests and with his cooperation and guidance, established the Hayachine Forest Commons Association. Using the donations collected from people from outside the area who gather mountain vegetables and mushrooms from the forests, the group raised money to implement environmental conservation measures for the forest commons established within the national forests. These measures included efforts such as setting up signs for improving visitor behavior, maintaining forest roads, collecting rubbish, establishing a forest market, and conducting patrols. Thus, the network of local residents in Tsukimoushi-machi—a combination of people born and raised in the area and new residents from large cities—has effected social change by implementing environmental conservation efforts for the forest commons and is leading the endogenous development in the local society (Okuda et al., 2003).

In Kaneyama-machi, Yamagata Prefecture, a network comprising residents, local carpenters, architects, sawmills, forest associations, forest owners, and the town office was formed with the common desire to preserve and nurture the beautiful townscape. They aim to effect social change wherein local people opt for Kaneyama-style houses that are built by Kaneyama carpenters. Thus, this network is leading the endogenous development in the local society (Okuda, et al., 2004).

The following chapter utilizes collaborative governance theory to consider the conditions required to engender endogenous development in mountain villages.

2. Endogenous Development Theory and Collaborative Governance Theory

2.1 Endogenous Development Theory

Grobally, newly developing countries are pursuing the same type of Western economic development achieved by countries such as Japan and South Korea, and the countries which have achieved such economic development have begun scrambling for resources. The quest for simple extrinsic economic development is bound to exacerbate the deterioration of the global environment and depletion of resources, and unless some action is taken, our nation and communities will be forced into a situation where they will have to pursue independent, polyphyletic, and endogenous economic development (Tsurumi, 1996). In the report "What Now?" submitted by The Dag Hammarskjöld Foundation at the 7th Special Session of the United Nations General Assembly held in 1975, two forms of development—endogenous development and self-reliance—were proposed under the title "Towards Another Development". The report stated that "if development is the development of man as an individual and as a social being aiming at his liberation and at his fulfillment, it cannot but stem from the inner core of each society".

In the article "Deployment of the Endogenous Development Theory", Tsurumi (1996) stated that "the processes to achieve endogenous development and lifestyles that result in the achievement of common goals of people in an area are created autonomously by those people using local resources such as water, wood, and minerals. At the same time, they are autonomously created by new skills in which traditional or local skills are integrated with outside knowledge, skills, and systems in accordance with the particular natural environment, cultural heritage, and history of the area. This process of social change by which the goal is achieved is endogenous development".

Tamanoi (1977) defined regionalism as a sentiment in which the residents of a certain area have a sense of unity regarding the local cultural environment and community, and seek political and economic self-sustainability and cultural independence. Moreover, the residents in such an area themselves determine their lifestyle and form of local development. However, Tsurumi (1996) stated that "what is common to various definitions of community is, 'common ties with a limited place'. Here, 'limited place' is a permanent dwelling place where residents live for a long time and 'common ties' are common values, goals, and concepts shared by the local residents".

In light of the abovementioned debate, the process of achieving endogenous development can be defined as, "a process of social change with the common goal of protecting and fostering that which is important to local residents who constitute various networks of people from a wide area, including cities and other municipalities outside the village, and align with outside knowledge, skills and systems". To revitalize communities and form new identities in areas that have hitherto held little attraction for outsiders, there are high hopes for endogenous development that makes local residents' activities more vibrant.

2.2 Collaborative Governance Theory

Inoue (2009) proposed "collaborative governance" as a "glocal (global + local)" and sustainable strategy for managing and using "local commons", which involves diverse organizations and persons—such as governments, citizens, NPOs, companies, and scholars—participating across borders and collaborating in the governance and independent and voluntary management of the commons. He also states that assuming that the activities are collaborative works by the local residents for protecting, developing, and utilizing the local environment and resources, the conditions for realizing "collaborative governance" are as follows: the "principle of involvement", which allows inclusion of nonlocal persons, opinions in the management and usage of the environment and resources; the "commitment principle", which allows nonlocal persons to be involved in verifying plans and designs of activities depending on the degree of their involvement; and "open-minded localism", which assumes that the main manager and user of the environment are the local people and organizations, and they allow utilizing by outside entities.

Inoue (2009) defined "collaborative governance" as an elementary idea for developing the commons theory into environmental governance for formulating resource (environmental) policies based on the actual states of the region by expanding the target of discussion from small local communities to more general themes on governing bodies, consensus building, and public participation, while abiding by localism. He also states that collaborative governance is a system of resource management in which various entities (stakeholders) such as the central government, town office, residents, enterprises, NGPs/NPOs, and global citizens collaborate.

3. The Network for Supporting Lives in Sawauchi-mura

3.1 Present State of Sawauchi-mura

On November 1, 2005, Sawauchi-mura was amalgamated with Yuda-machi to form Nishiwaga-machi. Sawauchi-mura is bordered on the west by Misato-machi in Daisen City, Akita Prefecture, on the north by Shizukuishi-machi in Iwate Prefecture, on the east by Hanamaki City and Kitakami City, and on the south by Yuda-machi. It is located in a basin in the Ou Mountains. The Waga River flows from north to south through this basin. The Morioka-Yokote Prefectural Highway runs through the center of the basin and spans approximately 60 km to Morioka City, Kitakami City, or Hanamaki City, and 50 km to Yokote City. It is possible to reach any of these destinations by car in approximately an hour.

The basic tenet of Sawauchi-mura is "respect for life". It was one of the first municipalities in Japan in 1960 that provided free medical care to people 65 years or older, and has continued with policies that emphasize health and welfare. Currently, it pays for all medical costs of persons aged 60 years and older (since FY1961) and of infants who are less than 12 months old. On the social welfare front, Sawauchi-mura has established the "Katakuri no Sono" welfare center for elderly residents, and is currently working with the Sawauchi-mura Social Welfare Association to provide employment for people with handicap. Thus, Sawauchi-mura is developing its own richly local policy of "warm living in a snowy region" and it is renowned throughout Japan as a "health and welfare village".

The total population of mountain villages in Japan fell dramatically to 59% of its former population between 1960 and 2000. Sawauchi-mura was no exception, though the decrease was not as drastic despite the fact that life in winter is difficult with more than 2 m of snow. The population dropped from 6,451 in 1960 to 3,974 in 2000—62% of the former count.

3.2 Survey Location and Method

In June 1997, a questionnaire survey was conducted in households in the Ryozawa and Shichinai settlements in Sawauchi-mura to explore the living conditions in the colonies and the relationships between brothers and sisters and parents and children. A total of 28 households, 14 of the 17 households in Ryozawa and 14 of the 16 households in Shichinai, agreed to be interviewed. Another questionnaire survey was also conducted in June 1999 to examine the utilization of *Furusato Takkyubin* by members and their ties to Sawauchi-mura. The questionnaire was included with *Furusato Takkyubin*, and of the 330 questionnaires distributed, 129 responses were received (a response rate of 39%).

On October 7, 2010, to learn about "*Furusato Takkyubin*" and volunteer activities, a questionnaire survey was performed with the staff of the community work center, a gathering place for people with handicap (the work center was renamed the Work Station Yuda/Sawauchi after the amalgamation of Sawauchi-mura and Yuda-machi in April 2006).

3.3 Present State of Surveyed Colonies

The 28 households involved in the survey represent a total of 126 people. This equates to 4.5 people per household, which is higher than the average number of people per household in Sawauchi-mura as a whole. The middle-aged segment ranging from 35 to 64 years of age accounted for 44% of the total number of people surveyed, whereas people aged 65 years and older accounted for 33%. Young people in the 20–34 age group—the ones who will support the colony in the future—accounted for a mere 7% of the population. Furthermore, on examining the age structure of each household, more than half the number of households (17) were noted to have a member belonging to the 34 years or younger age group. Of these, there is a high possibility that children aged 19 or younger in 10 of these households will leave to attend other schools, meaning that only seven households—a mere 25% of the total—had young people in the 20–34 age group who will support the colony in the future. The remaining 11 households have no members in the 34 years or younger age group, and among these households, there are three households with only one elderly person. However, parents in three of these 11 households are hopeful that their children will return, whereas those in three households were uncertain, and in one household, the parents deemed thought that their children would probably not return. The remaining four households have working-age successors in the village; however, those individuals are still unmarried (see Tables 2–1 and 2–2).

3.4 Current State of Interpersonal Ties within Ryozawa

There are strong interpersonal ties based on territorial bonding among the people living in Ryozawa. Nowadays, though it is rare to see the tradition of *yoe* (or *yui*) of neighbors helping each other with farm work, people of the settlement plant flowers as part of the "Ryozawa Marugoto Koen Jigyo", which is one of the environmental development activities of the colony. Furthermore, there are numerous opportunities for families living in the colony to get together to share their home cooking and mingle at events such as *Sanaburi* (a party given to honor those who have planted rice fields), New Year, and during the cherry blossoms or firefly viewing parties. The Seikatsu Kaizen Group (Eating Habit Improvement

Table 2–1. Age Structure of Ryozawa and Shichinai Colony Inhabitants

colony name	sex\age	–19	–34	–49	–64	65–	Total
Ryozawa	Male	4	5	8	3	13	33
	Female	5	1	7	8	11	32
	Total	9	6	15	11	24	65
Shichinai	Male	8	2	10	5	7	32
	Female	4	1	7	7	10	29
	Total	12	3	17	12	17	61
total	Male	12	7	18	8	20	65
	Female	9	2	14	15	21	61
	Total	21	9	32	23	41	126

Source: Questionnaire findings.

Table 2–2. Structure of the Households Living in Ryozawa and Shichinai colony

Structure\colony name	Ryozawa	Shichinai	total
1. (elderly +) middle + Early childhood	3	7	10
2. (elderly +) middle + Youth (+ Early childhood)	5	2	7
3. (elderly +) middle	4	4	8
4. only elderly	2	1	3
total	14	14	28

Source: Questionnaire findings.
Notes: "Elderly" includes those aged 65 years old or older, "middle" 35–64 years, "youth" includes those aged 20–34 years and "early childhood" includes who are less than 19 years old.

Group) run by rural women plays an important role in creating interpersonal ties through territorial bonding in the colonies. Since 1950, such groups were created under the guidance of the Ministry of Agriculture, Forestry and Fisheries. Though similar groups that were created under the guidance of the authorities have become a shell of their former selves, this group is still a vibrant organization in this colony. The Eating Habit Improvement Group, which works in league with Nagaseno—another colony—has 16–17 members and is mainly active in the winter during the non-farming season when they indulge in other activities such as making *kiriboshi daikon* (strips of dried Japanese radish) and *yuki natto* (soybeans fermented in the snow). The Eating Habit Improvement Group has continued to thrive because local families understand that it brings the community together. There is no pressure to be particularly involved, newcomers can easily enroll, and when members need to take a break for personal reasons, the other members makes it easier for them to rejoin the group.

3.5 Activities of Volunteer Groups

In Sawauchi-mura, various volunteer groups are active because of the interpersonal ties wrought by the above-described territorial bonding within the colony. There are 34 volunteer groups registered with the Sawauchi-mura social welfare association: 15 senior citizens groups; 9 "Snowbusters", members of which remove snow from the homes of the elderly, and 10 other groups. In addition, there are "House helpers" who help with simple repairs in homes of the elderly, *ominaeshi* who manufacture and sell small chests of drawers and boxes and donate the proceeds to Sawauchi-mura social welfare association; and "rindo" who assist in the community work center and nursing homes, among others.

Snowbusters, formed in December 1993, in Sawauchi-mura, comprises of 10 units, each of which is assigned to one of the village's 10 colonies to remove snow from homes of the elderly in teams. As of 2006, snowbusters had 126 regular members—mostly youth groups—and 121 members who were junior high school and high school students. The group performs this work in 3 joint events every year and when necessary, on Sundays. However, they are not forced to push themselves too hard and are certainly not asked to climb on roofs. This movement has spurred the establishment of Snowbusters chapters throughout Iwate Prefecture. Recently, individuals and volunteer groups from other municipalities visited Sawauchi-mura to support the activities of the Snowbusters. Thus, the network of people who help to support the lives of the elderly in Sawauchi-mura during the snowy winter season has expanded outside the village and reached urban areas.

3.6 *Furusato Takkyubin*

Furusato Takkyubin is a unique service in Sawauchi-mura. Launched as a special project by the community work center for people with handicap, it is part of a village planning movement for social welfare that included disabled and elderly people as participants since 1985.

Products harvested or processed within the village are sent four times a year to subscribers of *Furusato Takkyubin*, who are called "*Furusato* members". In addition to products that are produced or processed directly by the community work center, the service includes handicrafts and the harvesting and processing of mountain vegetables by groups such as the Eating Habit Improvement Group and Senior Citizens' Club. Through such territorial bonds, *Furusato Takkyubin* helps to support interpersonal ties.

Initially, there were fewer than 100 *Furusato* members, but their numbers increased every year, reaching a peak in 1993 with 537 members. However, after 1993 the annual membership fee was raised from 12,000 JPY to 20,000 JPY and membership declined to 230 people in 2006. Currently, there are 40 workers at the community work center and this number is considered appropriate for the smooth functioning of various activities.

According to a questionnaire completed by *Furusato* members in June 1999, the vast majority of *Furusato* members (82%) reside in the Kanto (East Japan) region, with 8% living in the Chubu (Central Japan) region or further west and 10% living in the Tohoku / Hokkaido

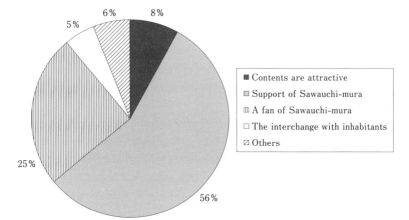

Figure 2–1. The reason that contracts "*Furusato Takkyubin*"
Source: Questionnaire findings.

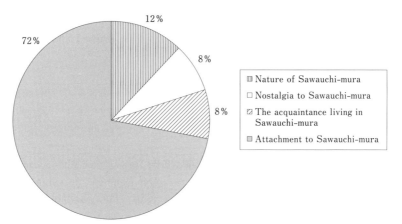

Figure 2–2. What joins you and Sawauchi-mura together?
Source: Questionnaire findings.

area. Many members who live in distant areas learned of the service either by visiting Sawauchi-mura, or by word of mouth. Results of the questionnaire revealed that the biggest reason cited for using the service was to "support Sawauchi-mura" at 56%, followed by "I'm a fan of the village" at 25%. Thus, it was apparent that almost all respondents used the service because it was a "project of Sawauchi-mura" (see Figure 2–1). Figure 2–2 shows that 72% of the members said that the connection between "*Furusato* members" and Sawauchi-mura was a common desire to support the efforts of Sawauchi-mura, 12% stated that it was "the natural environment of Sawauchi-mura", and 8% said that it was because

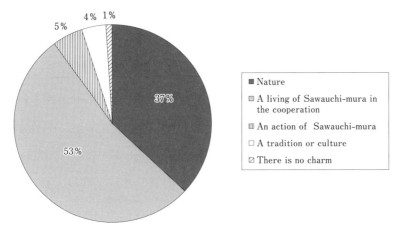

Figure 2–3. What is the charm of Sawauchi-mura?
Source: Questionnaire findings.

of a sense of nostalgia toward their home town or toward human relationships. Figure 2–3 shows a positive image of Sawauchi-mura in that 53% respondents liked the way the residents live their lives helping each other, 5% appreciated that the municipal government was making a concerted effort toward health and welfare, and nearly 60% of respondents had a positive image of Sawauchi-mura as a "people-friendly village". The above findings show that *Furusato Takkyubin* is supported by a broad network of people from cities and other outside areas who want to support Sawauchi-mura in its efforts to address health and welfare and implement its own unique and richly local form of village planning.

3.7 Endogenous Development and the Network for Supporting Lives in Sawauchi-mura

The municipal government of Sawauchi-mura adopts its own uniquely local approach toward issues (improvement of medical care, etc.) and is involved in a variety of activities based on territorial bonding and interpersonal ties to develop well-founded and lasting policies. Furthermore, a network of people from a wide area, including cities and other municipalities outside the village (see Figure 2–4), has been forming to support these activities (Okuda, 2001).

The stronger and wider such networks established in various forms by people from within and outside the village—become, the more vibrant will be the activities of the people living in these colonies. There are firmly established networks in the region to help people with handicap. Such volunteer groups and networks with city residents supporting the elderly and people with handicap have led to endogenous development. These networks are connected by a strong common desire of residents to protect life in the region.

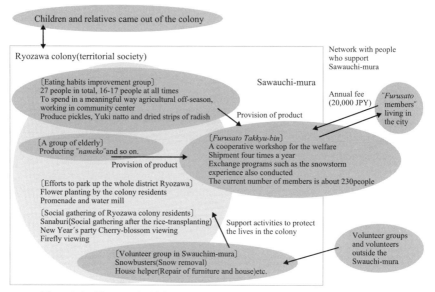

Figure 2-4. The network which supports a living in the mountain colony of Sawauchi-mura

Such activities are autonomouly conducted by the local residents and outside volunteer groups. *Furusato* members are only involved in supporting the activities and not in the planning and designing functions. Because of the small size of the local volunteer group, the support of outside volunteer groups is indispensable for protecting the "life of the elderly". Thus, the project for protecting the life of the people with handicaps could not have been realized without the support of *Furusato* members.

4. Network for Growing *Quercus serrata trees* in Tsukimoushi-machi

4.1 *Satoyama* has Supported Life in Mountain Villages in Tsukimoushi-machi

Kitao (2001) defines *satoyama* as "forests located near farms and mountain villages that have been used by farmers to make a living". In the Edo Period, such *satoyama* were indispensable for people in mountain villages as places to collect manure, forage for oxen and horses, and material and fuel to use at home and in lodges. They have wisely managed such areas by treating them as *Iriai* forests and setting limits on their exploitation. These forest commons were an important element of rural life and a symbol of the solidarity of village communities. *Iriai* forests resembled the rigid "local commons" for which certain rules and rights regarding management and exploitation (when, in which areas, with which tools, and to what extent to gather resources) were imposed within the community (Inoue, 2003).

However, after the 1868 Meiji Restoration, the government ignored such traditions and attempts to secure tax income; it tried to impose the concept of private land ownership by selling or transferring land to individuals. The Meiji government would not allow systems such as *Iriai* forests where the land was held in common; rather, they ensured that the land belonged to either private owners or public authority. As wood resources ran short and the purpose of forest exploitation changed from fuel production to material production, *Iriai* forests were increasingly being enclosed to convert them into national forests and to place forest resources under state control. Thus, by taking advantage of local farmers' lack of awareness regarding land ownership, the government rapidly redefined land ownership with regard to *Iriai* forests and converted them into national or private forests. The exploitation of *Iriai* forests by local farmers was gradually made impossible because of artificial forestation aimed at the development of forest resources. Consequently, during the Meiji Period, successive conflicts occurred between owners of large forests and those who sought to maintain the right of access to *Iriai* forests. Some of these cases continued until after World War II, such as the Kotsunagi Affair of Ichinohe-cho in Iwate Prefecture.

Iriai forests, which remained village-owned forests, became community-owned forests with the introduction of the municipality system in 1889 and the subsequent merger of different villages. Many such forests became the property of local authorities because of the regularization and integration process of community-owned forests that began in 1910. Later, with the Act for the Promotion of Merger of Towns and Villages in 1953, 40% of forests owned by old-system municipalities were handed down to the new municipalities; however, in regions where people did not want such succession of forest ownership, these forests once again became community-owned forests. The same 40% of forests then became property ward forests as determined under the Local Autonomy Act. Until 1955, it is estimated that approximately 2.2 million ha of such *Iriai* forests existed throughout Japan. The Act Concerning Revision of Rights for Common-Forest Use was authorized in 1966, following which *Iriai* forests have been allotted to rights holders, and Production Forest Associations have been established by entities with the right to use *Iriai* forests. During the 42-year period until 2008, 6,636 *Iriai* forests encompassing 574,175 ha were regularized (see Figure 2–5).

However, as the forest owners and members of the Production Forest Associations aged and the Japanese forestry that was stagnated by wood prices declined, interest in forestry decreased and they began to neglect forest maintenance.

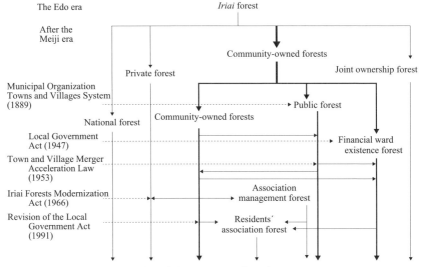

Figure 2–5. The change of *Iriai* forest

4.2 Survey Location and Method

Tono-City in Iwate Prefecture is known for "Tono Story", which was written by Kunio Yanagida, a folklorist. This city is located in the Tono Basin, the largest basin in the Kitakami Highland in southern Iwate Prefecture, and has prospered since ancient times as the castle town of the Tono Nambu Family. It is approximately 60 km southwest of Morioka city via route 396 and is strategically located position for transportation between coastal and inland areas.

Tsukimoushi-machi, the target of the survey, is located in the northernmost part of Tono-City at the foot of the Hayachine Mountains. The population of Tsukimoushi-machi peaked in 1955 with 3,417 people, then it started to decrease rapidly because of the period of high economic growth. In 2010, the population stood at 1,379 people—approximately half of the 1955 population. The number of households peaked at 586 in 1960 and fell to 428 by 2010 (see Figure 2–6).

The survey was completed with each household in October 2010 to gain an overview of the colonies of Oide and Onodaira in Tsukimoushi-machi. In June 2005 and June 2010, polls were conducted with people living in the colonies and working at *Kokuseikyo* (a cooperative supporting their lives by harvesting forests and selling them) regarding the growing status of *Quercus serrata trees* in the National Forest.

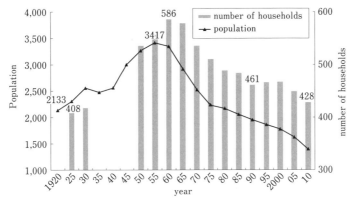

Figure 2–6. Trend in population and number of households in Tsukimoushi-machi
Source: Population Census Data (The Ministry of Internal Affairs and Communications)

4.3 Life in Tsukimoushi-machi

The colony surveys of June 2005 showed that households comprising solely of people aged 60 years or above accounted for nine of the total 28 households (32%), confirming that aging in these colonies has increased. Furthermore, 10 of the 28 households from the two colonies that were surveyed had people aged 39 years or under, of which two households were new residents (e.g., people from outside the surveyed colonies who had moved from urban areas). Eight households had children aged 15 years or younger who attended primary or junior high school, of which one household consisted of new residents. Nine households comprised solely of people aged 60 years or above (see Table 2–3).

These numbers indicate a difficult situation for the colonies' future survival. In addition, examining the age structure of the 88 people in the surveyed colonies, revealed that 39 were aged 60 years or above (the largest group at 44%) and nine were in the 20–39 age group, accounting for 10% of the total residents. When examining the 20–59 age group (the group of 35 people who will support the colony in the future), it can be noted that there were 11 returning migrants (people from the surveyed colonies who returned after temporarily leaving) and eight new residents. This makes a total of 19 residents, indicating that the proportion of returning migrants and new residents in the 20–59 age group was more than half of the total population (54%) (see Figure 2–7).

4.4 Hayachine Cooperative for Growing Shiitake Mushrooms

Industries that support the lives of people living in mountain colonies in Tsukimoushi-machi included agriculture and forestry. However, because of the difficult production environment, many households were engaged in the cultivation of shiitake mushrooms to supplement their income. In 1985, 10 of the 28 households within the colonies established the Hayachine

Table 2-3. Structure of the households living in the the mountain colony of Tono-City

colony	Households consisting entirely of members over the age of 40	Households consisting entirely of members over the age of 60	Households with one or more members of 39 years or less	with children	total
Oide	7	5	2[2]	1[1]	9[2](1)
Oonodaira	11[2]	4[1]	8	7	19[2](7)
total	18[2]	9[1]	10[2]	8[1]	28[4](8)

Notes: The number in parenthesis ( ) denote the number of the households with children under the age of 18.
The number in brackets [ ] denote the number of the households moving in from outside between 2005–1995.
Source: The result of interviews conducted in June, 2005.

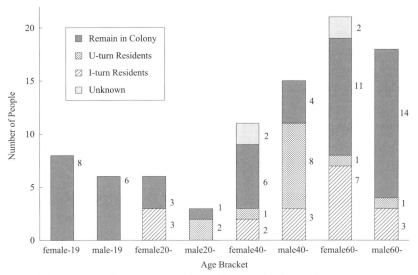

Figure 2–7. The number for every move experience of the mountain colony residents

Cooperative for growing shiitake mushrooms and process-sharing forest production for producing wood for cultivating shiitake mushrooms. This cooperative controls a drying facility for shiitake mushrooms that was established with the government's assistance and has a 65-year contract with the government for leasing national forest land to produce wood (oak, *Quercus serrata*) for cultivating shiitake mushrooms. The cooperative will make the first harvest 25 years after plantation; it plans to make subsequent harvests twice every twenty years and promote germination of the harvested trees. Thus far, 4,000 seedlings per ha have been planted, and the process-sharing proportion is 30% for the government and 70% for producers. Plantation has been implemented systematically since 1986, and in 2009, the planted area spanned 157 ha (see Figure 2–8).

Thus, residents in Tsukimoushi-machi borrow national forest land from the government to create a system that allows them to utilize the land by collaboration.

4.5 Endogenous Development and the Network for Growing *Quercus serrata* trees in Tsukimoushi-machi

In Tsukimoushi-machi, Tono-City, residents grow *Quercus serrata* trees in national forests to produce beds for growing shiitake mushrooms, which are the main product of the region, by obtaining assistance and advice from the prefectural government, the manager of national forests, and *Kokuseikyo*. In Tsukimoushi-machi, the network of the residents (see Figure

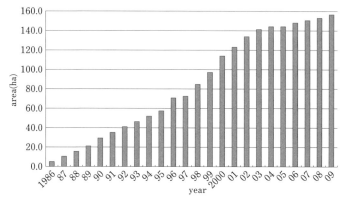

Figure 2–8. Transition of the forest area borrowed to produce the wood for cultivating Shiitake mushrooms

2–9) is bringing about social change by growing *Quercus serrata* trees in national forests, leading the endogenous development in the local society (Okuda et al., 1999).

The network is connected by the residents' strong desire of wanting to reside in the region. Growing *Quercus serrata* trees in national forests is an act toward restoring the former commons of *Iriai* forest, which have become irrelevant for the residents today, and using it to produce beds for growing shiitake mushrooms.

Though the residents have played a vital role in planning and designing the *Quercus serrata* forest commons, supports from outside regions has also played an important part, such as advice from the manager of national forests about where to plant the trees and grants from the prefectural government and *Kokuseikyo*.

5. The Network for Conserving the Environment of the Forest Commons

5.1 Survey Location and Method

The survey discussed in Section 4 was conducted in the colonies of Oide and Onodaira in Tsukimoushi-machi, Tono-City. In March 2001, the staff of the Tono Branch of the Iwate South Forest Management Office was interviewed regarding the state of management and utilization of national forests in Tono-City. In July 2001, both local and new residents were surveyed concerning the state of national forest management activities in common use being proactively undertaken in the forest commons.

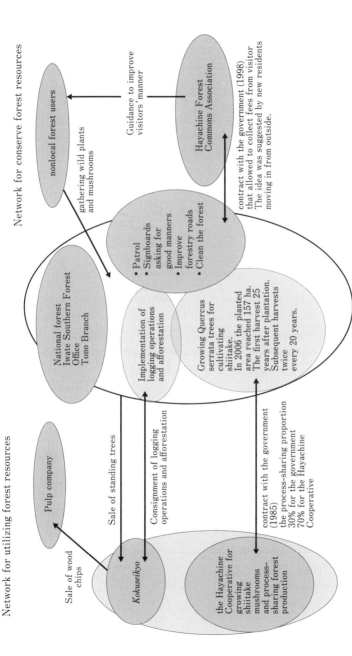

Figure 2-9. The network to conserve and utilizing forest resources in National forest of residents living in the mountain colony of Tono-City

5.2 Hayachine Forest Commons Association

Originally, the Oide Forest Commons Association had an agreement regarding the national forest at the foot of the Hayachine Mountains in Tono-City with the manager of the national forest of the Mogami district, which granted them the right to harvest mountain vegetables and mushrooms in the forest in exchange for bearing the responsibility for everyday conservation and protection, such as preventing forest fires. However, in recent years, the number of people coming to the forest from outside the area to harvest mountain vegetables and mushrooms has increased. This has caused overexploitation of mountain vegetables and mushrooms and illegal dumping of garbage, which has increased to such an extent that it can no longer be ignored. Several measures were implemented to deal with this issue, such as putting up signs, distributing pamphlets, and conducting patrols; however, limited financial resources were available, which caused difficulty in improving the effectiveness of such measures. At that time, the community came together in the hope of offering effective guidance to people outside their area. They wanted to inform people about the fact that these management activities have been supported by the manager of national forests and about measures for securing sources of funding, the lack of which was restricting such activities. At the suggestion of one of the new residents, the community decided to request permission from the manager of national forests to charge nonlocal forest users. Consequently, the Oide Forest Commons Association and the Koide Forest Commons Association amalgamated to form the Hayachine Forest Commons Association. In March 1998, an agreement was signed with the director of the Tono Branch of the Iwate South Forest Management Department. In FY1999, the charges collected from nonlocal forest users enabled signs for brush cutting (3), garbage collection (2), forest markets (3), and forest patrols (73) with the aim of improving behavior as well as forest road maintenance.

5.3 Endogenous Development and the Network for Forest Conserving the Environment of the Forest Commons

In Tsukimoushi-machi, the network of local residents (both those born in the region and those who moved in) (see Figure 2–9) has brought about social change related to conserving the environment of the forest commons and led to endogenous development (Okuda et al., 2003). What connects the network is the common desire of the residents who want to protect the forest commons from the dumping of garbage and excessive collection of mountain vegetables as well as the desire to live in a "beautiful forest".

Regarding the area's relation with outside regions, it was in fact the new residents who suggested the idea of conserving the environment of the forest commons, which was essential for the formation of the project. Thus, new residents can play a major role in planning and designing these measures. Requesting permission form the manager of national forests to collect harvesting charges is another example of the importance of relation with outside parties. However, of the 54 involved local residents approximately 12 or 13 usually participate further, and although 60,000–70,000 JPY was collected in 1999, this figure dropped to approximately 30,000 JPY in 2000. Thus, the project has been inactive of late, and further incentives are needed to reactivate the project.

6. The Network for Preserving and Nurturing a Beautiful Townscape in Kaneyama-machi

6.1 Survey Location

Kaneyama-machi is an old post town located on the old Ushū Highway, which connected Fukushima Prefecture and Aomori Prefecture in the Edo period. Kaneyama-machi is located in the Mogami area of the northeastern Yamagata Prefecture and is surrounded on the north and west by Mamurogawa-machi, on the south by Shinjo-City, and on the east by Ogachi-machi, Akita Prefecture.

Isabella Lucy Bird, who later became a member of the Royal Geographical Society, visited Kaneyama-machi on her way to the Tohoku region and Hokkaido and noted the following "After leaving Shinjo this morning, we crossed over a steep ridge into a singular basin of great beauty, with a semi-circle of pyramidal hills, rendered more striking by being covered to their summits with pyramidal cryptomeria, and apparently blocking all northward progress. At their feet lies Kaneyama in a romantic situation…" (Bird, 1880).

According to the population census data of the Ministry of Internal Affairs and Communications, the average mountain village population in Japan peaked in 1955 at 7,991 and started to rapidly decrease around 1960 with the onset of the period of high economic growth. It continued to decrease from 1975 until 2005, when it reached 3,903—49% of the peak population. Although the population of Kaneyama-machi followed almost the same trend as seen in mountain villages throughout the rest of Japan, the rate of decrease started to slow, and in 2005, the population was 6,949—67% of its peak population.

The number of households continued to increase until it reached 1,747 in 1960 and has remained at approximately the same level ever since, with 1,728 households recorded in 2005 (see Figure 2–10).

According to the Census of Agriculture and Forestry 2005, there were 647 agricultural management entities, 193 forestry management entities, and 746 management entities engaged in both agriculture and forestry. This revealed that Kaneyama-machi is a town with high social and economic pressure on these primary industries.

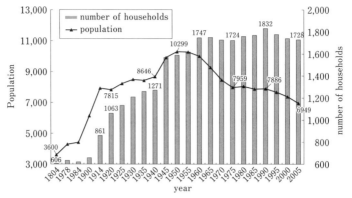

Figure 2–10. Trend in population and numbet of households in Kaneyama-machi
Source: Population Census data (The Ministry of Internal Affairs and Communications)

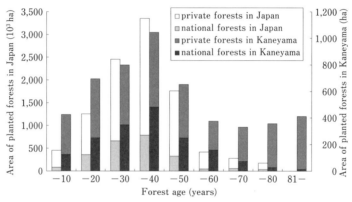

Figure 2–11. Area of planted forests in Japan and Kaneyama-machi
Source: Census of Agriculture and Forestry 2000

6.2 Forest Resources

According to the Census of Agriculture and Forestry 2000, there were 6,843 ha of national forests and 5,809 ha of private forests in Kaneyama-machi—a total of 12,652 ha. The town has a total area of 16,179 ha, 78% of which is covered in forests, which is 9% higher than the prefectural average of approximately 69%. The area of plantation forests in national forests is 1,720 ha, whereas that in private forests is 3,377 ha. There are more private forests than national forests in the countryside, and in private forests, which are smaller in terms of area than national forests, there is a much higher proportion of plantation forests (58%) than in national forests (25%). Compared with the proportion of plantation forests in private forests throughout the prefecture (41%), the proportion in Kaneyama-machi is 17%

Plate 2–1. Kaneyama-machi.
Photo by Hironori Okuda.

Plate 2–2. Typical Kaneyama-style houses.
Photo by Hironori Okuda.

higher at 58%. Furthermore, compared with the age-class location of plantation forests throughout Japan, there is no difference with the national figures of a peak age of 31–40 years. However, in contrast to the national proportion of 4.4% of total forested area consisting of trees 10 years or younger, the corresponding figure for Kaneyama-machi is 8.3%. Moreover, the national proportion of afforested area occupied by trees aged 51 years or older is 8.6%, whereas the corresponding figure for Kaneyama-machi is 37.6%. Thus, plantation forests have a greater proportion of both younger and older trees in Kaneyama-machi. This is because replanting has been steadily conducted since before World War II until the present; the result of forestry management aimed at long rotation large diameter tree production. In Kaneyama-machi, this age structure provides Kaneyama carpenters with a steady supply of various Kaneyama cedar products from construction materials to structural materials, processed by local sawmills (see Figure 2–11).

According to the Census of Forestry 2000, there are 286 owners with forest area less than three ha, 200 owners with forest area between three and 100 ha, three owners with forest area between 100 and 1,000 ha, and four owners with more than 1,000 ha. Therefore, although in terms of numbers most owners are small- to medium-sized forest owners, the four large forest owners with more than 1,000 ha own a total of approximately 4,500 ha each. Thus, approximately 75% of private forests are owned by large forest owners; because over the years, in the midst of fluctuations in the economy, small- to medium-sized forests were acquired by large forest owners through sale or foreclosure. This made it possible for forest management to aim at stable long rotation large diameter tree production in Kaneyama-machi with a rotation of 80 years or more.

6.3 Activities for Preserving and Nurturing Beautiful Townscapes in Kaneyama-machi

Kaneyama-machi is an old post town on the Ushū Highway. The upper parts of the outer walls of the Kaneyama houses are made of white earth, the lower parts are made of Kaneyama cedar, and the roofs of these houses are gabled. The houses blend in harmoniously with the greenery of the mountains, creating an atmosphere of serenity in the town. This townscape is the result of "town beautification campaigns" that have been undertaken since 1963.

These campaigns started after the then-mayor Eiichi Kishi visited European and American communities and decided to preserve and nurture a townscape for Kaneyama that would blend in better with the natural environment and culture in an appropriate manner (see Table 2–4).

A residential architecture competition that was held in 1978 promoted Kaneyama-style houses; it improved the skills of Kaneyama carpenters, introduced the concept of a beautiful townscape, and encouraged a groundswell of pursuit of beauty. Furthermore, in 1984, the Kaneyama municipal government implemented the New Kaneyama-machi Basic Design and launched the ambitious town beautification campaigns titled "100-year movement for preserving and nurturing a beautiful townscape".

Thus, preserving and nurturing a beautiful townscape in Kaneyama-machi has been continuously promoted since 1963, reaching a milestone with the enactment of the ordinance to form a beautiful townscape in 1986. This ordinance states that "preserving and nurturing the townscape and natural environment that is unique to Kaneyama-machi, which is an important asset shared by the townspeople, and enhancing it to pass on to future generations is recognized to be an important responsibility that is incumbent on the townspeople. Preserving and nurturing the townscape in Kaneyama-machi shall be carried out based on the following five pillars: preserving and nurturing a townscape that is rich in character; pre-

Table 2–4. The process of the action for preserving and nurturing the beautiful townscape in Kanayama-machi

1958–1982	The beginning of making of beautiful townscape in Kanayama-machi
1958	Then-mayor Eiichi Kishi has been European and American communities and decided to preserve and nurture a townscape in Kaneyama-machi that would blend in better with the natural environment and culture.
1963	"Town beautification campaigns" has been undertaken.
1978	The first "residential architecture competitions" has been held.
1983-present	Cultivate the atmosphere that encourages residents to select "Kaneyama-style houses" when building new houses in Kaneyama-machi
1984	The Kaneyama municipal government implemented the "New Kaneyama Basic Design"
1986	The town council issued "the Kaneyama-machi Landscape Ordinance"

Source: The basic design of the action for preserving and nurturing the beautiful townscape (Kanayama-machi, 2001)

serving and enhancing natural beauty; preserving and nurturing a beautiful townscape; creating a comfortable town; and creating a town of which people can be proud" (Kaneyama-machi, 2001). According to the ordinance, there is a system whereby residents notify the town when building new homes, and the town provides guidance to the residents. An advice program has been established wherein if a house's style meets the standards, it is called a "Kaneyama-style house" and subsidies are provided to cover 1/3 of the building or repairing expenses up to a maximum of ¥500,000 (until 1996, the maximum was ¥300,000). The ordinance does not impose penalties if a Kaneyama-style house is not built. The number of cases that received subsidies steadily increased until it reached 145 in 2002, after which there was a decrease in residential construction. This decreasing trend continued and by 2008, the number of cases dropped to 69, with subsidies of ¥6,057,000. A total of 1,312 houses had received the subsidy until that time, representing a total subsidy amount of ¥210 million (see Figure 2–12).

6.4 Survey Methods

To understand residents' opinions concerning Kaneyama-style houses, questionnaires were sent to the heads of 1,617 households of Kaneyama-machi in September 2002. There were 453 respondents (28% of the households). Moreover, in August 2003 and November 2004, interviews regarding wood business were conducted with 12 Kaneyama carpenters, three construction companies, and two lumber companies. Previously, in August 2002, the town of Kaneyama had conducted a survey to learn about the state and distribution of Kaneyama-style houses. However, to gain a more thorough understanding of this situation, in January 2007, Okuda and Komaki (Forestry and Forest Products Research Institute) conducted a survey of all Kaneyama-style houses in the town.

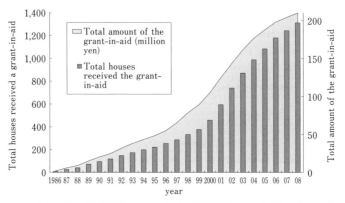

Figure 2–12. Trend in total houses received subsidies and amount of the subsidies in Kaneyama-machi
Source: Statistics document of Kaneyama-machi

To gain a better understanding of their "appraisal of the townscape", questionnaires were sent to the heads of 1,631 households in October 2008 for which, there were 447 respondents, 27% of the households.

6.5 Evaluation of Kaneyama-style Houses by Kaneyama-machi Residents

According to the results of the September 2002 questionnaire, 432 of the 453 respondents knew about the ordinance to form a beautiful townscape and 435 respondents knew that subsidies are provided by the town if a house's style meets the standards.

According to the 2002 survey conducted in the Kaneyama-machi municipal government, of the 1,678 houses in the area, 886 (52%) were Kaneyama-style houses, whereas in the 2007 survey, 1,006 houses (60%) were reported as Kaneyama-style houses. Though it is possible that there were differences in the ways that Kaneyama-style houses were identified, the number of such houses is undeniably increasing. Many residents appear to understand the policy of preserving and nurturing a beautiful townscape because the town's characteristic policy regarding the townscape comes from being constructed in this manner. The September 2002 survey.

Furthermore, 161 of the 237 respondents (68%) who lived in a Kaneyama-style house expressed satisfaction with their homes. When asked about the image of Kaneyama-style houses, 319 respondents, 70% of all respondents, said that Kaneyama-style houses suited them and 43% respondents considered such houses "traditional". The image of Kaneyama-style houses as suited to Kaneyama-machi had become firmly established. However, some difficulties regarding Kaneyama-style houses were pointed out through comments such as, "maintenance of the exterior requires effort and money" and "they cost more to build". Of the total respondents, 79% stated that they would consult with a Kaneyama carpenter or a professional architect when building a home. Thus, the popularity of Kaneyama-style houses depends on the attitudes of Kaneyama carpenters and architectural design offices. A poll surveying Kaneyama carpenters revealed that all of them actively promoted Kaneyama-style houses to the townspeople. Moreover, 215 (82%) of the 261 respondents who wanted to rebuild a house wanted to live in a Kaneyama-style house, and 182 (85%) of these 215 respondents wanted to build a Kaneyama-style house using Kaneyama cedar (see Table 2–5).

6.6 Flow Structure of Kaneyama Cedar Lumber

In FY2002, approximately 46,800 m^3 of Kaneyama cedar logs were produced in the town. Of this, 10,200 m^3 (22%) was supplied as building materials for local use, and the remaining 78% (36,600 m^3) was shipped out of the town to places such as Shinjo-City or the former Ogachi-machi in Akita Prefecture. Sawmills in such towns processed the logs into 6,700 m^3 of Kaneyama lumber, 2,000 m^3 of which was used locally. In FY2002, 29 building notifications were received for wooden houses in Kaneyama-machi. According to the interviews in August 2003 and November 2004, the flow of locally-produced lumber in Kaneyama-machi in FY2002 indicated floor area per house to be approximately 200 m^2, with 0.25 m^3 of lumber being used per 1 m^2 of floor space. Thus, approximately 1,500 m^3 of lumber has been used for building 30 houses in the town. Even if this local lumber were used in all home construction in Kaneyama, 500 m^3 would still remain unused. It appears that almost all construction materials where lumber is used are made from local Kaneyama cedar lumber.

6.7 Forming the Network for Preserving and Nurturing a Beautiful Townscape in Kaneyama-machi

In 1986, the town council issued an ordinance to form a beautiful townscape and to assist in the spread of Kaneyama-style houses. According to the ordinance, which does not impose penalties, residents are required to follow the guidelines stipulated in the "Townscape Formation Standards" when building or repairing a house in Kaneyama-machi. Many residents are aware that subsidies will be provided by the town if a house's style meets the standards. Further, they want to build Kaneyama-style houses that are in harmony with the townscape and the town's environment to preserve and nurture the traditional townscape that they care about. The Kaneyama-machi Chamber of Commerce and Industry has held residential architecture competitions since 1978 to improve the skills of Kaneyama carpenters, in which nonlocal architects are invited as judges. The interviews conducted in August 2003 and November 2004 revealed that when townspeople want to build a house and accordingly consult with Kaneyama carpenters or architectural design offices, they receive

Table 2-5. Kaneyama residents' evaluations of "Kaneyama-style house"

	Kaneyama-style house	Other wooden house	Non-wooden house		Total
Do you want to live in "Kaneyama-style house"? (of the 261 respondents who want to rebuild a house)	215 (82%)	40 (15%)	6 (2%)		261 (100%)
	Kaneyama cedar lumber	Other domestic lumber	Any lumber	No reply	Total
What material do you want to build a "Kaneyama-style house" of? (of the 215 respondents who want to live in a Kaneyama-style house)	182 (85%)	20 (9%)	12 (6%)	1 (0%)	215 (100%)
	Yes, satisfied	No, dissatisfied	Neither	No reply	Total
Are you satisfied with your Kaneyama-style house? (of the 237 respondents who live in Kaneyama-style houses)	161 (68%)	30 (13%)	40 (17%)	6 (3%)	237 (100%)
	Suitable	Traditional	Highly ranked	Popular	Total
What is your image of a typical Kaneyama-style house? (of all 453 of all respondents)	319 (70%)	195 (43%)	73 (16%)	105 (23%)	453 (100%)

Source: Questionnaire findings (September, 2002).

Table 2-6. Local residents' views of the Kaneyama townscape

	Excellent	Good	Average	Not good	No reply	Total
What do you think of Kaneyama's townscape?	120 (27%)	194 (43%)	108 (24%)	11 (3%)	14 (3%)	447 (100%)
	Form the overall townscape including mountains and fields	Continue with the current focus on residential areas	No special effort is necessary		No reply	
What kind of concerted effort do you want to make toward townscape development?	231 (52%)	171 (38%)	7 (2%)		38 (8%)	447 (100%)
	The municipal government	The community	Outside influences		No reply	
Who or what is the main player in forming Kaneyama's townscape?	204 (46%)	221 (49%)	17 (4%)		25 (6%)	447 (100%)

Source: Questionnaire findings (October, 2008).

Plate 2–3. *Soba* noodle shop in a closed school.
Photo by Hironori Okuda.

recommendations to build Kaneyama-style houses. Depending on such advice, the townspeople usually decided to construct Kaneyama-style houses. Kaneyama carpenters, on receiving orders from townspeople, order a set of the lumber required for building these houses from a manufacturer or forestry association in the town. For their part, local forest owners are asked to provide a stable supply of Kaneyama cedar materials, for which they focus on large-scale production of trees aged 80 years or older with strong and easy-to-process wood.

Thus, in Kaneyama-machi, building houses is connected with preserving and nurturing a beautiful townscape within the context of the town's traditions, skills, history, landscape, and resources. The townspeople's desire for a beautiful townscape leads to construction of Kaneyama-style houses by Kaneyama carpenters using an abundant supply of Kaneyama cedar. Therefore, a network is being formed that includes residents, Kaneyama carpenters, architectural design offices, sawmills, forest associations, forest owners, and the town office to preserve and nurture a beautiful townscape in Kaneyama-machi (see Figure 2–13). Preserving and nurturing a beautiful townscape of Kaneyama-machi can be seen as "a proactive movement to promote the development of the town through the collaboration of residents and municipal government" (Muramatsu, 2002).

6.8 Results of Preserving and Nurturing a Beautiful Townscape

The number of visitors frequenting to the town is increasing as a result of the beautiful townscape created by rows of Kaneyama-style houses. To increase awareness regarding the beautification of the environment, waterways are being stocked with carp and volunteers provide refreshments to the visitors and guide them to the town.

Each year, nearly 16,000 visitors come to Taniguchi, a small colony of 36 houses nestled in the mountains to the north of Kaneyama-machi. Visitors flock to the soba noodle

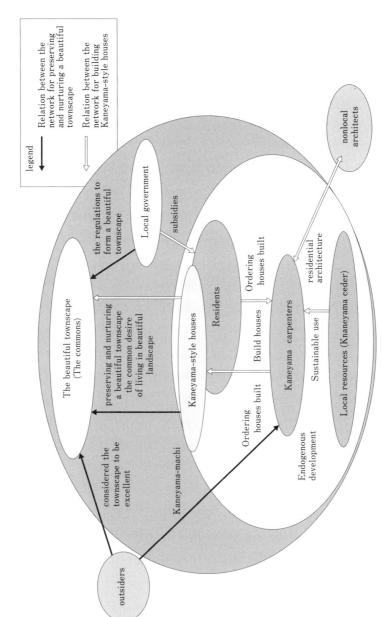

Figure 2-13. Relation between the network for preserving and nurturing a beautiful townscape and building Kaneyama-style houses.

shop that is run primarily by the residents of Taniguchi in an education extension campus that was closed in 1996. The residents of Taniguchi were able to accomplish this project as a result of numerous discussions in the local area, by receiving ideas from the town councilors and the staff of the town office, and with the cooperation of people from outside the town (residents of Miyagi Prefecture and Tokyo). Moreover, although they initially purchased buckwheat flour from a supplier in Tendo in Yamagata Prefecture, they now acquire it from farmers in Kaneyama-machi who started producing buckwheat as a crop conversion. Thus, the network of efforts of regional collaborations is rapidly expanding. This also illustrates the fact that products created in a certain place can lead to the creation of other products. For example, with the increase in the number of visitors seeking the beautiful townscapes of satoyama that have begun to be preserved and nurtured in Kaneyama-machi, the volunteer activities and agricultural production that targets such visitors become active. This type of sustainable and organic local involvement revitalizes mountain villages and increases the area's appeal.

6.9 Endogenous Development and the Network for Preserving and Nurturing a Beautiful Townscape in Kaneyama-machi

According to the results of the September 2002 questionnaire, 27% of respondents considered their townscape "excellent" and 43% rated it as "good", a 70% favorable rating. In addition, 52% of respondents wanted to make a concerted effort to form an overall townscape in which everything—including the mountains and fields—would be considered. When combined with the 38% of respondents who wanted to make a concerted effort to continue with the current focus on residential areas, 90% of respondents were interested in townscape development. Furthermore, 46% of respondents thought that the main contribution in forming a beautiful townscape of Kaneyama-machi should come from the municipal government, 49% opined it should come from the community, and 4% thought it should come from those from outside regions (see Table 2–6). Moreover, according to the ordinance to form a beautiful townscape, there are no penalties if a Kaneyama-style house is not built. The decision regarding the style of house is left to the residents. Though many respondents are of the opinion that the municipal government has a major influence on such decisions, it is in fact the residents who are responsible for creating the beautiful townscape. The unique townscape of Kaneyama-style houses in Kaneyama-machi is a common property resource that the residents should preserve and nurture.

To encourage residents to build Kaneyama-style houses, residential architecture competitions have been held since 1978, in which nonlocal architects are invited as judges. In 1986, the town council issued the ordinance to form a beautiful townscape to cultivate an atmosphere that encourages residents to select Kaneyama-style houses when building new houses in Kaneyama-machi. This led to a network among residents, Kaneyama carpenters, architectural design offices, sawmills, forest associations, forest owners, and the town office. The beautiful townscape project has been autonomously implemented in Kaneyama-machi by the local residents. This network has brought about social change by encouraging the local residents to select a Kaneyama-style house built by Kaneyama carpenters and thus engendered endogenous development (Okuda et al., 2004). The network is connected by the common and fervent desire of the local residents to "live in a beautiful townscape". However, some people, particularly the youth, are averse to living in houses that are similar in appearances and thus, have built houses different from the townscape, resulting in difficulty in preserving and nurturing the "beautiful townscape". Building Kaneyama-style houses as per the rules is an autonomous act of the local residents for preserving and nurturing the beautiful townscape in Kaneyama-machi that would otherwise be destroyed. The question of whether it is possible to continue the activities or not entirely depends on whether the network of local residents can maintain their shared desire of "living in a beautiful townscape".

In the residential architecture competitions, nonlocal architects provided opinions and advice, proposed Kaneyama-style houses (white-walled houses with cedar boards and gabled roofs), encouraged Kaneyama carpenters to improve their skills, introduced the concept of a beautiful townscape, and brought about a groundswell in seeking such beauty. Thus, outsiders have played an important role in deciding the direction of activities.

7. Endogenous Development and Collaborative Governance

"Endogenous development" is a process of social change in which local residents who foster a common desire to protect "important things" solicit outsiders for knowledge, skills, and systems and allow their involvement in planning and designing activities based on the degree of their involvement in "common property resources (the "commitment principle" in the collaborative governance theory)".

Although the activities to protect, develop, and use "common property resources" must be practiced mainly by local entities, involvement in these activities should be open to outsiders; this is called "open-minded localism (of the collaborative governance theory)". This is how the process toward endogenous development can be easily explained using the collaborative governance theory.

Based on the abovementioned four examples (two examples of Tsukimoushi-machi, and one each of Kaneyama-machi and Sawauchi-mura), endogenous development is based on the local network of residents connected by the common desire to protect "what is important" (see Figure 2–14), and these activities would not have led to endogenous development with-

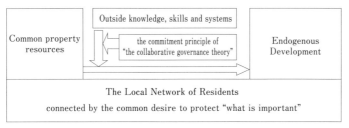

Figure 2–14. Structure of "Endogenous Development"

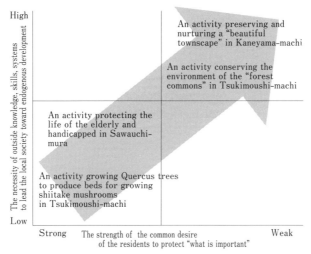

Figure 2–15. Relation between the strength of the common desire of the residents

out the involvement of outsiders. Tsurumi (1996) stated that, "the process toward endogenous development and lifestyles that result in the achievement of common goals of the local residents are created autonomously by those residents using local resources like water, wood, and minerals. At the same time, they are autonomously created by new skills in which traditional skills are integrated with outside knowledge, skills, and systems in accordance with the particular natural environment, cultural heritage, and history of the area".

Thus, endogenous development should be led autonomously in the region, and integration of outside and traditional/local knowledge, skills, and systems should be realized in the region by obtaining the assistance and advice from outsiders. To lead the local society toward endogenous development, it is necessary to involve outsiders in planning and designing activities based on the degree of their involvement in "common property resources". However, there is a tendency among local residents to seek deeper involvement of outsiders and ask for more knowledge, skills, and systems because the common desire of the residents to protect "what is important" is weaker and endogenous development is more difficult to achieve in the local society (see Figure 2–15).

References

Bird, I. L. 2000. *Unbeaten Tracks in Japan* (A translation into Japanese by Kenkichi Takanashi). Tokyo: Heibonsha.

Dag Hammarskjöld Foundation. 1975. *What Now?: The 1975 Dag Hammarskjöld Report on Development and International Cooperation* prepared on the occasion of the Seventh Special Session of the United Nations General Assembly (New York, 1–12 September 1975), Uppsala: Dag Hammarskjöld Foundation.

Inoue, M. 2003. "A Concept and the Effectiveness of Commons". *Encyclopedia of Forest*. Tokyo: Asakura Publishing, pp. 593–596.

Inoue, M. 2009. "A Design Guidance of 'Collaborative Governance' about the Natural Resources of the Global Era". In: T. Murota (ed.) *Gurobaru Jidai no Rokaru Komonzu [The Local Commons in the Global Era]*. Kyoto: Minerva Shobo, pp. 3–25.

Kaneyama-machi. 2001. "The Basic Design of the Action for Preserving and Nurturing the Beautiful Townscape".

Kitao, K. 2003. "Satoyama". *The Forest and the Forestry Encyclopedia*. Tokyo: Maruzen, pp. 347–348.

Muramatsu, M. 2002. "Characteristics of the Landscape Development Policy in Rural Area: A case study of Kaneyama Town, Yamagata Prefecture, Japan". *Journal of Farm Management Economics*, 33: 67–82.

Okuda, H. et al. 1999. "History and Prospect of the National Forest Utilization by Local People: A Case Study of a Mountain Village in Tono City, Iwate Prefecture". *Forest Economy*, 611: 27–34.

Okuda, H. et al. 2001. "Human Networks Supporting a Life in the Mountain Village: A Case Study in Sawauchi-mura, Iwate Prefecture". *Journal of Japanese Forest Society*, 83(1): 47–52.

Okuda, H. et al. 2003. "Development of Co-management Approach by the Local Residents of the Region Forest". *2002 Research Results Anthology*. Tsukuba: Forestry and Forest Products Research Institute (FFPRI), pp. 22–23.

Okuda, H. et al. 2004. "Factors Involved in the Formation of a Self-sufficient System for Timber Products in Kaneyama-town". *Journal of Japanese Forest Society*, 86(2): 144–150.

Tamanoi, Y. 1977. *Idea of Regional Decentralization*. Tokyo: Toyo Keizai.

Tsurumi, K. 1996. *Development of Endogenous Development Theory*. Tokyo: Chikuma Shobo.

Chapter 3
Collaborative Forest Governance in Mass Private Tree Plantation Management
Company-Community Forestry Partnership System in Java, Indonesia (PHBM)

Yasuhiro Yokota, Kazuhiro Harada, Rohman, Nur Oktalina Silvi, and Wiyono

1. Introduction

1.1 Arguments on Collaborative Forest Governance

Throughout Southeast Asia, a large proportion of forest management in terms of the overall area is performed by state and public institutions. Previously, forest management was primarily performed in a monopolistic manner, either by the state or designated institutions and groups recognized by the state. It is historically evident that sustainable forest management cannot be realized under such an elitist form of forest management. Furthermore, although entities, such as the state, have introduced various participatory approaches, in an effort to urge local people who utilize the forest to participate in forest management, many of these government-run programs entail only limited transfer of ownership and management rights (IGES, 2007).

This chapter will consider the possibilities and limitations associated with the feasibility of establishing collaborative forest governance in the context of the management and administration of forests in an environment with limited transfer of rights. The chapter will then examine, as an example, PHBM (*Pengelolaan Sumberdaya Hutan Bersama Masyarakat*), the company-community forestry partnership system run by the State Forestry Corporation of Indonesia (SFC) (*Perum Perhutani*) on the island of Java, Indonesia. This system, which was implemented under the guidance of the SFC (an outsider), could be an example of "adjustment strategy" formulated by local people (wherein the opinion of an outsider accounted for more than that of the local people). Furthermore, it could also be cited as exemplifying a situation where the main participants in forest management prior to the implementation of these efforts were—both legally and in reality—not the local people themselves but the SFC. In this example, it was not a case of the local people welcoming outsiders into forest management, but the opposite, i.e., local people were invited by outsiders.

A similar example in Indonesia is MHBM (*Mengelolaan Hutan Bersama Masyarakat*), a company-community forestry partnership program established by a concession holder of industrial tree plantation (private company) in southern Sumatra. MHBM jointly manages (state-owned) industrial plantation concession with local people of the surrounding area (Awang et al., 2005). This program is very similar in content to PHBM. As, at present, only one company has obtained a concession within the appointed time period, one might assess that it is rather limited spatially and temporally in comparison with PHBM.

Other examples of such government-run programs in state forests include community forests (HKm: *Hutan Kemasyarakatan*), village forests (*Hutan Desa*), and community plantation forests (HTR: *Hutan Tanaman Rakyat*) (e.g., Awang et al., 2005; Fujiwara et al., 2012; Nawir et al., 2003; Rohadi et al., 2010).

Under the HKm program, aimed at both forest conservation and improvement of the welfare of local people, 35-year usage rights (extendable and renewable) are provided to groups of local people. This program is applicable for areas in state-owned production forests and protected forests where no other rights have been established. This program was inaugurated in 1995 under Decision No. 622 by the Ministry of Forestry and was modified on six occasions until 2007, with the terms of such usage rights and categories of involved forests changed each time. When the 1998 amendments were implemented, local people were the main entity engaged in forest management for the first time in the history of Indonesian forest governance. This was a clear departure from the manner in which matters had been conducted until 1998. However, these repeated modifications resulted in confusion at the local level and inhibited the program's smooth implementation. Moreover, there were other problems including delays in processing at the central government level, issuance of usage rights at the sole discretion of the local government, unsustainable forest use, and issuance of rights only in degraded areas (Djamhuri, 2008; Shimagami, 2010; Van Noordwijk et al., 2007).

Under the *Hutan Desa* program, the village committee have an acknowledged right to manage and use production forests and protected forests for 35 years for the welfare and benefit of the village. The harvesting of non-timber forest products and timber production in production forests is also permitted. This program was defined in Article 41 of the Forest Law in 1999 and inaugurated through Decree No. 49 of the Ministry of Forestry in 2008. Similar to the HKm program, issues such as a tendency to prioritize the assignment of logging rights to companies and lack of coordination between central and local governments were pointed out (Zainuri, 2011).

Under the HTR program, to improve the welfare of the local people and promote effective utilization of low-quality production forests, permission is granted to families and cooperative associations living in and around the forests to engage in afforestation and production and sale of timber for a maximum of 60 years. This program is for low-productivity production forests in state-owned forests where no other management plans or concessions have been established (Obidzinski and Dermawan, 2010; Van Noordwijk et al., 2007).

HKm, *Hutan Desa*, and HTR can engage in a broader scope of activities within forests under their management than PHBM. On the other hand, they rarely have powerful partners such as the SFC, and it is not unusual to find that usage rights are issued only for forests that are not rich in timber resources.

This chapter focuses on PHBM as an example and examines the manner in which the local people have dealt with a forest management system run under the direction of outsiders and influence of this management. This chapter will then consider the degree to which forest management for local people is possible under outsider-controlled forest management systems that involve limited transfer of rights to local people and have inherent limitations, especially from the perspective of the benefits that can be obtained by local people. To recap the characteristics of this system, the legal position of the forests involved is that of state-owned forests managed by the SFC (a large forestry company), for the purpose of timber production (as set by the SFC). The purpose of collaborating with local people is to minimize the social risk in forestry management (e.g., illegal logging) through cooperation by the local people.

1.2 History and Management Scheme of PHBM

The history of tree plantations in Java stretches back to teak plantations during the Dutch colonial period. After Indonesia became independent in 1945, these plantations became state-owned forests; their management was assumed by the SFC. Java is divided into three forest units (UNIT) and 57 forest districts (KPH: *Kesatuan Pemangkuan Hutan*).

The boundaries of plantations are clearly recognized within local society because of their long history and the SFC's strict management. Access to these plantations by unauthorized entities has been officially prohibited. Local people have been involved in plantation management through participation in the plantation system of agroforestry, known as *Tumpang Sari*[1]. Unofficially, they have also harvested non-timber forest products such as firewood, fodder, herbs, and teak leaves.

However, as a result of the Asian currency crisis in 1997–1998, a vicious circle began in Indonesia: economic recession brought about social instability, which in turn led to political instability, ultimately resulting in deepened economic recession. This caused an increase in crime due to the accompanying deterioration of social order, which also affected

---

1) In exchange for farmers planting trees at no cost, they are allowed to cultivate between the rows of planted trees for a few years following planting.

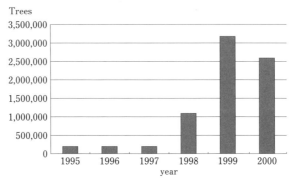

Figure 3–1. Total number of trees lost by the SFC through illegal logging.
Source: The SFC's internal data.

public institutions. Such deteriorating conditions were also observed in state forests under the jurisdiction of the SFC, where there was a dramatic increase in illegal logging by stakeholders both from within and outside local areas (see Figure 3–1).

To reduce damage from a massive increase in illegal logging and in response to calls for public institutions to contribute to local society that arose accompanying the rapid pace of democratization within the country, the SFC introduced PHBM system in conjunction with local people. This was based on successful implementation of Joint Forest Management in India. PHBM was inaugurated in 2001 based on Decision No. 136 / KPTS / DIR / 2001 of the SFC's board of directors[2]. PHBM was not only a program but was also made part of the SFC's overall system. Details regarding the content of such activities are designed and implemented by each forest district.

1.3 Overview of Madiun Forest District and Madiun Regency

This chapter examines PHBM implemented by the Madiun forest district (KPH Madiun) in the Madiun Regency (*Kabupaten Madiun*) in the western part of the East Java province (see Figure 3–2). The Madiun forest district started PHBM in 2002 and is one of the forest districts that have been actively implementing PHBM since then.

---

2) Prior to the introduction of PHBM, SFC was engaged encouraging the local people's participation and collaboration with them and had already implemented various programs such as *Tumpang Sari*, the prosperity approach, PMDH, and social forestry. This indicates that the SFC experienced the pressure of incursion into forest land by local people for many years, and that efforts were constantly being taken to improve these programs because it was not easy to achieve a reduction in that risk (Peluso, 1992; Shiga, 2012). The major difference between these approaches and PHBM is that all attempts until that time were an additional program to the SFC's forest management system while PHBM is a part of the system.

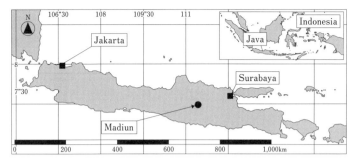

Figure 3–2. Survey area.
Source: Authors.

The Madiun forest district has jurisdiction over 31,221.8 ha, most of which (27,267.9 ha) consists of teak plantations. Within the district, there are two sub-forest districts (sub-KPH), four planning areas (BH: *Bagian Hutan*), and 11 branch offices (BKPH: *Bagian Kesatuan Pemangkuan Hutan*). Sitting astride Madiun Regency, Ponorogo Regency, and Magetan Regency, the northern sub-forest district (16,031.5 ha consisting of two planning areas and six branch offices) belongs to the Madiun Regency (Perum Perhutami KPH Madiun, 2009). This chapter will thus report on the status of PHBM started by the Madiun forest district in this sub-forest district.

The Madiun Regency is approximately 160 km (four to five hours by car) away from both Surabaya, the capital of the East Java province, and Yogyakarta in Central Java. It encompasses an area of 101,086 ha, most of which (85.3%) is at an altitude of less than 500 m. The terrain is basically flat and 88.8% of the villages in the area are located on flat land, while the remaining 11.2% are located in the hills (BPS Kabupaten Madiun, 2003). When PHBM was first implemented in 2002, the area consisted of 15 sub-Regencies and 206 villages with a combined population of 666,548 people, representing a population density of 659 people per km^2 (BPS Kabupaten Madiun, 2003). Approximately 70% of the population resides in rural areas (Faculty of Forestry, Gadjah Mada University, 2005).

As of 2002, there were 41 forest villages[3] (*Desa Hutan*) under the jurisdiction of the northern sub-forest district. The main economic activities in the area are agriculture and agriculture-related activities (such as agricultural labor), mainly focused around paddy fields (two to three crops per year). In places where paddy field cultivation is impossible, cassava or corn are cultivated. There is a general scarcity of agricultural land and the average area of

---

3) Forest villages refer to villages that are located in areas surrounding forests under the SFC jurisdiction and have SFC forest within their administrative boundaries.

Table 3–1. Status of Surveyed Villages

Name of the surveyed village	Level of engagement in PHBM in 2004	Demography in 2004		
		Population (people)	Household (HH)	Area (ha)
Da	active	3,848	972	561
Ba	middle	1,044	319	397
Bo	not yet active	2,766	712	1,113

Source: "Level of engagement in PHBM in 2004," Authors, "Demography in 2004," BPS Kabupaten Mediun (2005a; 2005b).

agricultural land per household is approximately 0.20 ha (KPH Madiun, 2009). Furthermore, going to work in the cities during the agricultural off-season is an important source of income, with some people (especially women) migrating overseas for a number of years for this purpose.

To survey the state of PHBM implementation at the village level, three of these 41 forest villages were selected[4] (see Table 3–1).

2. An Outline of PHBM in the Madiun Forest District

From the viewpoint of collaboration with local people, there are four characteristics of the PHBM system in the Madiun forest district: the Forest Resource Management Group; agricultural use in the SFC forest; forest utilization and management; and local people support[5].

2.1 Introduction of the Forest Resource Management Group

When implementing PHBM, the SFC requested that a volunteer group be formed to encourage autonomous and responsible activities by the local people. In Madiun, this group is known as the Forest Resource Management Group (MPSDH: *Masyarakat Pengelola Sumber Daya Hutan*) (hereinafter referred to as the "management group")[6] and one group is formed in each village through the initiative of the local people. As part of the promotion of PHBM, a contract is signed between the two groups (the management group and the forest district). The

---

4) When selecting villages for survey, people with knowledge of the program were asked to categorize villages into three groups (active, middle, not yet active) according to the degree of PHBM activity. One village was then selected from each group.
5) Though the characteristics of systems in other forest districts are similar, the details differ.
6) In other forest districts, such groups are generally called Forest Village Community Associations (LMDH: *Lembaga Masyarakat Desa Hutan*).

reason for forming a new group of interested parties is that, depending on the village, there may be people whose lives have little to do with the forest, making it difficult to motivate activity by the village as a whole.

Establishment of the Forest Resource Management Group
It could be said that the formation of a management group is encouraged by outsiders such as the SFC and Gadjah Mada University; however, it actually depends on the intent of local people who decide whether the group will be established and is by no means forced. Prior to the actual establishment of the group, local people meet frequently to discuss issues with the support of Gadjah Mada University and the SFC, but the pace of such discussions is determined by each village. Differences in the state of progress according to village were observed. Furthermore, although management group bylaws are similar because of the influence of outside support, each village determines the content of its own bylaws. Each management group has several subgroups, named Working Units (KKP: *Kelompok Kerja Pesangam*) at the sub-village level. The number of and member size of the subgroup varies as per the management group.

Members of the Forest Resource Management Group
The method of securing members differs from village to village; in some villages, everyone living in settlements near the SFC forest becomes a member, whereas in other villages, volunteers are recruited. However, in most villages, one person from each household or family usually becomes a member to represent his/her household or family, and people can join the management group after it has been established.

Contract between the Forest Resource Management Group and Forest District Office
A (10-year renewable) contract clearly outlining the rights and responsibilities of each party is signed between the management group and forest district office for joint long-term management of all the SFC forests within the administrative boundaries of that village. Forest compartments relevant to the management group are clearly outlined in the contract. Depending on the situation, certain forest compartments may be better accessed from a neighboring village. In such a case, that particular area of the forest can be designated as being under the jurisdiction of the management group of the neighboring village. Furthermore, where extremely few SFC forests fall within the administrative boundaries of a village are, such a village may join the management group of a neighboring village instead of forming an independent management group.

Forest Resource Management Group Activity Objectives and Organization Continuity
Although a farmers group (KTH: *Kelompok Tani Hutan*) existed under the old system, formed by farmers participating in the *Tumpang Sari* program, it was different in nature. The activities of the farmers group concentrated exclusively on planting teak trees through the *Tumpang Sari* program, whereas the activities of the management group include not only

those covered under the *Tumpang Sari* program but also all forest management and utilization activities, such as maintenance, harvesting and forest conservation management. Moreover, depending on the members of the management group, some may participate only in *Tumpang Sari* while others participate only in paid forestry work or forest conservation management activities. Furthermore, in terms of organization continuity, the farmers group was temporary and disbanded upon completion of the *Tumpang Sari* work (in reality, in many cases, groups went to a different location to participate in the *Tumpang Sari* program). Contrarily, the management groups are permanent organizations that exist as long as there is a contract with the forest district, regardless of whether the *Tumpang Sari* program is in operation. In addition, whereas each farmer of the farmers group and local forest district administrators signed *Tumpang Sari* contracts directly and individually, contracts are concluded between the management groups and the forest district as organizations under PHBM.

Role of Local People

Although under previous programs, local people were suppliers of labor, under PHBM, management groups hold the place of partners who jointly perform all forest management activities.

2.2 Agricultural Use in the SFC Forest

Tumpang Sari

Before PHBM, local people had agricultural lot in the SFC forest through *Tumpang Sari* program. Under the old *Tumpang Sari* program, agricultural intercropping was limited to two years. However, as this restriction has been lifted under PHBM, the possibility has now opened up for local people to use forest land on an ongoing basis for planting shade-tolerant crops and producing fuel wood—even after canopy closure—on the condition that they do not damage the planted teak trees.

Priority Given to Tumpang Sari Participants

In the old *Tumpang Sari* program, site supervisory staff sourced *Tumpang Sari* volunteers from the vicinity—some through personal contacts—and formed farmers groups. Participants in *Tumpang Sari* were not limited to the local people from settlements surrounding the plantations; local people from other villages could also participate. In contrast, after the establishment of the management groups, site supervisory staff first approached the management groups with jurisdiction over the planned *Tumpang Sari* plantation area. This meant that in terms of distributing opportunities, there was possibility for improvement in transparency and equality for *Tumpang Sari*, and that compared with the old system, management group members have been given priority for *Tumpang Sari* participation within the village.

Plate 3–1. Intercropping of *Kunir* (turmeric) under planted trees within a teak plantation.
Photo by authors.

Intercropping under Planted Trees
Planting within teak forests that have already been established is also permitted, and thus the amount of land for intercropping agriculture has been expanded. Plate 3–1 shows intercropping under planted trees within a teak plantation.

2.3 Expansion of Forest Management Utilization Opportunities
Profit Sharing from the Proceeds of Final Cutting and Thinning of Plantation Trees
When final cutting and thinning is conducted in forests under the jurisdiction of the management groups and the forest district office has obtained income, the management groups are paid a maximum of 25% of the profit share according to the number of years from the time of the conclusion of the contract. The method used to calculating the profit sharing rate is roughly shown below[8]. The profit sharing rate may be further reduced, depending on how much illegal logging occurs within the forest under the jurisdiction of the management group as a whole.

Final Cutting: $Pa = 25\% \times$ (the lesser of either the number of years since the contract was concluded or the stand age at the time of cutting) / (the stand age at the time of cutting)

---

8) For example, final cutting of teak is executed at the age of 80 years; thus, five years after the conclusion of the contract, the profit share rate would be ($25\% \times 5/80$), which is approximately 1.39%. If thinning is implemented 15 years after the previous thinning was performed, then five years after the conclusion of the contract, the profit sharing rate would be ($25\% \times 5/15$), which is approximately 8.33%.

Thinning: Pa = 25% × (the lesser of either the number of years since the contract was concluded or the number of years since the previous thinning) / (the number of years since the previous thinning)

The distribution of this profit share is applied from the moment the contract is concluded. Of course, because only a few years have passed since the contract was concluded, the profit sharing rate is very low.

Prioritization of Forestry Work Opportunities

Priority is now given to management group members for forestry work opportunities arising within forests under the jurisdiction of the management group. Although before PHBM, the site supervisory staff of the forest district made the effort to find people themselves, since the establishment of the management group, such decisions are made after consulting the group.

Official Authorization to Harvest Forest Products

When a contract is concluded with the forest district office, the harvesting of non-timber forest products such as firewood, fodder, herbs, and teak leaves is officially authorized not only for household consumption but also for sale. Furthermore, since trees logged during the first thinning are small in diameter, the entire volume is provided to management group members to be used as fuel.

Requirement to Cooperate with Forest Preservation

The management group is required to cooperate with protection and management activities of the forest district office. Patrols are performed to prevent illegal logging, recover illegally logged timber that is left within the forest, extinguish forest fires, provide information regarding forest management, and prepare reports, without receiving any remuneration from the SFC for performing such tasks.

2.4 Implementation Structures for Local People Support

Under PHBM, the Madiun forest district has formulated a system or a network to support local people and smoothly promote the system. Figure 3–3 shows the various stakeholders involved.

In conjunction with the introduction of PHBM, a PHBM division (*Sub Seksi PHBM dan Bina Lingkungan*) was established within the forest district office and a field facilitator was employed to provide onsite support for the local people (initially three facilitators and now four, all from the local area).

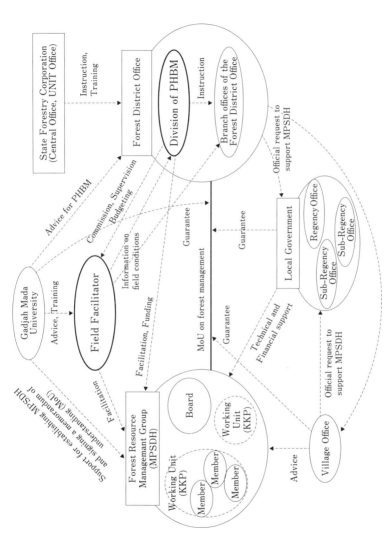

Figure 3–3. Major stakeholders in PHBM in the Madiun forest district.
Source: Field research.

Along with providing advice concern the design of PHBM in the Madiun forest district, Gadjah Mada University also trained the field facilitator[9].

Local people formed the management group, and Gadjah Mada University and the field facilitator provided guidance and support during the process of concluding the contract with the SFC. The sub-regency chief, village chief, village council, and the Gadjah Mada University School of Forestry acted as guarantors for the contract.

In management group operations, the forest district office and field facilitator work together to provide support, including advice and financial assistance. The field facilitators gather information regarding the opinions of local people and state of the site and provide such information to the forest district office.

To promote PHBM, the SFC is raising awareness among executives and staff, establishing curricula regarding PHBM for use in various training programs (including promotions) and strengthening efforts to change people's thinking.

3. Local People's Response to PHBM in the Madiun Forest District

3.1 State of Contract Conclusion and Registration

Conclusion of Contracts (State of Forest Resource Management Group Establishment)
In the entire northern sub-forest district, almost all forest villages had concluded a contract with the forest district as of February 2011, whereas some villages completed this process faster than others: 2 villages by November 2002, 18 villages by April 2003, 4 villages by September 2003, 1 village by March 2004, 3 villages by May 2005, 5 villages by June 2006, and 3 villages by August 2007. Three villages joined the management group of another village as the area of forest under their jurisdiction was small, and one village had yet to conclude a contract[10]. The three villages selected for the survey—Da, Ba, and Bo—concluded contracts with the forest district on April 30, 2003.

---

9) Prior to PHBM in the period between 1991 and 2001, Gadjah Mada University and the Madiun forest district implemented a pilot project for developing a desirable teak forest management system. At that time, Gadjah Mada University provided on-the-job training.

10) The reason for differences among villages in terms of concluded contracts can be explained as follows. First, factors concerning local people included social conditions (such as motivation of village leaders, level of interest among village leaders regarding PHBM, degree of agreement among local people), local people's attitude (such as degree of trust regarding the new system and the forest district, sense of hesitation/burden regarding forest management responsibilities), economic state (such as agricultural foundation, especially in terms of paddy fields), and forest resource state (distance from forest compartments outlined in the contract, state of forest resources). Other factors included the enthusiasm of both the forest district personnel and local government support.

Registration with the Forest Resource Management Group

This section examines households who registered with the management group as members in each of the three villages in the survey. The number of such households at the time of the conclusion of contracts with the forest district office were as follows: 60 households in Da village, (6.0% of the total households in the village); 75 households in Ba village (30.0% of the total households in the village); and 357 households in Bo village (47.9% of the total households in the village).

The results of a (multiple choice) questionnaire[11] conducted with 30 households each from the three surveyed villages from 2004 to 2005 revealed that the primary motivations for management group members to participate were economic reasons to "increase income" (approximately 68.9%), including responses such as "to secure agricultural land",[12] and "to obtain a share of profits from the sale of teak". Social reasons such as "my acquaintances and neighbors are participating" and "I was invited" did not feature as prominently (approximately 35.6%). Other reasons included "I want to contribute to sustainable forest management" and "I'm used to *Tumpang Sari* / forestry work". For Da and Ba villages, only those households who wanted to become members did so, whereas in Bo village, all members of households in sub-forest districts close to the forest became members.

The questionnaire was similarly conducted for 20 unregistered households; their responses were as follows. "I own enough agricultural land" (20.0%), "I'm engaged in other employment / jobs" (25.0%) (It could be said that these two replies indicate that these households did not consider PHBM sufficiently lucrative). Further replies included "health problems" (25.0%), "lack of labor capacity within the household" (10.0%), "I never received notification about it" (25.0%), and "I don't understand it so I'm scared or uncertain about it" (20.0%).

Furthermore, among management group members, although there are many households engaged in agricultural activities within the forest as their primary source of income, there are other households that are not engaged in such activities with the aim of deriving profit shares and households that have registered to cooperate with PHBM to fulfill their local public roles.

The number of management group members as of February 2010 was 104 households in Da village, 163 households in Ba village, and 427 households in Bo village. Reasons for

---

11) In the three villages surveyed, a random sample was selected of 30 households registered with the management group in each village living in areas close to the forest. A random sample of 10 households in each village was also selected from among households in the same areas who had not registered with the management group. In Bo village, it was not possible to carry out the survey on unregistered households as everyone in the area had already registered.

12) Reasons for local people's desire to secure agricultural land included "lack of agricultural land to own or use", "inability to find jobs other than in agriculture", and "the only work while living in the village is in agriculture".

increasing participation among registered households included an increased level of understanding regarding PHBM; a realization of the merits of PHBM such as the distribution of profit share and support from outside; return of people working outside the village; and the fact that sharecropping contracts elsewhere had come to an end and people were seeking agricultural land.

3.2 State of Activity Implementation

Agricultural Activities within the Village

The *Tumpang Sari* program accompanies reforestation after final cutting. However, there are few opportunities each year for individual villages due to the uneven age composition of the forest and the long rotation period of 80 years. For Da and Ba villages, there were no opportunities during period between 2003 and 2010. Nevertheless, no real problems could be observed overall with regard to this lack of opportunity. This is because after the conclusion of the contract with the forest district, priority was given to village management group members if the opportunity for *Tumpang Sari* arose within the village, and there was a high degree of satisfaction regarding the content of PHBM. However, in some places, soil fertility is poor because of which some did not want to participate even if there were opportunities. Furthermore, in the case of Bo village, when other land for cultivation was no longer available for use, the local people desired to cultivate even in areas with poor soil fertility where they initially did not want to farm. This indicates that the desire to participate changes according to factors such as soil fertility and the possibility of utilizing land for agricultural use.

Intercropping under planted trees (*Lahan di bawah tegakan*) was mainly used for growing highly shade-tolerant root crops (presently, the *Porang* plant from the arum root family or turmeric known as *Kunir*). In Da village, *Porang* was actively grown (approximately 60 ha, with another approximate 2 ha of the herb *empon-empon*). In Ba village, cassava (approximately 6 ha) and *Kunir* (approximately 4 ha) are planted (a couple of hectares of *Porang* was planted previously, but is no longer cultivated as the attempt was unsuccessful). In Bo village, people did not attempt growing these crops after several failed attempts; however, around 2010, they were inspired by examples of success in other villages. Efforts were once again under taken to grow these crops (approximately 2 ha and 1 ha of *Porang* and *Kunir*, respectively) and to grow grass for livestock (approximately 5 ha), bananas (approximately 10 ha), and cassava (approximately 5 ha). In this case too, the desire to participate changed according to land productivity factors, such as soil fertility and drainage, as well as other agricultural land use possibilities. In some areas, some management group members were unaware of the intercropping under the planted trees system.

Although cultivation in gaps created from the death of plantation trees is not defined in the SFC's forest management system, the SFC neither encourages nor prohibits the spontaneous use of such areas by local people. Generally, this is implemented in small gaps in areas, and though some people deem that this is intercropping under planted trees, in the case of Ba village, there are large areas of non-performing plantation land cultivated by many management group members. Furthermore, in Ba village, there were more than a few management group members engaged in the cultivation of vacant land surrounding power lines. Such land is not particularly attractive as agricultural land because it has low soil fertility. However, when no other land is available for cultivation, it is valuable as arable land. Thus, the desire to participate can be considered as changing according to factors such as soil fertility and the possibility of land utilization for agricultural use.

Opportunities for Forestry Work

After concluding a contract with the forest district, priority is given to management group members to participate in opportunities for forestry work. However, similarly to *Tumpang Sari*, such opportunities do not arise every year. Furthermore, though forestry jobs may include activities such as creating service roads, thinning, final cutting, and transportation, it is not possible to conduct thinning or final cutting without tools such as a chainsaw. The thinning and final cutting of large trees also requires specialized techniques, which may lead to some limitations. In addition, there is not much work in terms of volume, probably merely a few days of work for a few people.

In the case of Da village, there is a forest district nursery where many management group members work that provides ongoing work opportunities[13]. However, this type of employment is not available in all villages.

Profit Sharing

When there was logging within the forest under the jurisdiction of the management group and the forest district office received income from sales, part of this income was distributed to the management group after contract conclusion with the forest district office. However, when final cutting is conducted, it is based on the SFC's long-term forest management plans, and there is no guarantee that there will be ongoing profit share income available for each village every year. The history of profit sharing until now, as shown in Table 3-2, reveals that it has been received twice by Da village, five times by Ba village, and five times by Bo village. There was no real dissatisfaction about the fact that there is no profit share income

---

13) Results of the 2004 survey showed that wages were 10,000–15,000 rupiahs/day, which was approximately the same level as for other labor.

Table 3-2. Distribution of profit sharing

		Year						
		2004	2005	2006	2007	2008	2009	2010	2011 (Forecast data)
profit-sharing distribution to villages surveyed (thousands of rupiah)	Bo	0	4,364	0	45,346	19,388	0	1,801	311
	Ba	1,605	5,934	0	0	141	2,920	2,130	0
	Da	0	899	0	0	0	0	20,959	0
profit-sharing distribution in the north sub-KPH	number of MPSDHs participating in profit-sharing (groups)	9	9	15	16	24	22	13	23
	number of MPSDHs in the north sub-KPH (groups)	24	25	28	33	36	36	36	36
	proportion of MPSDHs participating in profit-sharing (%)	37.5	36.0	53.6	48.5	66.7	61.1	36.1	63.9
	total amount of shared profit (thousands of rupiah)	15,477	163,993	204,746	283,792	337,099	195,168	346,622	292,060
	average shared profit (thousands of rupiah/group)	1,720	18,221	13,650	17,737	14,046	8,871	26,663	12,698

Source: Madiun forest district internal data.

every year and that the dividends received also fluctuate greatly every year; there seems to be an opinion that, unlike before the introduction of PHBM, at least some amount is received.

Profit share income obtained from the forest district is redistributed among management group members, management group board members (honorarium), management group budget, and the village office budget. The proportions are determined by each management group and clearly outlined in the group's bylaws. In many cases, 80% of the income goes to management group members, 10% to management group board members, 5% to management group accounting, and 5% to village office. Although dividends are directly paid to management group members in some cases, dividends are deposited in savings that are used to fund management group collaborative activities or to develop local infrastructure (this will be discussed later in detail). In the case of Ba village, some amount is used to partially fund efforts such as intercropping under planted trees; however, such decisions are carefully discussed to ensure transparency of the decision-making process regarding usage of funds, most of which is deposited in a bank.

Cooperation with Forest Protection Administration

By concluding a contract with the forest district office, management groups become obliged to cooperate with forest district office forest protection and management. The main task, which is also the aim of introducing PHBM system, is to cooperate with measures to combat illegal logging. Although forest patrols are one of the specific activities that are expected to be performed, primary patrol participants are administrative staff rather than management group members. However, management group members going to agricultural land within the forest or going into the forest to harvest forest products, such as firewood and fodder, essentially have the same effect as patrols. Intercropping under planted trees could make people think twice about cutting down trees in a particular area, according to the forest district office. In the case of Da village, the management group has been carrying out forest fire patrols since 2006[14]. This is because they planted *Porang*, a plant from the arum root family that is susceptible to fire, as a management group collaborative activity and want to reduce fire damage to the crops. This functions as patrolling to combat illegal logging too.

In addition, management group members respond to requests to recover timber from illegal logging that has been left in the forest and to extinguish forest fires promptly.

On the basis of the contract with the forest district office, management group administrative staff provides a variety of information regarding forest management and regularly prepares reports.

---

14) These were not performed in 2010 as the fire risk was low because of rainfall even in the dry season.

Activities as Part of Village / Settlement Development Activities

As previously mentioned in "Profit Sharing", the management groups use profit share income to fund collaborative activities of the group and to develop local infrastructure (such as construction of meeting houses, road improvement, and simple water system development). The collaborative activities that the management group engaged in include not only production activities such as the cultivation of cash crops but also the provision of microcredit / mutual finance (*arisan*[15]) to management group members, education, and vocational training. These activities are funded by the profit share income derived under PHBM as well as by subsidies from the government and voluntary organizations and loans from investors. Furthermore, this does not apply exclusively to activities involving management group members; rather, it includes activities involving local people as a whole. The management group plays a vital role both within PHBM and in the development of the villages and settlements of the area. In reality, the provincial government and the regency government have begun considering management groups as capable local groups that can form the basis for local policy implementation.

The 36 management groups within the northern sub-forest district are engaged in a variety of activities including intercropping under planted trees (23 groups in arum root cultivation; 16 groups in cassava cultivation; 15 groups in herb cultivation, and seven groups in grass cultivation for livestock), loans (13 groups), seedling production (seven groups), fertilizer production (seven groups), construction of public facilities (seven groups), and road improvement (six groups). Da village is the most proactive of the surveyed villages in terms of collaborative activities and is engaged in activities such as intercropping under planted trees, teak seedling production, fertilizer production, loans, fish farming, and manufacturing and selling wooden products. Ba village is primarily engaged in intercropping under planted trees, whereas Bo village is engaged in chainsaw rental, loans and mutual finance, and intercropping under planted trees. Intercropping under planted trees features prominently in all three villages, which might have led to the increase in the desire to use the forest and forest ownership according to the forest district office and field facilitators.

Meetings

Management groups hold meetings to further their activities. Although the topics discussed in these meetings vary according to the management group and the time of the year, important issues usually include such as activity evaluation, formulation of plans for the follow-

---

15) Every person attending the meeting contributes money, and one person from among those in attendance receives the entire amount. This system of finance is widely used throughout Indonesia. Though there are various ways to select the person who receives the collected amount, selection was accomplished by drawing lots in the survey villages, with only people who had yet to receive funds included in the draw the next time.

ing financial year, and financial reports (especially the distribution of income from final cutting and thinning). There are meetings exclusively for the board members and those that all management group members can attend. In addition to being held regularly, the administration meetings are also held whenever it is necessary to discuss urgent matters (such as when requests are received from the forest district office for cooperation with certain work or offers have been received from outside investors regarding joint projects). In the case of Ba village, such occasional board members meetings have been held several times a year. In 2008, these meetings were held 10 times in Da village, twice in Ba village, and twice in Bo village.

The way in which meetings for ordinary (non-executive) members of the management groups are held differs according to the management group. Some groups, such as that of Bo village, hold them at the same time as religious activities or local meetings, whereas other groups, such that of Da village, hold meetings solely to discuss matters relating to the group. In villages with many members, such as Bo village, discussions are held in each subgroup, and a meeting for all members is normally not conducted. The frequency of such meetings also varies, with some groups such as that of Da village meeting more than once a month, whereas other groups such as that of Ba village meeting roughly once a year. Concerning the attendance rate, attendance in Da village is high at approximately 80% on average because mutual finance activities are also conducted at the time of the meeting. Similarly, attendance is high in Bo village because the meeting is held at the same time as the monthly meeting of the sub village. However, when discussions are being held with outsiders during the day or when discussions are limited to specific topics, the meeting is attended only by those who are interested and have spare time to attend. On the other hand, when Ba village held a meeting of all members in 2010 to engage in intercropping under planted trees, only 20 members including the board members attended.

Contact between the board members and ordinary members is maintained through the abovementioned meetings, subgroup leaders, everyday individual communications between the board members and ordinary members, and when people meet for management group collaborative activities.

4. Strategy Analysis from Collaborative Governance Perspective

4.1 The Strategy Adopted by Local People

If the strategy adopted by local people with regard to PHBM were to be classified as a response, resistance, or governance strategy, as indicated in the introduction, it would be classified as an "adjustment strategy". Forest that is the subject of PHBM is state forest that, over the past 100 years, has been managed in a monopolistic manner by the colonial government, state, and SFC. Local people no longer possess any customary or legal rights, and local society has accepted the exclusive management of the SFC. When the SFC proposed a new program proposal for collaborative management of the forest, local people were unable to implement a resistance strategy to eliminate outside influence or implement a governance strategy to collaborate with outsiders while maintaining resident leadership. The only choices left for local people were to respond or oppose (or ignore).

The following is a summary of the proposal from the forest district office. The forest district office formulated a system with little burden to obtain the local society's cooperation in reducing social risk (specifically to combat illegal logging) in forest management while providing many benefits to local people. PHBM provides the local people with long-term (at least 10 years; renewable if there are no specific problems) opportunities for agricultural activities within the forest, such as official authorization for harvesting forest products, opportunities for forestry work, and profit share income from the proceeds of teak sales. On the other hand, local people have few responsibilities or burdens, they are not forced to establish a management group, and the establishment and operation of management groups is left to be self-governed by the local people themselves. Though the responsibilities of management groups are determined in the contracts that are concluded, there are no penal provisions for occurrences of infringements (it could be said that a reduction in the profit sharing rate from teak sale income that depends on how much illegal logging occurs is a positive incentive). Concerning cooperation to combat illegal logging, there are no specific responsibilities for individual members. Furthermore, for individual local people the introduction of the group system signifies a decrease in workload and an increase in negotiating power, because they represent their own interests in negotiations with the forest district office.

In terms of the operation of the system, Gadjah Mada University provided a facilitation service for local people for the establishment of management groups and introduction of field facilitators and, later, for the operation of the management groups. Furthermore, on suggestion by the head of the forest district office, the office has endeavored to change the staff's way of thinking and has actively promoted PHBM, which favors local people as partners. This is also the policy of the SFC as a whole, and the SFC's internal regulations regarding PHBM state that efforts regarding PHBM will begin by first changing the way of thinking of the SFC staff. Building a good relationship between local people and the forest district office seems to be going comparatively smoothly in the Madiun forest district compared with other examples, though there is frequent mention of the lack of progress in building a relationship of trust (Fujiwara et al., 2012).

In response to such proposals from the forest district office, most local people do not consider it necessary to actively or aggressively oppose these. They have chosen to either adopt an "adjustment strategy" or passively oppose (ignore or reserve their opinion) such proposals. People with little involvement with the forest, such as company employees and farmers who live far from forest land, chose to passively oppose the proposals. The remaining local people chose the "adjustment strategy" (as the authors have already mentioned, the proportion of all households in the villages that had registered with the management group as members was 6% for Da village, 30% for Ba village, and 48% for Bo village). Since PHBM was implemented, a number of positive things occurred—advantages such as the distribution of profit share dividends have become obvious; awareness regarding the content of PHBM has increased; the forest district office continues to actively promote PHBM; and the management groups receive support and subsidies from outside. Therefore, people who had ignored PHBM or had reserved their opinions have changed their strategy to "adjustment" to PHBM, and the number of group members has increased.

On the other hand, on considering stakeholders who thought that they would be adversely affected by PHBM (specifically people who profited from illegal logging and secured opportunities for *Tumpang Sari* or forestry work by networking with local forest district supervisors), it may have been necessary for them to actively "resist" in order to maintain such positions. However, in the end, they accepted the local people's decision as a whole.

4.2 Issues regarding Strategy

Possibilities and Limitations for Local People: Impact on Livelihoods of Participants

Local people chose "adjustment strategy" toward PHBM because they decided that the relative advantages were greater than any disadvantages / costs. When evaluating PHBM, it is important to identify the impact and limitations on improving the livelihoods of participating local people.

Regarding the effect on improvement of livelihoods, it was possible for individual management group members to enjoy increased opportunities for agricultural activities within the forest, such as official authorization for harvesting non-timber forest products, profit share dividends, and financial and physical benefits from group activities. From the results of the questionnaire conducted by the authors in 2004–2005, the surveyed households acquired approximately 53.8% of their arable land within the forest and nine households secured their arable land only within the forest. Furthermore, according to the results of the survey, approximately 10.8% of the total cash income (gross income) of these households was derived through PHBM, and two households derived cash income solely through PHBM. This income was derived from activities such as selling agricultural products produced within the forest, forestry wages, selling harvested forest products, and group activities. Psychologically, a relationship of mutual trust has been built between the forest district office and local people. In addition to obtaining the right to officially enter the forest and harvest non-timber forest products, local people no longer fear encountering forest district staff when entering the forest.

With regard to the impact on the management group as a whole and the local society, sources of funds for collaborative activities have been secured through profit share dividends and outside support, including the forest district office. Village and settlement development has progressed through the management groups, which have become the recipients of outside support.

On the other hand, there have been no real burdens or costs on an individual basis since the introduction of PHBM; however, on a management group basis, there has been an increase in the workload of the board members. Although in one village a lack of clarity on the redistribution of profit share income within the village caused some temporary conflict, it was quickly resolved. There has also been no real change in social relationships between the members and other people in village. In the case of other forest district offices, it is reported that a lack of trust in the relationship between the board members and ordinary members that was not resolved resulted in a widening economic gap in the local society (Shiga et al., 2012).

However, limits were observed with regard to the effect on improvement of livelihoods and in terms of agricultural activities within the forest. With *Tumpang Sari*, the crops that can be cultivated in teak plantations are generally limited because of lack of sunshine after crown closure. It is also not possible to continuously provide different areas for *Tumpang Sari* within villages because of the uneven age composition of the forest, nor is it possible to supply adequately large areas of land to all those who want it. Thus, it is difficult to utilize *Tumpang Sari* as a stable way to secure arable land. According to the results of the 2004–2005 survey, 46.1% of the households that were surveyed did not possess arable land within the forest. The same was true of intercropping under planted trees regardless of *Tumpang Sari*, and soil fertility was low in some areas. Though there are various influencing factors such as people and the areas in question, intercropping is not necessarily always seen as more attractive than sharecropping or work outside the forest. In terms of cash income, similar to *Tumpang Sari*, it was not possible to provide profit share dividends and opportunities for forestry work on a regular basis every year because of the forest age composition issues, and in many cases the volume was inadequate.

Therefore, it is difficult to say that there are no qualitative, quantitative, or ongoing issues with regard to the benefits provided through PHBM. Similar instances are also being reported in other forest district offices (e.g., Maryudi and Krott, 2012; Fujiwara et al., 2012). Therefore, it is difficult for all management group members to rely on PHBM for their livelihood, and to some extent, some type of foundation for their livelihood is required outside PHBM.

Problems Arising from Limiting the Transfer of Rights to Local People
As mentioned above, considering the causes of the limitations of the PHBM's ability to

improve the livelihoods of local people, problems identified include maintaining a balance between forestry land use and agricultural land use and the long-term nature of land use cycles due to the long rotation periods associated with teak forestry. This is partly a technical issue and there are many points that can be addressed using the present PHBM with a little ingenuity. However, when considering what varieties to plant and how to distribute land for forestry use and agricultural use, issues arise that are determined by the SFC headquarter. Even at present, although it is possible for management groups to voice their opinions regarding forest management, they only have the right to suggest whereas the right to decide lies with the forest district office (depending on the issue, decisions may need to be advanced further up the chain of command to a UNIT or headquarters).

This instance illustrates the limitations of PHBM. If adjustments could be freely made with consideration of the local people's opinions at the local level, it would lead to a huge undertaking regarding how to achieve the sustainable forestry management required by the SFC; how to secure sustainable income as a corporate entity, and how to establish and adjust forest management goals. Furthermore, on the national level, this could lead to an issue of how to reconcile national land development plans with people's desires at the local level.

In addition, under the present system where many decision-making rights lie with the forest district office, it is feared that the enthusiasm of the top people and the attention to detail in these efforts will disappear because of the short transfer cycle for both the head of the forest district office and staff charging in PHBM.

Applicable Local Conditions

There are a number of prerequisites for PHBM to achieve and maintain good results. This chapter will list the four main requisites below. Whereas there are some conditions that were originally satisfied by the forest district office, there are other conditions that have required some effort. If these conditions collapse in the future, it is possible that they will interfere with the operation of PHBM.

(a) **Management Goals Agreement**
- **Local people accept company goals of forest management**: PHBM was able to commence with relative ease even though teak plantation management had been determined *a priori* as a forest management policy. This was because the management policy for forests under the jurisdiction of the SFC in local society was recognized as a teak plantation management policy, and no discussions were required when management policy was determined.
- **Land ownership is clear and stable**: Although this condition is related to the above-mentioned condition, the reason for the local society's easy acceptance of the priority rights of the forest district office in relation to policy decisions on land use is that the

geographical boundaries of land under the jurisdiction of the forest district office are recognized within local society.

(b) Management Group Operation by Local People

- **Relationships among local people are good**: As the operation of management groups is left to be governed by the participating local people, social relationships of the local society are incorporated in the operation of management groups as mentioned by Maryudi and Krott (2012) and Shiga et al. (2012). It is vital that local people share good relationships to avoid hindering the operation of management groups. Furthermore, when a system is introduced that will benefit local people, good relationships among local people are important to avoid conflict regarding the distribution of benefits and costs.

- **Presence of motivated leaders with an interest in forest management**: As this involves entrusting the introduction of a system that has not existed before under the initiative of local people, whether this can be introduced and utilized depends on the leaders and a core group who will promote this system. In examples from Madiun and from those of PHBM from the Bogor forest district office in the Western Java province and the South Keduh forest district office in the Central Java province, which were also surveyed by the authors, the importance of the existence of motivated leaders and a central core support group has been suggested.

(c) Effectiveness of Incentives

(c-1) Provision of Arable Land within Forests and Profit Share Income

- **Areas with a shortage of land**: For the effective functioning of the incentives provided, there must be a demand within the local people for such incentives. From examples within the Madiun area, the incentives provided were not considered attractive by households who already owned enough agricultural land or those who were company employees. They showed little interest regarding PHBM. Furthermore, not only land area, but also soil fertility and agricultural productivity are important.

- **Areas with a shortage of cash income**: Similar to the abovementioned agricultural land, the incentive of cash income did not apply for households where people worked in companies in the city, had a certain degree of stable cash income, or whose family had emigrated overseas to work and were sending money home.

- **Few people with a drastic shortage of agricultural land**: As mentioned in 4.2 above, it is not possible to satisfy everyone in terms of the quantity of agricultural land provided or the regular provision of such land. Thus, if there is a drastic shortage of agricultural land in the Madiun area, it is conceivable that cooperation could not be obtained from local society because of the inability to provide an adequate volume of incentives.

- **Few people with a drastic shortage of cash income**: Similar to agricultural land, if there is a drastic shortage of cash income in the Madiun area, it is conceivable that cooperation could not be obtained from local society because of the inability to provide adequate volume of incentives.

(c-2) Distribution of Income from Teak Sales to Local People
- **High income tree species such as teak**: PHBM was able to provide various incentives for local people because the forest district office was able to earn a profit even when it provided the incentives. Furthermore, because a profit share of up to 25% of profits from sales was provided, it is important for the product to generate a high profit to be able to provide local people with an attractive incentive.
- **Quantities of a resource that can be sold immediately**: When PHBM was first introduced many local people doubted that it would really benefit them. What may have swept aside this doubt and increased trust in the system was the fact that the forest district was able to secure adequate resources and immediately pay some amount of profit share.

(d) Good Partnership between the Company and Local People
- **Good relationship between company and local people**: The local people cooperated with PHBM when it was introduced as well as on an ongoing basis because it is supported by a good relationship between the company and local people, although there are also incentives and few requirements.
- **The company is making serious efforts**: Many local people doubted that the content of the system shown to them could be implemented as written when PHBM was first introduced by the forest district office. To dispel this doubt, the forest district office continued and implemented the system, modified the staff's way of thinking, improved the system and its method of operations, and held discussions with local people. By doing so, a closer relationship was formed between the forest district office and local people, trust in the system increased, and cooperation with the system was strengthened.
- **Internal company order is maintained**: Initially, in the local level of the Madiun forest district, there was a lack of understanding that PHBM was different from previous concepts, and there was backlash from some forest district branch offices. In areas where such branch offices were located, cooperation from local people could not be obtained; there was a delay in organization of the management group and local people voiced a degree of dissatisfaction.
- **Presence of advocates and advisors such as field facilitators**: In the Madiun forest district, field facilitators and Gadjah Mada University played a major role in supporting the building of good relationships between the forest district and local people, conducting extension activities for local people, and training the staff of the forest district. It was necessary to secure adequate numbers of such people and activity budget to ensure that these advocates and advisors were able to appropriately utilize their skills.

Future Possibilities

This section considers the future possibilities of PHBM from the perspective of collaborative governance. In terms of forest management itself, there is a minor possibility that there will be a transition from "adjustment strategy" to *"kyouchi* strategy" (see the Introduction in this book) because of the limitation in the transferring ownership and decision-making rights.

At present, in terms of the utilization of benefits obtained through forest management and avoidance of disadvantages, considerably broad decision-making rights are given to the local people. In addition, because of the forest district office's attitude of favoring local people as partners, local people's cooperation with forest management has been obtained, and local people participate in a system of environmental and resource management through solidarity and collaboration with a diverse range of stakeholders. Local people contribute as important stakeholders; furthermore, they also enjoy benefits from the established system. Whether this will continue in the future depends primarily on the SFC and the Madiun forest district office as the entity promoting and responsible for PHBM. In the future, given the present rise of democracy in Indonesia, the possibility of a change in attitude of both the SFC and forest district office or the cessation of PHBM is not high. However, there is a possibility of a gradual decrease in the benefits to local people and increase in costs due to rising costs, which depends on the management state of the SFC and on whether there is a decreased risk of illegal logging and clearing of land. The presence of this risk could be seen as a weak point of the "adjustment strategy".

To prevent unilateral decision making by the SFC and / or the forest district, several measures are implemented, such as ensuring that a contract is concluded between the forest district office and the management groups and guaranteed by local authorities, and that the field facilitators act as an intermediary between both parties. It is necessary to ensure that these measures are more than a façade, in order to reduce risk and amplify the local people's voice.

Acknowledgement

The cooperation of various people in the survey villages, the SFC, Gadjah Mada University, Forest and Nature Conservation Research and Development Center, the Indonesian Ministry of Forestry, and Bogor Agricultural University was invaluable for conducting our field investigations. We would like to especially thank Dr. Mohammad Na'iem, Dean of the Faculty of Forestry at Gadjah Mada University; Dr. Taulana Skandi and Ms. Sri Suharti, Forest and Nature Conservation Research and Development Center, the Indonesian Ministry of Forestry; and Madiun field facilitators Agus Widiyant, Didik Purnomo, Sutoyo, and Nasarudin Latif for their invaluable assistance in designing, implementing, and analyzing the survey. The authors would like to take this opportunity to express our heartfelt appreciation.

References

Awang S. A., H. Purnomo, W. Wardhara, P. Guizol, P. Levang, S. Sitorus, N. Murtiyanto, and Y. Susanto 2005. *LPF Project South Sumatra Case Study*. Bogor, Indonesia: CIFOR.

Badan Pusat Statistik (BPS) Kabupaten Madiun 2003. *Kabupaten Madiun dalam Angka 2002 (Madiun Regency in Figures 2002)*. Madiun, Indonesia: BPS Kabupaten Madiun.

BPS Kabupaten Madiun 2005a. *Kecamatan Kare 2004*. Madiun, Indonesia: BPS Kabupaten Madiun.

BPS Kabupaten Madiun 2005b. *Kecamatan Dagangan 2004*. Madiun, Indonesia: BPS Kabupaten Madiun.

Djamhuri, T. L. 2008. "Community Participation in a Social Forestry Program in Central Java, Indonesia: the Effect of Incentive Structure and Social Capital". *Agroforestry Systems*, 74: 63–96.

Faculty of Forestry, Gadjah Mada University 2005. *Survey on Community Involved Forest Management in Madiun Forest District, East Java, Indonesia*. Bogor, Indonesia: Forestry Research and Development Agency (FORDA) and Japan International Cooperation Agency (JICA).

Fujiwara, T., R. M. Septiana, S. A. Awang, W. T. Widayanti, H. Bariatul, K. Hyakumura, and N. Sato 2012. "Changes in Local Social Economy and Forest Management through the Introduction of Collaborative Forest Management (PHBM), and the Challenges It Poses on Equitable Partnership: a Case Study of KPH Pemalang, Central Java, Indonesia". *Tropics*, 20(4): 115–134.

Institute for Global Environmental Strategies (IGES) 2007. *Decentralization and State-sponsored Community Forestry in Asia*. Hayama, Japan: IGES.

Maryudi, A. and M. Krott 2012. "Poverty Alleviation Efforts through a Community Forestry Program in Java, Indonesia". *Journal of Sustainable Development*, 5(2): 43–53.

Nawir, A. A., L. Santoso, and I. Mudhofar 2003. *Towards Mutually-beneficial Company-community Partnerships in Timber Plantation: Lesson Learnt from Indonesia*. CIFOR working paper 26. Bogor, Indonesia: CIFOR.

Obidzinski, K. and A. Dermawan 2010. "Smallholder Timber Plantation Development in Indonesia: What is Preventing Progress?" *International Forest Review*, 12(4): 339–348.

Peluso, N. L. 1992. *Rich Forest, Poor People: Resource Control and Resistance in Java*. Berkeley, CA: University of California Press.

Perum Perhutani KPH Madiun 2009. *Ringkasan Publik (Public Summary)*. Madiun, Indonesia: Perum Perhutani KPH Madiun.

Rohadi, D., M. Kallio, H. Krisnawati, and P. Manalu 2010. "Economic Incentives and Household Perceptions on Smallholder Timber Plantations: Lessons from Case Studies in Indonesia". Paper presented at Global Conference on Agricultural Research for Development, Montpelier, France.

Shiga, K., M. Masuda, and N. Onda 2012. "Jawa ni okeru Ringyoukousya no Chiiki Taisaku no Hensen oyobi Jumin Kyoudou Sinrin Kanri Sisutemu no Kadai – Seido to Unyou no Jittai" [Changes in People's Involvement Measures by the State Forestry Corporation and Challenges of Joint Forest Management System in Java: Policy Design and Implementation Process]. *Ringyo Keizai Kenkyu* [*Journal of Forest Economics*], 58(2): 1–13.

Shimagami, M. 2010. "Indonesia ni okeru Komyunitirin (Hkm) Seisaku no Tenkai – Ranpun Syu Butun Sanroku Syuhen Chiiki wo Jirei to site" [History of Community Forestry Policy in Indonesia (Hkm)]. In: M. Ichicawa, F. Ubukata, and D. Naitou (eds.) *Nettai Ajia no Hitobito to Sinrin Kanri Seido—Genba kara no Gabanansu Ron* [*People and Forest Management in Tropical Asia: Local-Level Impacts of Diverse Governance Systems*]. Kyoto: Jimbun Shoin, pp. 128–147.

Van Noordwijk, M., S. Suyanto, S. Budidarsono, N. Sakuntaladewi, J. M. Rosbetko, H. L. Tata, G. Galudra, and C. Fay 2007. *Is Hutan Tanaman Rakyat a New Paradigm in Community Based Tree Planting in Indonesia*. ICRAF Working Paper 45. Bogor, Indonesia: ICRAF.

Zainuri, H. 2011. Hutan Desa (http://www.jatan.org/wp-content/uploads/2011/11/2011.10.21-Zainuri-Mitra-Insani.pdf) (accessed May 24, 2012).

Chapter 4
Legitimacy for "Great Happiness"
Communal Resource Utilization in Biche Village,
Marovo Lagoon in the Solomon Islands

Motomu Tanaka

1. Collaborative Governance and Dynamics of Local Societies

1.1 Dynamics of Local Societies

This chapter examines the dynamics of local societies formed through communal utilization of local natural resources that are the foundation of people's lives and focuses on the concept of legitimacy that has been formed in relation to the communal utilization of such resources. Further, it also describes the fluctuations that have occurred as local societies have accepted the involvement of outsiders.

Various local societies have been formed globally by people living in and utilizing a wide range of natural environments. However, because of increasing globalization and the spread of urban societies that rely on many external resources, lifestyles based on local natural resources, the bonds developed among people through the communal utilization of natural resources, and the diversity of local society are slowly being lost. The collaborative governance theory has arisen from discussion in the midst of changes regarding what people perceive to be the ideal state of local society.

The diversity of local society in the Solomon Islands (see Figure 4–1) has been formed as a result of people's joint utilization of the area's rich natural resources—the sea, forests, and rivers. The communal use of traditional resources and sharing of land by tribes are recognized both legally and practically, and local tribes have the strongest rights in matters regarding development. All the more reason, therefore, for the presence of collaborative governance at work in societies that are formed through some of the most fundamental human activities—the communal use of local resources. Inoue (2009:11) not only affirms the involvement of outsiders but also places importance on considering the legitimacy of outsiders' involvement from the perspective of the people in the local society. In the Solomons, local people consider legitimacy to be formed within the framework of the communal utilization of local resources. The logic of the people in the local society regarding

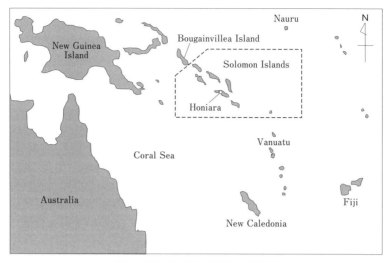

Figure 4–1. Location of Solomon Islands

the pros and cons of developing collaborative governance and their views regarding outsiders' involvement in local society is considered herein.

1.2 Local Network in the Solomon Islands

The Solomons features 2.213 million ha of forest that constitutes 77% of national land (FAO, 2010: 9) and the sea, which is blessed with an abundance of marine resources, such as skipjack tuna; this forest area is the subject of various development initiatives. In addition to commercial logging[1] mainly performed by Southeast Asian companies, fishing and mining activities are engaged in by Japanese, Chinese, and Australian companies. Indeed, this richness of natural resources is the foundation that supports local society, and the various development initiatives effect local people's self-sufficient lifestyle and interpersonal relationships.

The people of the Solomon Islands have no real sense of national identity; they derive their sense of belonging from the sharing and communal use of various resources and through

---

1) Commercial logging, which is actively performed in Southeast Asia, began in the Philippines. From the 1970s, regulations on the export of raw wood were consolidated, and commercial logging increased in Indonesia as well as in Sabah and Sarawak states in Malaysia. During the 1990s, logging volumes rapidly increased in Papua New Guinea and the Solomons (Tanaka, 2004). The Solomons is one of the last places where commercial logging of tropical natural forest is still performed, and since regulations were tightened in Malaysia, Indonesia and the Philippines, the Solomons has become the location for behind-the-scenes commercial logging.

their interpersonal relationships. In the Solomon, 87% of the state land is referred to as customary land, and the communal utilization rights of the tribes in each area are recognized with regard to such land (Statistics Office, 1995). When activities such as commercial logging are performed, disputes regarding contracts and land use arise, and the rights of tribes are not ignored. In fact, local tribes themselves are a part of the entities involved in introducing such developments.

Involvement with outsiders is not limited to involvement in large-scale developments; it also includes developments trialed by international NGOs, including small-scale logging by villagers, eco-tourism, the sale of timber, and beekeeping (Miyauchi, 1998; Tanaka, 2002). In 1998, ethnic disputes broke out when tensions escalated regarding the number of people from Malaita Island coming to the capital Honiara, resulting in increased foreign government involvement aimed at peacekeeping and the restoration of law and order.

The Regional Assistance Mission to the Solomon Islands (RAMSI) was deployed to the country in 2003. RAMSI comprised personnel from other countries in the South Pacific, with Australia playing a central role, and as of 2014, personnel are still stationed there. RAMSI has contributed to the country not only in terms of restoring law and order but also in aspects such as politics and justice. Nature conservation projects have also been actively implemented by foreign research institutions and NGOs.

As previously mentioned, communal use and the ownership of customary land is recognized legally as well as practically in the Solomons, and the Solomons is one of the few places where customary land ownership groups have a strong influence in development and local resources. On the basis of local natural resources and relationships, they have adopted a "collaborative governance" strategy where they accept—or, at times, reject— foreign governments and corporations, research institutions, and NGOs, etc. In other words, as a group, the local people seem to play a proactive role in decision making.

Resources in local society are not limited to natural resources—the people living in the local area themselves and their interpersonal relationships are also important resources. Mutual assistance within groups engaged in communal utilization of natural resources is common. In the Solomons, this assistance is not limited to sharing labor, such as that involved in the communal harvesting of crops, assistance in raising children, and the sharing of harvested items, but also includes sharing knowledge and techniques regarding how to utilize natural resources. Not only is there communal utilization of local natural resources by networks of people who share a common bond of trust, there is also the mutual sharing of techniques, knowledge, and labor acquired through such utilization. These networks are hereafter referred to as "mutual networks" (see Figure 4–2).

Mutual networks that support natural resources and local society are a resource that transcends the geographical boundaries of what is defined as "local" and can be better

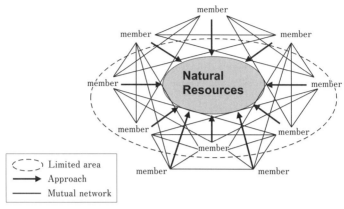

Figure 4–2. Local Network

referred to as "human resources." Furthermore, cultural resources are formed through the action of human resources on natural resources. With the exception of migratory birds and fish, which range over a wide area, natural resources are, to some extent, fixed in a certain geographical location. However, mutual networks can transcend the geographical boundaries of specific areas with natural resources.

Local society in the Solomons is formed on the basis of the communal use and ownership of resources and associated geographical and blood relationships. Furthermore, there is a concept of right and wrong within such mutual networks—the concept of "legitimacy." The term "member legitimacy" is used herein to refer to legitimacy within the local society regarding who among the members of the local society with the resources, will use resources and in what way.

Though member legitimacy associated with the communal utilization of a diverse range of resources changed due to development, the Solomons continue to be a legally and practically recognized region. Gatokae Island and Biche village (see Figure 4–3 and Plate 4–1), featured in this chapter, are areas that have accepted commercial logging and the involvement of international NGOs and researchers in resource management. This chapter clarifies concepts regarding the use of resources in Biche and ascertains member legitimacy. Then, the possibility of the formation of collaborative governance is considered on the basis of how local people considered member legitimacy in terms of outsiders' involvement with respect to local resources.

1.3 Abundance of Natural Resources

Gatokae Island is a volcanic island approximately 10 km in diameter located at the southeastern tip of Marovo Lagoon in the Western Province of the Solomons. In the center of

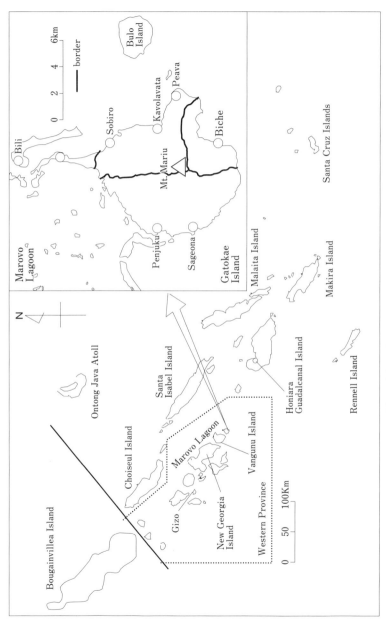

Figure 4-3. Solomon Islands and Gatokae Islands

the island is Mt. Mariu, an active volcano, approximately 800 m in height. According to the local people, there was a small-scale eruption in the 1930s.

The people of Gatokae Island speak Marovo and derive their livelihoods from shifting cultivation (*chigo*) and fishing (*chinaba*). As of December 2013, there were approximately 1,500 people living in the area, comprising the Mategele tribe and their wives (hereafter, the "M tribe"). When missionaries first visited the area in 1915, the total population of the island was recorded as approximately 200 people (Nuefeld, 1976: 196–197). There are currently seven villages on the island: Biche, Penjuku, Peava, Bili, Sobiro, Kavolavata, and Sageona.

As of July 2014, there were 25 households with 138 people, all belonging to the M tribe, living in Biche and engaged mostly in shifting cultivation and fishing. The leaders (*bagara*) of the M tribe are from this village. Though the northwestern side of the island is on the shores of Marovo Lagoon and has quiet seas, Biche is exposed to the open seas on the south side of the island where the seas can be rough.

There are two theories regarding the origin of the name *Gatokae*. One maintains that the word is derived from the words *nga* (myself) and *tokae* (help), meaning "the island that helps me (in my life)" or the "island where I can be self-sufficient." The other theory states that it is derived from *gato kale*, which means, "a place left on the edge (of Marovo Lagoon)." As evident from these theories, Gatokae Island is a small island on the edge of the Marovo Lagoon blessed with an abundance of natural resources, such as forests, rivers, and the sea, from which people can obtain the resources they need to live.

2. Rights, Ownership, and Utilization in Biche

Rights (*Nginira*)

In the Marovo language, *nginira* is a concept that is similar to rights. *Nginira* means a type of "strong power" for claiming something. For this strong power to be acknowledged by other people as rights, it is important for it to make sense (*noro*). The common awareness among villagers that it should be so then becomes *noro*.

Torigoe (1997: 38–41) stated that while it is "impossible to understand peoples' minds," it was possible to ascertain the "logic behind the how they justify their behavior." *Noro* could also be described as a common awareness regarding what is deemed to be "just" in the "minds of people" in local society. Although *noro* is the basis for recognizing claims and attitudes related to resource utilization, it does not guarantee absolute "rights" that would eliminate others' claims. Such clashes of various "rights" are not a part of everyday village life, and when such clashes do arise, depending on the situation, either a better "logic" is sought in the midst of diverse claims or actions or there is interaction even through tacit eye contact.

Ownership (*Hinoho*)

In the Marovo language, the concept closest to ownership is *hinoho*. *Hinoho* means the state of "holding" something or "enjoying" it. A "*hinoho* person" signifies someone who enjoys the "prosperity" that comes from holding some resource.

On Gatokae Island and the surrounding uninhabited islands, villagers state that the M tribe has enjoyed the "prosperity" of possessing the resources of the surrounding seas. However the term *chakei* normally is used to describe the M tribe. *Chakei* means "to protect" or "to manage" and is also used in reference to "looking after a child" or "managing money."

Villagers who have "worked at" (*tavete*) natural resources are acknowledged as protectors of resources. In addition to cutting down trees for shifting cultivation or growing crops, "working at" transplanting wild plants, branding to avoid mistaken cutting, and weeding to prepare a growing environment for beneficial wild crops refers to semi-domestication activities.

Leaders of the M tribe are the representative owners and managers of Gatokae Island and the surrounding uninhabited islands. When outsiders seek to be involved with the resources of the M tribe, the leaders engage in negotiations as representative owners. The hinoho of the M tribe, specifically of the leaders, is emphasized to outsiders and endeavors are made to reach a *noro* conclusion. However, in terms of daily life in the village, what is important is not the "ownership" of resources but how the resources belonging to the M tribe can be "communally utilized" within the tribe, and there is constant awareness regarding this standpoint.

This chapter defines the "common awareness regarding the claims, attitudes, and actions of members that are considered *noro* in the communal utilization of resources within the tribe" as the "*noro* concept." The scope of people deemed to be members under the *noro* concept will change according to the resource in question, and in some cases, there will be "fluctuations" where elements that form the *noro* concept are partially ignored due to logic and systems outside the tribe.

Miyauchi (1998: 134–136) reported that in the Solomons, there were societies where something similar to "communal utilization rights" was formed that was vaguely recognized within local society and separate from the ownership rights that are recognized as absolute rights. In Biche too "ownership" is emphasized to outsiders such as logging companies, and something similar to "communal utilization rights" that form the basis of the *noro* concept within local society are part of the villagers' everyday awareness.

Utilization (*Tatavete*)

The Marovo word for "utilization" is *tatavete*. However, *tatavete* means the general "state of something being used." There is no specific Marovo word for "utilization rights." However, on attempting to ascertain how everyone becomes aware of how people are recognized as people who can "utilize" resources, aspects similar to "utilization rights" are noticed that can be broadly divided into two categories—"member utilization rights" and "priority utilization rights."

As a rule, wild plants and animals are acknowledged as a resource that can be utilized by all members of the M tribe. In this chapter, "member utilization right" refers to the "communal utilization right acknowledged as an innate right if one is a member of a certain group" People from other islands who are adopted by (*pinausu*) or are married to (roroto) members of the M tribe have member utilization rights while they are in Biche.

Furthermore, villagers and their families who cut down the trees and prepared the land for shifting cultivation were seen as managers of land for shifting cultivation and later had priority for utilization. This "priority utilization right whereby working on some resource is acknowledged" is hereinafter referred to as a "priority utilization right." Priority utilization rights can be inherited by descendants and are also recognized for semi-domestication of coconut palms (*Cocos nucifera*), sago palms (*Metroxylon* spp.), and canarium nuts (*Canarium* spp.).

Villagers acknowledged as having resource priority utilization rights face no problems as long as the rights are considered *noro*, and they may use the resource to obtain benefits or may dispose of the resource. The next section, provides a detailed explanation of the various elements involved in the formation and maintenance of the concept of *noro*.

3. The Concept of *Noro* Underpins the Legitimacy of the Communal Utilization of Resources

3.1 What is the Concept of *Noro*?

The concept of *noro* comprises four main elements: (1) "Generosity (*hinoho*)" in distributing harvested goods; (2) "Tolerance" (*vinamagua*) in recognizing communal utilization of resources and not strongly accusing others or strictly applying penalties for mistakes; (3) "Mutual assistance" (*vinari tokae*) for not seeking only one's own benefit; and (4) "Action (*tavete*)" that can protect resources to enable all to enjoy the "prosperity" of the village and also lead to claims for priority utilization rights.

Furthermore, "sufficiency" (*isiri*) and "taboo" (*hope* or *vinangira*) are concepts that can help maintain balance among these elements (see Figure 4–4). Actions that fall outside the scope of these six areas of common awareness are regarded as shameful.

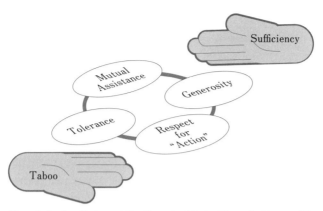

Figure 4–4. Four Factors and Two Concepts of the Legitimacy for Resource Utilization

3.2 *Vusivusi* and *Pela* Involved in Developing and Maintaining Generosity in Biche

To make generosity *noro*, the awareness of shame at being deemed *vusivusi* (stingy) also plays a part. *Vusivusi* is also used when accusing people of acting selfishly. In contrast, *hinoho*, meaning "generosity," also means "prosperity" and the "state of holding something." "People who hold something" or "prosperous people" are also considered as "generous people," and "generous" behavior is expected from them. People who are unable to behave in such a way are considered *vusivusi*. Furthermore, it is taboo to refuse the generous behavior of others. Thus, seeking generosity in each other, continuing to create such chances, and not hindering such actions are given importance.

In addition, in Biche, a fear of evil spirits (*pela*) continually promotes generous behavior. *Pela* are the evil spirits that the people of Biche feared prior to converting to Christianity in 1915. They believed that pela possess those who are extremely jealous of other people and help them murder people of whom they are jealous. Villagers fear that those possessed by pela will poison them through mediums such as food, manipulate wild animals using the power of the *pela*, and attempt to kill other people (Tanaka, 2006: 114–115).

The fear of *pela* leads people to avoid being the subject of someone's jealousy and being jealous of others. *Pela* have also been used as a means of restraining *vusivusi* behavior that could be the subject of someone else's jealousy and are a means by which a state of continual awareness is achieved wherein generous behavior is seen as *noro*.

3.3 Religion and Tolerance

The conversion of the villagers to Christianity in 1915 strongly influenced their tolerance toward outside society. As discussed later, the people of Biche regularly went headhunting on other islands[2]. When they were headhunting, Biche inhabitants considered the people of other islands their enemies and lived in fear of invasion. This shows that as a society, they were far from tolerant of outside societies. However, with their conversion to Christianity, headhunting ceased and the people of Biche became more tolerant toward religions of outside societies and of outsiders in general.

Evil spirits and reliance on psychic abilities are considered to have helped develop and maintain tolerance in local society. The fear of being possessed by *pela* helped people refrain from being jealous and being critical of others' behavior, which developed tolerance. Powers such as being able to calm peoples' anger (*poda vamanoto*) and to make them stop speaking abusively (*poda vadoma*) are examples of the psychic power that the villagers believed in prior to their conversion to Christianity. Even when people were infuriated with someone, if they were unable to express this to the other person, they would proffer the excuse that it must be because someone was using their psychic powers against them. The existence of psychic powers is used as an excuse for why people are unable to strongly criticize others in person, and this helps to maintain "tolerance."

There is also an awareness among people engaged in the semi-domestication of sago palms and canarium nuts that utilization by others should be tolerated because, for example, "bats increase the number of trees." Bats eat the hard flesh of fruit, which assists in germination and the growth of trees. Thus, the excuse that people advance for tolerating utilization by others is that they are merely managing trees that have grown naturally.

There are plants like sago palms that spread not only from fruit but also from stumps, and as bats are not the only means by which germination is propagated, the understanding of the villagers is not ecologically correct in light of modern science. However, the important role played by bats is a common environmental awareness shared by all villagers. It is widely accepted by villagers that certain kinds of trees are multiply merely on account of bats. Hence, even if some villagers claim priority utilization rights, there is a tolerance that refrains from excluding utilization by others and from criticizing them.

---

2) In some cases, people who were jealous of other people and judged to be possessed by *pela* were beheaded. It is said that similar to the psychic Monaka, who beheaded many Gatokae islanders in the 19th century, some villagers headhunted villagers who were possessed by *pela* within the island and did not go headhunting to other islands. Furthermore, villagers were able to solve the hostile relationship with the *bagara* tribes on Vangunu Island through marriage.

3.4 "Mutual Assistance" as the Foundation of Society

In Biche, in addition to the former headhunting expeditions, villagers helped each other by assisting in the construction of houses; a number of people would paddle canoes in pursuit of skipjack tuna, etc., and there were many jobs that required people to work together. Mutual assistance in the form of communal labor and the provision of labor is an important foundation of Biche society. Until the 1950s, there was only one kitchen in the village where everyone would collect their harvest and cook together. The catch of skipjack tuna was divided among the villagers, including the elderly who were unable to go fishing, and shared with people from other villages.

One of the teachings of the Seventh Day Adventist (hereafter, SDA) church, the branch of Christianity that the villagers believe in, is that ten percent of the harvest should be given to the local church. Such contributions can be utilized by people such as the elderly. This SDA doctrine has helped maintain mutual assistance in the form of giving the harvest to other people.

3.5 "Work" for Village Prosperity

The people of Biche express an aspect of the village prosperity by saying "We're blessed with resources that provide us with food, medicine, and sources of income if we only work." Only lazy people who do not work experience poverty (*malaga*) and people who do not work despite having resources that can be utilized are considered poor. In other words, the prosperity of having resources that can be communally utilized is connected to the emphasis on enjoying the fruits of one's labors.

To tolerate the utilization of resources by others and give generously of the harvest, it is necessary to develop an environment that can be used for obtaining wild plants and a good harvest. When something has been received from someone, unless one works, one is unable to harvest anything to give in return. Not behaving in a generous and tolerant manner with others or not participating in mutual assistance but always receiving things from others shows one's poverty and is acknowledged as a shameful condition.

Villagers accord so much respect to work because it is through work that one acquires the resources required to display generosity, tolerance, and mutual assistance. Work is also valued because it is necessary for increasing and protecting resources and supporting the other three elements.

Furthermore, setting boundaries in the forest to be cleared of trees and to claim such utilization rights is not customarily allowed. For example, the boundaries of priority utilization rights regarding land for shifting cultivation are unclear only when people continue to clear or utilize such forests. Moreover, as previously mentioned, semi-domesticated plants and wild plants that are acknowledged to have increased in number due to bats are communally utilized as a resource formed without the villagers' involvement. Because work is emphasized, resources that have been formed, to a great extent, by a force other than work are considered to be usable (should be allowed to be used) by anyone.

This type of awareness has made it difficult to divide the benefits obtained from things such as the aforementioned commercial logging and the forests that have been made into resource conservation areas. Regarding forest that has neither necessarily been formed through villagers' work nor actively protected by them, the benefits gained by outsiders from its use led to the awareness that anyone can benefit from such utilization (or engage in embezzlement).

3.6 *Isiri* and *Hope* Controlling Each Concept

The concept that keeps the four elements described above in balance and controls the excessive activities associated with each element are the customs of sufficiency (*isiri*) and taboos (*hope* and *vinangira*).

For example, *isiri* acts to control excessively generous behavior. Behaving generously can be linked with demanding repayment from the other person. If someone shows generosity that cannot be adequately repaid by the other person, it is not *isiri*. *Isiri* could be said to be an underproduction awareness that takes into consideration the generous behavior to ward or from others and increases or decreases harvests accordingly. In the midst of village life, harvesting more than the amount required for your own use and behaving generously toward others is considered degradation of resources and should be avoided[3].

The same thing can be noted with mutual assistance. When asking other villagers to participate in communal labor for shifting cultivation or housing construction, it is difficult to ask for help unless one has *isiri* skill or strength. In addition, when asked to help in communal labor, whether one is able to adequately help is important. If others do not consider the reason for not helping adequate, such as due to age or sickness, the person may be mocked for pretending to be weak (*kebokebo*).

There are two types of taboo: *hope* and *vinangira*. *Hope* is used for things or people worthy of respect and places that are forbidden to be approached or destroyed because they

---

3) However, this awareness of underproduction is thought to have developed due to the fact that the long-term storage of items such as fish and potatoes is difficult as there are limited tools. Since the 1960s, fish and potatoes have been a source of income with the arrival of new tools, such as outboard motors. Thus, this awareness that *isiri* is a good thing has in some ways worn thin. It is thought that the ability to exchange goods for money, which does not rot; the introduction of tools such as outboard motors that can greatly exceed the limited labor available for catching fish; and the fact that using such tools requires money have changed awareness regarding underproduction.

Chapter 4 Legitimacy for "Great Happiness" 111

are sacred. Specifically, parents and the elderly, leaders, cemeteries, and the Bible (*buka*) are all referred to as *hope*. Actions that damage or threaten these entities or actions lacking in respect toward hope should be avoided. People do not tolerantly overlook such behavior, and sometimes may even directly criticize such disrespect. After the conversion of the villagers to Christianity, actions contravening what is written in the Bible also became *hope*.

Vinangira means exercising a strong power (*nginira*) over some resource. In some cases, when overfishing has occurred and fish stocks have become depleted in a certain area, certain measures, which are also called *vinangira*, such as managing the resources of the island and establishing no-fishing zones and periods are implemented by the authority of the leaders. In Biche, one of the sources of income comprised button shells (*Trochus niloticus*), which are used for accessories such as buttons. However, because of the depletion of button shells in the area surrounding the Magotu moat, the area was declared a no-fishing zone until the resource recovered. When preparing for a party such as a wedding, fishing would be forbidden in a certain moat for approximately a month, after which people would fish together and catch a large number of fish. Such no-fishing areas and periods were commonplace until the 1950s, and this system was used to recover resources by placing a taboo on things such as fishing when the resource was drastically depleting. Once the stocks were sufficiently recovered, excessive resource utilization would be controlled.

4. Christianity and Commercial Logging and Conservation

4.1 Establishment of Boundaries on the Island Based on Communal Utilization

Although details regarding the exact dates are unclear, from the beginning to mid-18th century, people of Gatokae Island were reported to have been headhunted on a large scale three times and were almost entirely wiped out. Gatokae Island is said to have been attacked by people from New Georgia Island or Rendova Island. Those who survived moved to a lookout on Mategele ridge, which is approximately a two hours walk and 2 km in a straight line from where they currently live. After they recovered their strength, the M tribe moved back to the coast, and from mid-19th century to the beginning of 20th century, they were renowned as headhunters, repeatedly attacking places such as Vangunu Island and New Georgia Island.

Since moving from Mt. Mategele ridge, they relocated seven times until they arrived at their present location in Biche. As their power as headhunters increased, they moved further down the mountain. Though the number of years they spent in each location is unknown, it has been at least 150 years since they inhabited Biche, which is the oldest village on Gatokae Island[4].

In 1893, the islands of New Georgia, Guadalcanal, Makira, and Malaita became a British protectorate, and by 1900, all islands became part of the protectorate. Then, they started to seize land from among customary land that they determined to be unutilized by residents. Headhunting posed problems in this respect. The colonial government commenced "pacification" of the Solomons to attract investors, converting them to Christianity and burning down villages that repeatedly engaged in headhunting (Bennett, 1987: 106–112, 147, 148). One of the villages that was burned down was Biche.

In 1912, the people of Biche, led by the brothers Paka and Vagolo, who were *bagara*, paddled their war canoe (*magoru*) over 100 km from their village to Lavi village on Guadalcanal and took the heads of missionaries and villagers. Angered by this action, the colonial government sent a warship to strafe and burn the people of Biche. The villagers fled and were safe. It is said that Paka and Vagolo were hiding among the rocks on the beach intending to take over the warship, but when they saw the awesome firepower, they were shocked and gave up their resistance.

In 1915, the people of Biche accepted the SDA and became Christians. However, Paka and some other villagers, did not accept the church, and although they continued obstructing mission activities in the village, most of the villagers—including Vagolo—became Christians, and headhunting met its demise.

Through converting people to Christianity, not only the people of Biche but those of many other villages also abandoned headhunting. This "pacification" by the colonial government was a success, and many villagers returned to coastal land that had been abandoned and left unused. However, because of the seizure of such coastal land by the colonial government and the development of coconut plantations by white colonists, many land disputes arose with villagers, who had originally utilized such land, and among the villages with regard to such developments. In Biche, although no land was seized by the colonial government, Gatokae Island was divided into four areas, with boundaries established by the church and government.

Prior to their SDA conversion, *bagara* exerted overall leadership over the M tribe society on Gatokae Island and the surrounding area, managed land and sea resources, negotiated with outside society, and arbitrated any disputes within the tribe. Those who became leaders, such as *bagara*, were generous toward others regarding the use of village resources, knowledge, skills, and harvests, and were respected as *tinoni getegetena* (big men); their position was secure.

---

4) At the end of the 19th century, almost no one lived on the eastern side of Gatokae Island. The present Sobiro village was formed when villagers, such as Apa, were pursued there by Paka because they were said to be possessed by *pela*.

In 1915, the *bagara*—Vagolo and Paka—became known as paramount chiefs, and representative owners of not only Gatokae Island but also the surrounding uninhabited islands, such as Bulo, Malemale, Kicha, and Borokua.

From 1915, church-related personnel, such as missionaries, began using the resources of Gatokae Island to build churches. The first church was built in Penjuku facing the lagoon with its gentle waves where it was easy for ships to dock and was the base for church's activities on Gatokae Island.

When using the resources on Gatokae Island, people from the church or colonial government always had to first seek permission from the *bagara* of Biche. However, it was difficult to land on the rocky shore of Biche, which faces the rough seas of the open ocean, and by land a round trip to Biche on bad roads across swamps, boulders, and steep slopes took an entire day. Therefore, wanting to avoid the inconvenience of having to go all the way to Biche, church people sought to divide Gatokae Island into four areas and have chiefs appointed to act as proxy *bagara*.

In 1922, a congress with missionaries was held in Penjuku Village to divide Gatokae Island into four separate areas. Frances and Gray from the colonial government also attended the congress. The congress began after Rupa, who was hosting the meeting, declared that Vagolo would act as chairman.

At the congress, it was decided that Vagolo and Paka would be chiefs of the Biche area and, as representatives, would manage Chubuil, Isu, and the four uninhabited islands of Bulo, Kicha, Borokua, and Malemale.

Furthermore, the area from Isu to Tetekarovo was under Penjuku and Sageona villages, and Salele was appointed chief. The area from Komabusi to Chubuil was under Kavolavata, Peava, and Sobiro villages, and Siana was appointed chief. The area from Varea to Bili was under Bili, and Hila and Tasa were appointed chiefs. Silver coins were given to the appointed chiefs in each area certifying their authority. When the members of the church or colonial government wanted to use resources, they simply needed to secure permission from the appointed chief.

Vagolo and Paka, who were the *bagara* at the time, led the chiefs as paramount chiefs and representative owners of the natural resources of Gatokae Island and the surrounding uninhabited islands, mainly managing the resources of Biche and uninhabited islands, such as Bulo. The new clan formed by the descendants of Vagolo and Paka and their wives (VP clan) live mainly in Biche and Peava villages. Since 1915, the *bagara* of the M tribe have been appointed from among the members of the VP clan.

For the sake of the church, the people of Gatokae Island allowed the island to be divided into four areas. However, it was merely the establishing of boundaries for outsiders, in the form of the church, to utilize the resources of Gatokae Island. For the people of the M tribe living in Gatokae Island, the customary leaders managing all resources were the *bagara*, and under them, they communally used the resources within the island and in the surrounding area. The chiefs appointed within the boundaries were merely proxies of the *bagara*.

Though the customary leaders, the *bagara*, were acknowledged as paramount chiefs who interacted with outsiders, it was assumed that the chiefs were actually managing each of the areas, which was a source of major confusion in contracts related to subsequent development, such as commercial logging. As discussed later, there were several cases in commercial logging contracts involving corrupt chiefs agreeing to procedures with logging companies without the consent of the *bagara*.

Paka died in 1928 and Vagolo in 1934, and they were succeeded as *bagara* by Birei and Papae. As *bagara*, these two individuals continued to be a strong voice as chiefs of the island; the four established areas were not emphasized and communal use of resources by the entire M tribe continued within local society.

During the 1940s, copra (dried coconut kernel) became a source of income. Members of the M tribe who had planted coconuts in Biche were acknowledged as having priority utilization rights, regardless of them being members of the VP clan. Even when members of the M tribe moved to other villages, they continued to maintain their priority utilization rights.

Furthermore, as mentioned above, until the 1950s, there was only one kitchen in Biche and meals were collectively cooked using harvested food. In addition to harvesting coconuts and canarium nuts with other M tribe members living in different villages, people worked together on a daily basis, fishing and practicing shifting cultivation together. There were close relationships among the villagers, and people worked together and shared the harvest. Various resources were communally utilized by the people of Biche, the VP clan, and the M tribe.

However, with the arrival of passenger ships and the increase in population in the 1960s, the introduction of cash crops in the 1970s, foreign fishing boats in the 1980s, commercial logging in the 1990s, and the establishment of resource conservation zones, the four divided areas became emphasized not only to outside society but also within the M tribe (Tanaka, 2007; 2008). The following sections explain the changes that occurred in the latter half of the 1990s that impacted the society of Gatokae Island as a whole, especially with regard to the four divided areas.

4.2 Opening the Door to Commercial Logging
Establishment of Boundaries and Misunderstanding the Power of Chiefs

After the death of Papae in 1956, Tesi became the second-in-command *bagara* when he returned in 1964 from working as a teacher on another island. When Tesi died in 1991, Penpio succeeded Tesi. Birei, who became a *bagara* in 1934, received much knowledge from Paka, had an outstanding ability to negotiate agreements within local society and with the outside world, and was highly esteemed and held in awe by the people. When managing land within Biche, he had a strong voice among the chiefs of the island and repeatedly refused logging contracts with Malaysian and Indonesian logging companies[5], such as Earth Movers and Silvania Products in 1994 and Golden Springs in 1995.

At a height of 800 m, Mt. Mariu towers over the center of Gatokae Island and is very steep in many places. Therefore, Birei and the other villagers were well aware of the fact that large-scale logging would probably cause disasters such as landslides. In the 1980s and again in 1994, in the area adjacent to Biche, shifting cultivation was attempted on a steep area called Toroso, but the crops were badly damaged because of landslides. Such failures made the villagers aware of the danger of logging steep areas.

To procure logging contracts, logging companies acted generously by giving or promising to give cash to the chiefs of villages, such as Peava. However, many villagers were concerned about the depletion of forest resources within the island through commercial logging, and in light of this, Birei did not allow any commercial logging when asked by some of the other chiefs on Gatokae Island.

Because of Birei's repeated refusal of logging, the logging companies set their eyes instead on Bulo Island, which is offshore from Peava village. In addition to being uninhabited and a sacred place for the M tribe, Bulo Island was a place where the hunting of coconut crabs and harvesting *naginagi* (*Cordia subsordata*) used for carvings was carried out. The Peava chief was allowed to manage Bulo Island under the *bagara*.

At the time, chief P of Peava and his father Y lived in the capital Honiara. A busy lawyer, P visited the island only once every few years. Commercial logging contracts require the signature of a tribe representative with ownership rights over customary land. However, seeking to monopolize royalties, P and Y secretly proceeded with a contract in Honiara with

---

5) Earth Movers is a subsidiary of Lee Ling Timber, Co. based in Sarawak, Malaysia, and Silvania Products is a subsidiary of the Malaysian company Kumpulan Emas Berhad Co. The president of Golden Spring, Co., is the Indonesian behind Sumber Mas Timber Group in East Kalimantan, Indonesia. Since the mid-1990s, commercial logging by such logging companies was actively performed and frequently caused friction with local residents in various areas, with some companies being ordered to cease operations due to major environmental damage (Bennett, 2000: 247, 287, 295, 344).

the Malaysian company Pan Pacific. Birei and the people of Biche were opposed to logging. However, they were unable to refuse the tearful demands of P who was a member of the VP clan and great-grandchildren of Vagolo, and thus, Birei eventually allowed operations on Bulo Island. With regard to the logging contract, the respect of the chiefs toward Birei as *bagara* continued, and emphasis on the consent of the *bagara* in logging contracts was, to some extent, still emphasized.

In 1997, the Porere land overlapping the boundary between Peava and Biche villages became the focus of logging. P went ahead with the Porere logging contract without obtaining the approval of Birei. As a lawyer, chief P was well versed in legal matters, had experienced many lawsuits regarding development, and was connected to logging companies. Although Birei had abundant knowledge regarding resources within local society and their utilization, P had knowledge about the law and developments, such as commercial logging, as well as the ability to negotiate. Through his relationships with outsiders, and by strengthening such relationships, this chief, who had left local society and lived in Honiara as a lawyer, was able to overwhelm the *bagara*.

Immediately before logging commenced, P held a meeting in Peava. Four people from Biche attended this meeting, but Birei was sick and unable to attend. Despite the fact that the people of Biche were opposed to logging, P defiantly said that it was impossible to stop the activity because he had already signed the contract. The logging company also told the villagers that they would pay a generous royalty of 12% of the free-on-board (FOB) log price to the VP clan and obtained approval for the logging. Birei died in 1998 after instructing the villagers to prohibit logging in the Biche section of the Porere land.

As a *bagara*, Birei was a major presence, managing a diverse range of resources. He was the personification of legitimacy in the use of resources, showing generous behavior toward the villagers and tolerance in allowing the use of resources. However, on his sick bed, he was unable to adequately exercise his power to control the self-serving behavior of the chiefs. After his death, Vagolo's grandson, Harron, became *bagara*.

In 2000, P signed a logging contract with the Australian company Emmett Logging in the Porere area. The people of Biche—including Harron—knew nothing of the conditions of the contract beforehand; a man from Peava, acting as an intermediary between P and Emmett Logging, merely informed them that logging would commence. Though the people of Biche were promised that reforestation would be implemented, absolutely no trees were planted.

The commercial logging contracts in which chief P showed disrespect to the *bagara* became the means by which other chiefs on Gatokae Island proceeded with commercial logging. As of May 2013, many logging companies have been in operation on Gatokae Island in villages such as Penjuku, Sageona, Kavolavata, Sobiro, and Bili. In every case of logging, the chiefs, and not the *bagara* were the central agents. Thus, by deepening their relationship with outsiders, the chiefs—who as proxies under the *bagara* were merely entities to *chakei* the resources in their designated area—took the liberty of introducing commercial logging. Consequently, the trees customarily used for construction have been depleted in many villages through such commercial logging.

Miyauchi (1998: 178) suggested that communal utilization rights may become unstable and environmentally-damaging development may occur through individual registration of land. In the example discussed in this chapter, as there is no individual registration of land the customary leaders, appointment of chiefs by outsiders to manage customary land, and establishment of boundaries dividing the island into administrative areas have been major causes for the promotion of commercial logging despite the presence of *bagara*.

In fact, the establishment of the boundaries and existence of the chiefs made it easy for resources to be utilized by outsiders. If the *bagara* and chiefs had shared a common concept of legitimacy regarding the use of nature as the foundation of their lives, development that disrespected the opinions of the *bagara* would not have proceeded, and no major problems would have emerged.

As is evident from this example, chiefs going to live in the capital maintaining no contact with the natural resources of Gatokae Island and being agents involved in logging contracts combined with a major difference in the involvement of chiefs and villagers regarding natural resources opened the door to commercial logging. As for the relationship with outsiders—in this case, logging companies—chiefs were appointed who disrespected the *bagara*, the customary leaders within local society. The establishment of boundaries dividing the island into four areas and the appointment of chiefs by outsiders, though strengthening the relationship between local and outside society, created a chance for promoting development such as commercial logging, in local society that was not in line with the wishes of the *bagara* or the local people.

4.3 Necessity of New Sources of Income
Expectation of Generous Logging Contracts and Tolerance of the Acceptance of Logging Companies

P and Y seized leadership with the aim of acquiring cash income from royalties and introduced commercial logging. Fearing the depletion of natural resources, villagers were opposed to commercial logging, despite being attracted by the thought of acquiring cash income.

Until the 1960s, cash income on Biche Island comprised the sale of stone bowls and trochus shells to Chinese merchants and the small-scale sale of crops. In 1969, the sale of copra increased under the guidance of the Agriculture Bureau and became the main source of cash income until the latter half of the 1980s. However, the international price of copra

Table 4–1. Annual Income of Biche Villagers in 2001 (SID)

Source of income	Selling to other villages	Employ-ment of logging	Selling in village	Royalty of logging	Lodge	Fish selling	Wood carving	Stone bowl	Shell selling	Total
Total	14,083	11,226	8,817	4,474	2,346	1,857	1,835	1,550	167	46,355
Average	880	702	551	280	147	116	115	97	10	2,897
Percentage	30.4	24.2	19.0	9.7	5.1	4.0	4.0	3.3	0.4	100.0

Source: Field Research.
Note: Author selected 16 of the 23 village house holds and visited them to ask their income and expense.

Table 4–2. Annual expenditure of Biche villagers in 2001 (SID)

Category	Schooling costs	Donation	Foods	Petrol	Soap	Seasoning	Kerosene	Battery	clothes	Medical costs	Others	Total
Total	9,708	9,271	8,664	2,344	1,678	1,244	837	780	453	141	2,037	37,157
Average	607	579	542	146	105	78	52	49	28	9	127	2,322
Percentage	26.1	25.0	23.3	6.3	4.5	3.3	2.3	2.1	1.2	0.4	5.5	100.0

Source: Field Research.
Note: Author selected 16 of the 23 village house holds and visited them to ask their income and expense.

Chapter 4 Legitimacy for "Great Happiness"

reached a peak of 710USD/t in 1984 before dropping to 309USD in 1987. By 1994, sales volumes for copra on Gatokae Island had dropped to 1% of the volume in 1984 (Hviding and Bayliss-Smith, 2000: 211). This was partially due to the fact that it needed to be fumigated in drums rather than sun dried and that, as a source of income, copra cultivation required considerable effect while providing little reward, and was thus disliked by villagers. In 2000, only two families in Biche sold copra; in 2001, no copra sales occurred.

Education is one of the biggest expenses for the people of Biche. The average annual income per household as of 2001 was 2897 Solomon Island dollars (hereafter referred to as SID, 1SID = 25 JPY as of 2001, see Table 4–1). The average schooling cost per household was 607SID (Table 4–2), which accounted for 26% of total cash income. Though the village had a population of 56 people in 1980, the number had increased to 109 people by 1995, indicating that per household education expenses had also increased. Prior to the introduction of commercial logging in 1996, villagers with many children struggled to pay education expenses.

Therefore, the new source of income revealed by chief P in the form of generous royalties from logging companies was attractive to the villagers. Moreover, as P had formed a relationship with outsiders in the form of logging companies, he was able to display off his power and promote logging contracts while controlling the opposition of the *bagara* and villagers.

With the drop in the price of copra and the increase in population, which meant an increase in education expenses, the villagers were faced with problems in terms of both income and expenditure and were searching for new sources of income. The generous royalties revealed by the logging companies were considered *noro* when signing the logging contracts; the villagers had high expectations. The reason that they allowed the promotion of the logging contracts by the chiefs and decided to see what type of generous behavior would result from the commercial logging promoted by the chiefs is connected with their concept of tolerance.

4.4 Royalty Distribution Expectations and Betrayal
Tolerance of Embezzlement and Rising Discontent

The royalties provided by commercial logging to Biche, including royalties from logging areas other than Porere, totaled 94,310SID or 4,100SID per household (as of 1996, 1 SID = 30 JPY). Compared with the average annual household income of 2897 SID in Biche, it would appear that commercial logging provided a significant amount of money to the villagers. However, villagers were largely dissatisfied with the royalties, as they felt not only that the amount itself was small but also that distribution had not been effected fairly.

The main royalties were distributed five times until 2001, with the average distribution per household being 699 SID in each instance. The villagers knew that at that time, logs were being sold in Honiara for 200 SID/m^3, and timber was being traded at a price of approximately 900SID. Hence, despite the fact that several ten thousand cubic meters were commercially logged, each household only received an amount equivalent to 3.5 m^3 of logs. Thus, it is only natural for the villagers to state that loggers had stolen their trees through commercial logging[6].

Commercial logging on Bulo Island was performed for approximately one year. The rumored royalties of tens of thousands of SID were pocketed by Y and P who lived in the capital and experienced a cozy relationship with the logging companies from the time the contracts were signed. Though the people of Biche were extremely dissatisfied, expressing this dissatisfaction directly to Y and P would be seen as a lack of tolerance and would not be considered *noro*. Thus, even though they were dissatisfied and tolerant, the villagers attempted to pocket the royalties by strengthening their own relationship with the logging companies (Tanaka, 2002).

Subsequently, when royalties were distributed, repeated cases were observed concerning part of the funds being pocketed by villagers entrusted with the job of receiving royalties from logging companies in the capital. By finding opportunities to be involved with outsiders or utilizing such opportunities, villagers tried to obtain greater benefits for themselves. Though not directly criticized in the village for such embezzlement (due to the tolerance of the people), many villagers were dissatisfied with being unable to obtain much in terms of royalties.

Starting with the chiefs who were agents for introducing logging contracts, people who maintained relationships with outsiders had several opportunities for embezzlement, which led to dissatisfaction among the villagers. Thus, they all started to seek opportunities of embezzlement for themselves. Trying to obtain *isiri* benefits themselves was seen as legitimate work and was tolerated without being strongly dissuaded. Commercial logging produced no benefits that everyone could be satisfied with as being *isiri*, and dissatisfaction continued to build within local society.

---

6) The volume logged was estimated from the volume carried by each ship and the number of ship loads as remembered by villagers engaged in measuring logs, etc. The actual volume was 4,000 m^3 for Emmett Logging and 20,000–30,000 m^3 for Pan Pacific.

4.5 Active Participation in Commercial Logging
Depletion of Useful Trees by Seeking Adequate Returns

The people of Biche, who were unable to obtain royalties from Bulo Island, sought partial embezzlement of royalties through commercial logging on Porere. They also actively engaged in logging jobs and sought an *isiri* return for themselves for the trees that they gave to the logging company.

An investigation of the work history of 48 men and women prior to the introduction of commercial logging revealed that 10 men and 21 women had no work experience. Of the 10 men who had no work experience, six obtained their first experience through logging work in Porere.

Of the 41 Pan Pacific Corporation workers in Porere, 11 workers from Biche were engaged in logging, measuring, and transportation. There were also two from Malaysia, 16 from Guadalcanal or Malaita, one from the Gilbert Islands, and 11 from Peava village, revealing a local labor ratio of 53% (workers from Gatokae Island as a percentage of total workers). Eight people from Biche were employed by Emmett Logging, which had a local labor ratio of 76%. Of the 23 households in Biche, 13 house holds had members who were participating in commercial logging, but only men actively participated.

What prompted the active participation of villagers and caused logging to continue was the labor situation whereby they could get paid according to the volume of timber they cut. This invited the depletion of useful trees through overcutting.

With regard to commercial logging, Birei designated the Biche zone of Porere as a protection area, and after attempting to persuade the villagers to prohibit commercial logging and make it taboo, he died. However, as villagers engaged in logging were paid according to the amount that they cut, they continued logging. This produced areas where the supply of trees that were important to the villagers for construction materials, such as *buni* (*Calophyllum* spp.) and *vasara* (*Vitex cofassus*), were depleted.

In the past, with the exception of certain types of trees, members of the M tribe were free to log native and secondary forests. Apart from the crops that they were cultivating, trees in fallowed areas and coconut plantations were not owned by any individual. Under lax rules, whereby it is preferable for users to obtain permission from individuals with priority utilization rights, communal utilization of trees was allowed for the entire M tribe. However, when commercial logging was performed in Porere, some villagers cut down all trees for construction purposes to increase their earnings.

Communal utilization of resources meant that having adequate volumes of construction timber for each house to use was *noro*. However, the *isiri* control mechanism did not function with regard to the work of logging resources such as trees for construction timber and converting such resource to money. Though *isiri* was valid as a control concept, with regard to the communal utilization of resources in local society, when such utilization was connected to sales to the outside world or when there were people obtaining even more benefits using connections with outsiders, the control did not seem to function.

4.6 Distribution of Royalties from Commercial Logging in Surrounding Villages
Emphasis on the Boundaries Dividing the Island into Four Areas

Though commercial logging proceeded in each village on Gatokae Island, it was rare for royalties to be distributed to the people of Biche. The VP clan monopolized royalties from Porere, and though they were distributed to members of the VP clan living in the capital, they were not distributed to other members of the M tribe. This shows that the boundaries dividing Gatokae Island into four areas were emphasized in the distribution of royalties.

On the other hand, the people of Biche claimed that as the VP clan was the only clan with a *bagara* fulfilling the role of managing the resources of Gatokae Island and the surrounding uninhabited islands, they had the right to receive royalties from commercial logging implemented in other villages on Gatokae Island. That claim continues to be ignored and the people of Biche are frustrated with the *bagara* who do not assert their leadership over the M tribe as a whole.

The people of the VP clan in Biche succeeded in receiving royalties only from the commercial logging performed by Golden Springs in Kavolavata and by Omex in Penjuku.

With the development of commercial logging, Gatokae Island society began emphasizing the boundaries that divide the island into four areas, which were merely established by outsiders such as the church who wanted to utilize resources. This has also become evident in the distribution of royalties within local society. Royalties are also distributed to members of the VP clan not living in Biche; however, these members' relationship with the M tribe as a whole is seen to be moving in a divisive direction because of the emphasis on the boundaries. These boundaries have become a tool for limiting those with communal utilization rights for resources to each village or the VP clan rather than the M tribe as a whole.

4.7 Establishment of a Conservation Zone in Biche
Acceptance of Generous Researchers and its Failure

Around 2004, a group from the University of Queensland in Australia visited the area around Gatokae Island to survey the marine resources in Marovo Lagoon. Though they surveyed the influence of fishing activities, commercial logging, and the effects of oil palm plantation development on marine resources such as sea cucumbers and coral, their focus shifted to Biche as much of the forest had still survived.

With its rich ecosystem and natural beauty, Marovo Lagoon stretches over 700 km^2 and has been nominated as a prospective world natural heritage site. However, this ecosystem's richness and beauty is threatened because of the acceptance of commercial logging by the villages surrounding the lagoon. In the midst of this predicament, researchers from the University of Queensland personally supported Biche, where much of the forest was still intact though commercial logging had been accepted in some areas such as Porere, that were under the control of the village.

The researchers suggested the establishment of a natural and cultural resource conservation zone for the sustainable use of resources, and in December 2006, a meeting was held in Biche and was attended by 24 men, three women, and six teenagers. It was decided that approximately 75% of the village area would be turned into a resource conservation zone in exchange for annual financial assistance of 50,000 SID (approximately 750,000 JPY), renewable on a five-year basis.

Areas excluded from the conservation zone included an area where people originally from Biche had once practiced shifting cultivation and areas for which ownership was claimed. Despite the fact that this was nothing more than an acknowledgement of priority utilization rights through shifting cultivation, many people mistakenly acknowledged them as ownership rights and, of course, such rights were neither acknowledged by the *bagara* nor the people of Biche. There were also people originally from Biche who sought to proceed with commercial logging in that area.

Prior to the meeting in September 2006, a Malaysian logging company, which had been operating in the neighboring village until May, approached the villagers for a logging contract for the area on the western side of Biche. At the meeting, the people of Biche decided to refuse the logging contract. However, through mediation by a person originally from Biche, one of the *bagara* was persuaded out for 5,000 SID and signed the contract; further, in December, another *bagara* and a spokesman were bought out for 2,000 SID and 1,000 SID, respectively, and two other villagers who were each paid 500 SID also signed. Even though they wanted to stop the logging, the villagers did not harshly rebuke the *bagara*. Furthermore, as there was no significant friction between the villagers and the people claiming ownership rights and seeking to proceed with commercial logging, the villagers decided to establish the resource conservation zone excluding the area involved in the logging contract.

Of the 50,000 SID received from establishing the conservation zone, during the first year, approximately 20,000 SID was spent on education expenses, 20,000 SID was spent on an outboard motor, and 10,000 SID was spent for repairing the church. An oil extraction machine was also supplied for producing coconut oil for sale, which was meant to provide a new source of income.

The researchers suggested establishing a nature conservation zone in Biche to ensure that commercial logging was not performed on village land. This claim emphasized the "universal concept" that the richness of nature and ecosystems should be protected and used sustainably.

In Biche, in addition to establishing no-fishing periods and areas to increase and sustainably use resources, the local people also adhered to the custom that *isiri* use was good. Moreover, the people wanted to protect the forest, especially the areas surrounding water sources where there were waterfalls and the land was steep, because they knew that if surrounding areas were logged, soil would permeate the rivers. Therefore, members were able to connect with the researchers' intentions in seeking to establish the resource conservation zone.

However, this connection was inadequate to proceed with establishing the conservation zone, and the situation transformed into a battle against the generosity of the logging company. The logging company did not only buy out the village elders but also gave the villagers rice and canned goods. However, the establishment of the resource conservation zone provided financial assistance of 50,000 SID, which was approximately 20 times the average annual household income as of 2001. This provision aimed at countering the generosity of the logging company.

Since 2008, there were plans to transform the entire Gatokae Island into a nature conservation zone. Though the researchers continued to promote the establishment of the zones, exhibited generosity to counter the logging companies, and obtained the cooperation of international NGOs such as Conservation International, things did not go as smoothly as planned. This hurdle arose because the villagers did not have any reason to accept this generous behavior.

The people of Biche slowly realized that researchers and tourists were visiting the forest that they had protected by refusing commercial logging because they discerned its value. They understood that these researchers and tourists would be generous and help them in continuing to protect the forest.

However, in Kavolavata and Penjuku villages, people began opposing the conservation zones. Both villages had introduced commercial logging throughout the area under their control; trees such as those used for construction had been depleted, and it was an area to which tourists rarely came. The villagers doubted the researchers and did not understand their generosity in wanting to conserve the forest. They were suspicious that by creating resource conservation zones, they would lose their land. In Kavolavata, despite their concerns, the villagers still expected to receive 20,000 SID in support each year, and though they did decide to accept it, this acceptance was only temporary.

In Penjuku, the researchers tried to convince the villagers by saying that they would give them a lot of money if they accepted the conservation zone. However, as some people were suspicious that such generous behavior would lead to loss of their land, it was easier for them to understand the arrangement with the logging companies who were prepared to pay them for logging; therefore, they did not establish the conservation zone.

Simultaneously, in Biche, the generous behavior of the researchers caused another problem, which was envy regarding the distribution of assistance to others both within and outside the village. Although 40% of the annual 50,000 SID assistance was spent on education expenses, 13 of the 29 households in the village either had no children or had no children of junior high school or high school age who had to attend expensive schools. Education expenses for primary school children were fully subsidized and amounted to approximately 40 SID per year. Initially, there was no opposition to this practice, as people thought that children who received higher education would in time contribute to the village in some manner.

Until now, 30 children have received support for educational expenses amounting to 1,500 SID to 1,700 SID in full or partial support for attending junior high school, high school, or vocational training school. However, as 18 were expelled for having relationships with the opposite sex, an increasing number of people considered this support for educational expenses a waste. There was also a case where a villager who received 1,000 SID for support to attend vocational training school was expelled for producing alcohol[7]. Though some villagers state that children do not fear being expelled because their school expenses are not paid by their parents' hard-earned money, some female students have had abortions for fear of being expelled if they were found to be pregnant.

Support for education expenses was limited to villagers who were involved on a daily basis in the resource conservation in Biche, but people who had moved out of the village were also seeking support for education expenses. In the conservation zone, though geographical boundaries were established using GPS, blood relationships cannot be severed by boundaries. Furthermore, there was jealousy regarding assistance with education expenses, such as in cases where assistance was refused to children from the first marriages of women who were not from Biche but who married someone in Biche. Because of such discord, several people left the village and some others stated that they wished to leave.

---

7) In 2010, at this vocational training school, 68 people—including children from other islands—were expelled for stealing, smoking, or drinking. Furthermore, in the same year, 23 of the 45 people who enrolled in the carpenter training course were expelled for drinking or smoking in the first year, and only 17 graduated after completing the full three-year course. The problem of children being expelled is not unique to Biche.

In July 2009, the conservation project began with financial assistance from the EU and technical support from the University of Queensland and Conservation International. The aim was to survey and protect forest resources, biological diversity, and cultural assets; the project, which operated for 17 months, had a large budget of 1.01 million SID. As part of the project, oxygen discharge volumes and carbon dioxide absorption volumes were measured, timber volumes and tree species surveyed, roads and rest areas were created, maps of sacred areas were created, and business guidance was provided regarding the sale of items such as palm oil.

Many villagers were employed in activities such as surveying forest resources, road maintenance, and building offices. However, depending on the work and the people involved, there were some cases wherein wages were not paid leading to dissatisfaction among the villagers. Eventually, because of some villagers' behavior, the project was aborted due to the degree of embezzlement and unaccountability of funds.

During the time in which the conservation zones were established and the forest protection project was in operation, considerable funds flowed into Biche and there were disagreements regarding their use. Furthermore, this has resulted in villagers finding themselves in the situation where, to borrow the words of villagers, "we lost our ability to be independent." Before starting an activity, villagers would discuss how they could obtain money from outsiders and how they could use the project. Though the children of these adults considered their parents spoiled and pathetic, they still sought employment in activities such as forest resource surveys and road maintenance in projects, similar to the adults.

Moreover, the project strengthened the villagers' tendency to seek paid employment rather than providing mutual assistance, such as the provision of labor and joint labor in work such as building houses and shifting cultivation for people in unrelated villages. Multiple groups were formed within the village, and people were employed for work such as shifting cultivation both within and outside the village. Though there were villagers who had experienced employment for the first time through commercial logging implemented until 2001, few people had experienced employment for everyday activities such as shifting cultivation and building houses within Biche. However, since the establishment of the conservation zones in 2006, many activities within the village have now become paid employment. It is also rare for villagers to demand for labor at no cost for activities such as building houses or to find opportunities for people to engage in communal labor.

4.8 Resumption of Commercial Logging in Biche

A Temporary Victory for Logging Companies in the "Generosity Competition"

In December 2011, commercial logging recommenced in Biche for the first time in approximately 11 years. Logging that extends to the conservation zone where villagers have many semi-domesticated nut and coconut groves and water source forests was restrained to a small area. However, operations still proceeded in the area to the west of Biche—an area that researchers had initially hoped to convert into a new conservation zone. This area spans approximately 400 ha, and ultimately, 68,000 m^3 of trees will be logged; logging of approximately 27,000 m^3 was expected to be completed by the end of November 2013.

The *bagara* and outside members of the M tribe, who had not greatly benefited from the establishment of the conservation zone, promoted the logging contract. The *bagara* possess considerable knowledge about various resources (including how to use them), they utilized or allowed others to utilize the resource, they were generous toward others in terms of harvests, and their position was acknowledged. However, when the conservation zone was established and money for the village budget was distributed among villagers who were employed, no consideration was accorded to the *bagara* who managed the resources, and no special payment was made. Thus, the establishment of the conservation zone was not used as an opportunity for expressing generous behavior toward the *bagara*.

In contrast, logging companies recognized the opportunity and bribed the *bagara* into signing logging contracts. This action of the logging companies acknowledged the position of the *bagara* and the role they play in local society. The *bagara*, whose authority was reinforced through this action, viewed commercial logging rather than the establishment of conservation zones as something that could provide them with legitimacy. With the money they had received as bribes, the *bagara* bought things such as rice and cooked and shared it with the villagers. However, this was not enough to suppress the dissatisfaction of the villagers, and the actions of the *bagara* were instead seen as acts of embezzlement to benefit themselves while ignoring the intent of the villagers. This served to further increase the villagers' desire to receive some kind of benefit from commercial logging.

Members of the M tribe in the surrounding villages, too, did not receive any particular benefits from the establishment of the conservation zone. The only benefit received were scholarships provided for children of those living in the surrounding villages whose ancestors were from Biche and who had nut and coconut groves. This was not because they did not live in Biche or were not engaged in resource management on a daily basis, but because if blood relationships were emphasized, the number of eligible people would be several times the population of Biche, and there were not adequate funds for distribution. However, this turned out to be a difficult problem as there are many people who could not be ignored even though they were no longer involved with the nature resources of Biche on a daily basis because they are related to people from the village.

In Biche, when establishing the resource conservation zone and distributing the benefits, it was emphasized that people living in Biche who have managed the village's resources have rights, even though they may have no blood relationship. This led to denial of social relationships with people outside Biche and caused friction in the local society. It was also associated with moves by residents of the surrounding villages to promote commercial logging contracts in Biche that were aimed at destroying the resource conservation zone. People from villages surrounding Biche engaged in secret intermediary activities between the logging companies and *bagara*, and in addition to obtaining many benefits, the company continued to bribe the *bagara*.

Before logging began, most villagers criticized the logging contracts and talked discussed to stop such companies. However, after operations commenced, some villagers were happy with the generous behavior of the logging companies when there was increased opportunity for employment and when things such as rice, instant noodles, canned food, sugar, and biscuits were delivered. The author a companied some of the young people of the village to assess the damage to the rivers at the time of logging and gather information and monitor boundaries in preparation for the lawsuit to claim damages for illegal actions, such as logging useful species of trees, including canarium nuts, that were important to the villagers, and make such companies cease operations. However, when the logging contract was established, the young people of the village also sought work from the logging company.

The establishment of a conservation zone in Biche was temporarily accepted due to normative sharing—adequate and sustainable utilization of resources—and generous behavior. However, villagers were pressured by the generous behavior and continuous confrontation with outsiders, such as logging companies, because of which they ceased resisting.

Seeing the villagers wavering in the direction of accepting the logging company, some villagers wept in anger while other villagers left the village. Some children said they would no longer believe the *bagara* and even started drinking and smoking. There were women who prostituted themselves to the Malaysians, who worked for the logging company, to obtain money and timber. The primary school that needed to be rebuilt for to continued functioning deteriorated because there were not enough people to participate in its building. The village lost its unity.

Since June 2011, in addition to the five households (20 people) who left the village until February 2012, three households (15 people) who had intended to return to the village decided against returning, and some villagers started to send their children to school in a different village. One of the villagers who had promoted the conservation project became greedy for money and together with his relatives, embezzled the equivalent of approximately 4,000 USD of the logging money and left the village.

The villagers began to consider both the logging company and the researchers as entities that would eventually leave and their involvement as something temporary, and they sought to obtain generous behavior from them. They deliberated among themselves regarding which outsiders they would seek to obtain generous behavior from and shamelessly sought help from outsiders while pretending to be weak, for which they were criticized by the young people. The researchers who wanted to establish the conservation zone had not returned to Biche as of December 2013, and it is unclear whether they will continue to be involved with Biche in the future as the village seemed to have accepted commercial logging.

At the beginning of June 2013, the logging company finally decided to terminate the operations. The primary school reopened in the conservation office and some people returned to Biche. However, the people in the villages surrounding Biche, who did not benefit from the conservation zone, are still secretly trying to proceed with more commercial logging contracts to cancel the establishment of the conservation zone itself, which is expected to continue to upset local society.

5. Is Collaborative Governance Able to Produce "Great Happiness?"

By using generosity (*hinoho*), which has legitimacy in the eyes of members, logging companies and researchers were temporarily accepted. With regard to establishing the conservation zone, because there was a shared awareness among the villagers and researchers regarding the conservation and sustainable use of resources, there was the possibility that their long-term contribution would be acknowledged. However, it was not possible to control the generous behavior of the logging companies. Moreover, because of the dissatisfaction of the chiefs and people in the surrounding villages caused by the establishment of the conservation zone, they decided to proceed with the introduction of commercial logging. The emphasis on the boundaries formed through the conversion of the villagers to Christianity and the boundaries established through the conservation zone was connected not only to the introduction of commercial logging but also to the destruction of blood relationships, which are the basis of local society.

The villagers say that this is the age in which they are overrun with selfish, stingy (*vusivusi*) "small happiness" (*minado kiki*). They say this in terms of comparison with the "great happiness" (*minado gete*) that they shared not only within Biche but also through their relationships with people in the surrounding villages by working on the communal utilization of many resources and sharing of harvests. The reason they could generously share harvests was only because there were people with the skills and knowledge required for planting crops, transplanting useful plants, and fishing. The joy of being able to share the harvest with others and to have their ability of behaving generously toward others acknowledged was their "great happiness." However, both commercial logging and the establishment of the conservation zone caused the villagers to lose their "great happiness."

The funds and scientific information possessed by the logging companies and researchers are huge and extremely powerful when compared with those of the villagers. It is probably very difficult for such outsiders to bring funds, information, and technology into the local society without causing a great deal of confusion. Whether the funds and scientific knowledge of outsiders are able to suppress and control the emotions of the people in local society and legitimize their involvement is another issue.

Regardless of how much knowledge considered "universal" or "scientific," skills and information, are emphasized, the attraction of money and material things is strong, and in a local society where generosity is legitimized, such things are perceived as all the more attractive by villagers. Through their generosity, it was easy for outsiders to gain temporary acceptance by the villagers. Even people who disapproved of the involvement of outsiders in resources did not forbid their "generous behavior."

However, through the emphasis on boundaries and embezzlement by some villagers, villagers who thought that the generous behavior they should have received was hindered felt extremely jealous. This jealousy threatened the cooperation that was at the heart of their small island society, which was based on the communal utilization of resources and everyday mutual assistance. This is the main reason that the presence of evil spirits, which suppress such things in Biche society, called *pela* is spoken of and that there were villagers who had the role of killing people possessed by such evil spirits[8].

From a different perspective, this society, where cooperation is a prerequisite, is a society where the creation, utilization, harvest, and sale of resources is constantly being monitored by people to assess who is involved the most, who obtains the most, whether they are sharing, and whether they are being generous. The emphasis of geographical boundaries and protection zones destroyed relationships with people outside the village—members of the M tribe with whom they have a blood relationship but live in other villages—and relationships with people within Biche society through jealousy.

Using collaborative governance in a local society, which is based on the communal utilization of resources, when seeking outsiders' involvement with the establishment of boundaries and the "commitment principle," whereby the people with the strongest involvement in the resources are given the strongest right to speak, is not always a good idea. Blood

---

8) Commercial logging and the development of oil palm plantations is proceeding both within and outside Gatokae Island; jealousy is caused regarding the resulting benefits, and there is talk in the village of the number of *pela* increasing. In many cases, when people die, it is said that they were cursed and killed by *pela*.

relationships that have continued for hundreds of years are the foundations of society, and through involvement with outsiders with overwhelming funds, technology, and information, people are inevitably confronted with the basic human emotion of jealousy. In other words, as long as collaborative governance investigates how a society should be, in terms of the relationship between nature and the people involved, it will probably always encounter problems. As with the example of Biche, the unity of local society, which should be the core of collaborative governance, is threatened by outsiders' involvement in forest utilization that is acknowledged temporarily on account of generosity.

How to overcome such a problem is one of the many challenges confronting collaborative governance. The type of approach that uses generous behavior to enable short-term involvement is used not only in commercial logging but also in international assistance and research, which is conducted in waves of several years and may be a convenient method. However, with forests, a long-term cycle is a prerequisite for creating, nurturing, utilizing, sustaining, harvesting, distributing the benefits, and recreating. Furthermore, for resources communally utilized by many people, the short-term involvement of outsiders, even in a small area, has the potential to distort this cycle and the relationships among people and cause confusion.

So, what type of collaborative governance has potential? The repeated short-term involvement of a diverse range of agents is important as it provides people in local societies with a wide range of potential opportunities. The local people then consider what to choose, act on, and gain experience—sometimes through failure—so that they may find some type of a hidden key to their particular situation. However, in the midst of repeated outsiders' involvement, it is essential that appropriate attention be focused on the ruin of local society and loss of the core value of local society, such as shifting cultivation, fishing, and harvesting, which is the communal utilization of resources that forms the foundation of their way of life. Without such consideration, collaborative governance is nothing more than the imposition of a "fake society" forced by outsiders on a confused local society, rather than the ushering in of people's "great happiness."

References

Bennett, J. A. 1987. *Wealth of the Solomons: A History of a Pacific Archipelago, 1800–1978.* Honolulu, HI: University of Hawaii Press.

Bennett, J. A. 2000. *Pacific Forest: A History of Resource Control and Contest in Solomon Islands, c. 1800–1997.* Isle of Harris, UK: The White Horse Press.

Food and Agriculture Organization of the United Nations. 2010. *Global Forest Resources Assessment 2010, Country Report, Solomon Islands.* Rome, Italy: Food and Agriculture Organization of the United Nations.

Hviding, E. and Bayliss-Smith, T. 2000. *Islands of Rainforest: Agroforestry, Logging and Ecotourism in Solomon Islands*. Burlington, VT: Ashgate Publishing.

Inoue, M. 2009. "Shizen Shigen 'Kyouchi' no Sekkei Shishin: Rokaru kara Gurobaru he" [The Design Guidelines of Natural Resources 'Collaborative Governance': Connecting the Local and the Global]. In: T. Murota (ed.) *Gurobaru Jidai no Rokaru Komonzu* [*The Local Commons in the Global Era*]. Kyoto: Minerva Shobo, pp. 3–25.

Miyauchi, T. 1998. "Jusoteki na Kankyo Riyo to Kyodo Riyoken: Solomon Shoto Maraitatou no Jirei kara" [Mixed Use of Environment and Collective Usufruct: A Case Study in Malaita, Solomon Islands]. *Kankyo Shakaigaku Kenkyu* [*Journal of Environmental Sociology*], 4: 125–141.

Neufeld, D. F. (ed.) 1976. *Seventh Day Adventist Encyclopedia*. Hagerstown, MD: Review and Herald Publishing Association.

Solomon Islands, Statistics office, Ministry of Finance. 1995. *Solomon Islands 1993 Statistical Yearbook*. Honiara, Solomon Islands: Statistics office.

Tanaka, M. 2002. "Solomon Shotou ni okeru Shogyo Bassai no Dounyu to Arata na Kaihatsu Kan no Keisei: Westan Shuu Marovo Raguun Gatokae Tou Biche Mura no Jirei" [Formation of the View on Development after Commercial Logging Introduction in Solomon Islands: A Case Study in Biche Village, Gatokae Island, Marovo Lagoon, Western Province]. *Kankyo Shakaigaku Kenkyu* [*Journal of Environmental Sociology*], 8: 120–135.

Tanaka, M. 2006. "Rito Murison Chiiki ni okeru Minkan Iryouyaku no Yakuwari no Doutai: Solomon Shoto Marovo Raguun Gatokae Tou Biche Mura no Jirei" [Movement in the Role of Folk Medicine in the Isolated Island without a Doctor: A case Study of Biche Village, Gatokae Island in Marovo lagoon, Solomon islands]. *Ecosophia*, 17: 104–120.

Tanaka, M. 2007. "Shigen no Kyodo Riyo ni Kansuru Seitosei Gainen ga Motarasu 'Yutakasa' no Kentou: Solomon Shotou Biche Mura ni okeru Shigen Riyo no Dotai kara" [A Study of Legitimate Communal Resource Utilization for Well-being: A Case Study of the Resource Utilization Movement in Biche Village, Solomon Islands]. *Kankyo Shakaigaku Kenkyu* [*Journal of Environmental Sociology*], 13: 125–142.

Tanaka, M. 2008. "Solomon Shotou ni okeru Rokaru Komonzu to Chiiki Hatten" [Local Commons and Regional Development in Solomon Islands]. In: M. Inoue, (ed.) *Komonzu Ron no Chousen: Arata na Shigen Kanri wo Motomete* [*Debate of the Commons: In Search of Various Modes of Resources Management*]. Tokyo: Shin-yo-sha.

Torigoe, H. 1997. *Kankyo Shakaigaku no Riron to Jissen: Seikatsu Kankyo Shugi no Tachiba kara* [*Theory and Practice of Environmental Sociology: From the Standpoint of Life Environmentalism*]. Tokyo: Yuhikaku Publishing.

Part II
Sharing Interests, Roles and Risks
The Process of Collaborative Governance

Chapter 5

Task-sharing, to the Degree Possible
Collaboration between Out-migrants and Remaining Residents of a Mountain Community Experiencing Rural Depopulation[1]

Mika Okubo

1. Rural Depopulation and Problem of Underuse

This case study examines a small mountain village in Japan. Located at an altitude of 900 m, Mogura village is surrounded by mountains. People first settled in the area during the Sengoku (Warring States) Period and earned their living by shifting cultivation. During and after the Pacific war era, the village endured hardships—many people were sent to war and the village was required to supply food. However, unlike the towns, they managed to produce enough food to survive.

As of 2009, 37 people lived in this particular village. With the exception of one person, all people were aged 60 or above. The villagers sometimes voiced their concerns: "After I die, what will become of the village?" "After I die, can I be buried in the village cemetery?"

A similar scenario is observed in villages throughout Japan, especially in mountain areas. According to a survey conducted by the Ministry of Internal Affairs and Communications (MIC), there are 10,091 communities wherein more than 50% of the population is aged 65 years or above (MIC, 2011). Odagiri (2006) described the process through which the number of people living in rural villages is decreasing in Japan, stating that although many members of the younger generation left the rural mountain villages during the period of high economic growth that occurred since the 1960s, their parents remained in the villages and continued to maintain agricultural and forestry land. Since the latter half of the 1990s, as the older generation has aged and started to pass away, the population of such villages has noticeably decreased. Challenges such as medical and welfare problems that accompany an increased number of elderly people who live alone, the desolation of fields and forests,

---

[1] This chapter has been translated and expanded from Okubo, Tanaka, and Inoue (2011).

and the weakening of community functions are being pointed out in rural mountain villages, where the aging of residents and depopulation is progressing (Ohno, 2005; Odagiri, 2006).

This situation can be considered as a "problem of underuse" of the commons (Kawada, 2009). Overuse has been the central theme of the commons theory since *The Tragedy of the Commons* (Hardin, 1968). On the basis of the premise that the commons will be overused if there are no adequate management systems, much research has been conducted adequate systems to sustainably manage the commons. In contrast, many of the problems in Japan's rural mountain villages arise because people are not implementing suitable actions. It is difficult to determine whether a resource is being overused or underused. In some cases, determining whether resource utilization is appropriate relates not only to the matter of quantitative sustainability, but also to how to actually determine value[2]. The increase in abandoned arable land, the desolation of artificial forest, and the decline in traditional arts in rural mountain areas are considered a problem, and the cause is thought to be lack of utilization and management. That is, underuse—not being able to take action regardless of intent—in Japan's rural mountain areas is considered a problem.

It is sometimes difficult for residents to solve such problems by themselves in depopulated areas. There are areas that explore collaborative governance-type measures, whereby people from outside, such as residents of urban areas, try to be involved in the management of terraced paddy fields or water conservation forests (Maeda and Nishimura, 2002; Ohno, 2005; Oura, 2008). There are also movements to promote a new type of residency—dual-area residency—where people live in both the city and the rural mountain area (Dual Area Residency Population Research Association, 2005). However, it has also been highlighted that the participation of new entities from outside does not necessarily solve the problems confronting rural mountain areas. Tamura (2001: 70–71) suggests that the reason for this is that urban residents, who are consumers seeking leisure, have a different perspective from people who live in rural mountain areas, who are producers. With regard to managing the local environment under the guidance of outsiders, Yoshida (2007) saw possible problems such as historical perspective being degraded and local residents being undervalued as administrators of the local environment. Bearing in mind the differences in the values of outsiders and

---

2) Suppose that a certain wild vegetable is no longer being utilized. Under environmental conditions where the volume of the resource is not reduced even though it is not being utilize, on considering only the volume of the resource, it cannot really be said that it is a problem of underuse. However, considering aspects such as the economic benefits obtained when appropriately utilizing such wild vegetables, or the passing on of folk knowledge such as where and how to harvest them and how to cook them, it can be judged whether lack of utilization is actually a problem.

local residents, respecting the opinions of local residents can be considered to be highly compatible with collaborative governance theory.

On the other hand, it should be noted that not all rural mountain communities experiencing depopulation are conducting some type of project with outsiders. In light of the limitations of urban/rural interaction by strangers, the importance of out-migrants from such communities who now don't live in the communities while being involved as supporters of the communities has been highlighted (Tokuno, 2008). This chapter considers the state of collaboration between out-migrants and residents and the method of realizing collaborative governance in mountain communities experiencing rural depopulation.

The community of Mogura in Hayakawa town, Yamanashi Prefecture, Japan, which is the subject of this research, is a mountain community experiencing rural depopulation and population aging. Residents sometimes express their fear of being the only person left in the community in the near future and their uncertainty of what will happen to the community after their death; however, there are no significant projects being implemented involving outsiders. Given the state of these communities 20 years in the future, it seems that out-migrants are the only hope. How will out-migrants be involved with the community in the future, if at all? Considering these issues, this research endeavors to ascertain how the relationship of out-migrants with their home community has changed and considers the social factors that have enabled the collaboration of out-migrants and residents.

Thus, the abovementioned social factors are shared by both out-migrants and residents. Clarifying matters shared by people who live in separate places will help to understand what should be shared by the various stakeholders involved in collaborative governance under the situation of underutilization.

2. Research Challenges and Focus

Existing research about out-migrants from rural mountain communities in Japan can be broadly divided into two categories. One category incorporates research conducted by Takahashi (1983), Matsumoto and Maruki (editor) (1994), Matsuzaki (editor) (2002), and Ajisaka (2009) that examined people from rural areas who now live in cities, the process of how they came to stay in the cities, and their relationships with others in the cities. In the research, people from rural areas were mainly referred to as "village leavers (*Risonsha*)" or "urban migrants (*Toshi-ijusha*)". In contrast, the other category comprises research where people from rural areas were mainly termed "out-migrant children (*Tashutsu-shi*)" or "out-migrant family members (*Tashutsu-kazoku*)".

According to existing research, out-migrant children have helped their elderly parents, who stayed in the rural areas, with farm work (Araki, 1994; Ishizaka, 2002; Ashida, 2006). These children also looked after their parents' welfare by visiting, taking care of them when they were sick, and calling them on the phone (Tachibana et al., 1998; Ishizaka, 2002; Ishizaka and Midorikawa, 2005). By focusing on the role of out-migrant children, the research succeeded in revealing out-migrant children as supporters of their aging parents in rural depopulated areas. In other words, he importance of reviewing the "human relationship resources that are able to support rural mountain communities now and in the future" rather than the traditional viewpoint that views out-migrant children as "simply people who have left their homes and communities" was clarified (Tokuno, 2008: 62–63). Such research has clarified the ways in which out-migrants are involved with their home communities. However, an analysis regarding factors such as why out-migrants were involved—or able to be involved— is as yet inadequate. Such research has only gone so far as to discover the following factors: (a) relationship with the parents living in the home communities depends on how far away the out-migrant children live (Ishizaka, 2002; Ishizaka and Midorikawa, 2005; Ashida, 2006); (b) marriage of out-migrant children or aging of their parents tend to strengthen the children's contribution to their home community (Ashida, 2006); (c) presence of friends in their home area increases the interest and participation of out-migrants (Sugawara et al., 2006); and (d) agricultural land and festivals act as a medium for human relationships (Yamamoto et al., 1998: 84). Therefore, this research will examine out-migrants who still maintain some relationships in their home village and the social factors that have enabled them to maintain such relationships.

This research differs from existing research in two ways. First, it includes out-migrants whose parents no longer live in their home community. In existing research, terminology such as "out-migrant child" and "out-migrant family" indicates that such research did not include out-migrants whose parents are no longer living in their home communities. Concerning people whose parents no longer live in their home community, there has been research describing the links among urban migrants from the same home area (Ajisaka, 1994; Yamamoto, 1994; Ajisaka, 2009). As noted later, only Yamashiro (2007) and Sugawara et al. (2006) considered the relationship of these people with their home village. However, as it is clear that there are owners of vacant houses who want to continue managing them without any help (Yusa et al., 2006), out-migrants whose parents no longer live in their home village are also expected to maintain some kind of relationship with their home area. As the aging parents' generation living in rural mountain areas is now starting to pass away, it is essential for such areas to understand the relationship that out-migrants have with their home communities after their parents no longer live there.

Second, the focus of this research is the involvement of out-migrants in matters regarding not only family aspects but also communal aspects in home community, such as systems of self-governance, collaborative activities, and festivals. Existing research (Araki, 1994;

Chapter 5 Task-sharing, to the Degree Possible

Ishizaka, 2002; Ashida, 2006; Tachibana et al., 1998; Ishizaka, 2002; Ishizaka and Midorikawa, 2005) has assessed the involvement of out-migrant children with their aging parents in their home community. On the other hand, research analyzing out-migrants' involvement in matters concerning local communal life in their home community, such as structures for self-governance and festivals, has been fairly limited. Yamashiro (2007) revealed the role played by Okinawan home village associations (*Kyoyukai*) as bodies through which the traditional culture of home areas can be passed on to future generations. A survey of home village groups in cities was conducted by Sugawara et al. (2006). Results of this survey revealed that some out-migrants were interested in matters such as preparation for festivals and management of local associations. However, this research did not go into any specific detail regarding their involvement. In rural mountain communities, the issue of decline in community self-governance functions accompanying the decrease in population is evidenced by the difficultly in securing leaders and finding people to perform collaborative activities (such as community work and traditional performing arts) (Ono, 2005:107–108). In the midst of a decline in the number of residents and resident households, it is important to see out-migrants not only as entities who can support their family in rural mountain areas but also as entities who can support the maintenance of local communal life and local resources in rural mountain area.

As part of matters regarding local communal life, this research specifically focuses on festivals and ascertains changes in the out-migrants' involvement. This is because significant changes have been observed in the out-migrants' involvement regarding festivals in the target community. Existing research that has examined the continuity of traditional events includes Shibuya (2000, 2006) and Ueda (2007). It is noteworthy that previous research emphasized people's understanding in rural communities that only local people should be involved in traditional performing arts. According to Ueda (2007), in the example of the *Taiko Odori* (drum dance) of Kajiwara in Itsuki-Village, Kumamoto Prefecture, despite it being difficult to continue the dance because of rural depopulation, people of the community refused an offer to work toward its continuance with people of nearby communities. Furthermore, according to Shibuya (2006: 156–157), in the example of *Hayachine Kagura* (a type of Shinto ritualistic dance) in Hayachine, Tohno city, Iwate Prefecture, though the number of performers is decreasing and in-migrants are carrying on the traditions, people from the community are of the opinion that, "We really should do it (perform) ourselves…" Such participation by in-migrants is seen as only a temporary situation until local people have been trained as successors. Even in Mogura, which is the subject of this research, it is often said that people who are not parishioners of the community shrine should not participate in carrying the portable shrine during the spring festival. With regard to the example of the drum dance, Ueda (2007) noted that the dance is essentially performed by the community as an act of thanksgiving to their ancestors and to provide comfort to the spirits of the dead on the occasion of their first ritual visit to the earthly world. Therefore, merely holding the dance as a means of preservation was not acceptable to the people of the community. Rural festivals are closely linked with tutelary gods and ancestors (Matsudaira, 2008), and it is said that there are restrictions on conditions for successors within the community. As a postscript to her thesis, Ueda (2010) mentions the case of a young man who migrated from the community and returned for the drum dance. As it is difficult to find successors for traditional performing arts throughout Japan due to rural depopulation and population aging (Hoshino, 2009), out-migrants are being looked to as successors of traditional performing arts in the context of communities. The present research will ascertain and illustrate this issue.

In commons research, Suga (2010, 2013) pointed out that cultural commons theory considers the management of cultural resources in the context of global commons as a way of managing culture that is modern and highly diffusible with marketability and market value. He also mentioned the necessity of local cultural commons theory as actual local commons. Thus, the festivals managed and enjoyed by the people of the community in this report are closed commons and examples of local cultural commons.

The subject of this research is the community of Mogura, Hayakawa town, Yamanashi Prefecture. According to the Basic Resident Register, in 1970, there were 288 people in a total of 67 households in Mogura, but by May 2009, the number decreased to 43 people in 26 households. However, a person having a residence certificate in Mogura may not necessarily be living in Mogura on a daily basis. As of April 2009, 37 people in a total of 23 households actually lived in Mogura[3]. Although a quick glance around the village revealed a noticeable number of vacant houses, out-migrants return to the village for *Obon* (a Japanese folk custom to honor the spirits of one's ancestors observed in mid-August), *Ohigan* (a Japanese holiday celebrated during both the spring and autumnal equinox), and other opportunities with their spouses and children to use and maintain the house. Many out-migrants also participate in community events, such as festivals and community collaborative labors.

Some people spend approximately half the week in Mogura, and it is difficult to ascertain whether they are residents or out-migrants. Therefore, based on the residents' perceptions, this research considered 37 people in a total of 23 households in the community to be residents. The out-migrants of Mogura are defined as people who have lived in

---

3) Specifically, in addition to interviewing a shopwoman in the community, It was ascertained through observation whether people were to be considered residents of Mogura.

Mogura and who are presently based outside the area. These people were treated separately from out-migrant spouses and children who had never lived in Mogura.

The research involved a survey of 34 residents, 38 out-migrants, 6 out-migrant spouses, and 17 out-migrant children and observed their activities in Mogura. These out-migrant, out-migrant spouses and out-migrant children are all currently involved in Mogura. At this stage, this research will exclude out-migrants who currently have no involvement with Mogura. The survey was concleted 13 times over 120 days during June 2008–December 2009. Moreover, the research also involved a questionnaire survey of 30 out-migrants who had returned to the village on December 6, 2009 for community collaborative labor[4]. Furthermore, on November 8–9, 2009, the author stayed in Showa town, Yamanashi Prefecture, in which many people from the community have migrated. This town provided an opportunity to observe the *Mujin* men's group meetings. *Mujin* is a type of traditional group. In this case, *mujin* group members comprised male out-migrants from Mogura in their 30–40s. These members meet for drinks once a month at a Japanese-style pub near Kofu city, the capital of Yamanashi Prefecture, and take a trip together once a year. This group is still active, and it provides a venue for out-migrants to keep up with the developments in each other's lives.

3. Overview and Migration History of Mogura

Located approximately two hours by car from Kofu station, the center of Yamanashi Prefecture, Mogura is a community on a winding stretch of road at an altitude of approximately 900 m. Until the mid-1950s, shifting cultivation was practiced in the area, after which forestation became more prominent. Paddy field cultivation is difficult on sloping land. Because the people of the area wanted to cultivate and eat rice, around 1950, approximately 20 households cut lumbers from their mountain and sold it in order to purchase land for paddy fields in Showa town or Kofu City in Yamanashi Prefecture (Hayakawa town Board of Education, 1980, and interviews with residents). With the closure of Kofu Industrial High School's Misato campus and Mogura mine in 1965 and 1968, respectively, the number of people migrating from the area after graduating from junior high school increased after the late

---

4) With regard to the questionnaire, it is necessary to consider that the subjects were people who had come home on December 6, 2009, for community collaborative labor. Community collaborative labor is a community's event for which large numbers of out-migrants return to Mogura. The questionnaire was not aimed at all out-migrants, and it is necessary to interpret it as a survey of only out-migrants who were participating in community events at that time.

Figure 5–1. Out-migrant's present place of residence.
The circle shows the location of Mogura. The numbers behind the municipality names indicate the number of out-migrants living in that municipality.
Source: Questionnaire (Number of valid responses (26). One out-migrant living in Saitama city in Saitama Prefecture is not shown on this map.)

1960s. Households that had paddy fields in areas such as Showa town used the sheds built adjacent to their paddy fields as new bases to be used as schools for their children.

As of April 2009, there were 37 residents in 23 households in Mogura, of which 33 residents were 65 years old or above. Of these households, nine comprised individuals living alone, accounting for approximately 40% of the total number of households. Of these households, 19 had places outside Mogura, where they could stay such as with their children (the situation was unclear for the other four households). Some residents were based outside Mogura for approximately half the week, whereas others were considering moving to live with their children outside the area.

Chapter 5   Task-sharing, to the Degree Possible   143

According to the results of the questionnaire survey, 24 of the 26 out-migrants migrated to places within Yamanashi Prefecture, one moved to Fujinomiya City in neighboring Shizuoka Prefecture and one moved to Saitama Prefecture (see Figure 5–1). The reason that a significant number of people (nine) moved to Showa town was that many households that bought paddy fields in Showa town also built houses on the land and were living there. Furthermore, except the one person who moved to Saitama Prefecture, the other 25 people lived within 45–100 minutes by car, and most of the out-migrants surveyed in this research were those who lived a medium distance away (Ishizaka, 2002).

Concerning the age of the out-migrants, 13 of the 29 out-migrants were in their 50s and 10 were in their 60s, revealing that more than 90% out-migrants aged the 40–60 year (see Table 5–1). Of 27 people, six replied that their father or mother or both parents lived in Mogura, while approximately 80% of out-migrants said that their parents were not living in the community (see Table 5–2). However, although their parents are no longer living in Mogura, many out-migrants have relatives who are from Mogura[5]. On examining the frequency with which out-migrants returned to Mogura, it was noted that 14 of the 25 people mentioned above returned to Mogura at least once a month. On the other hand, 11 out of 24 people replied that they only stayed there overnight once or twice a year, with two out-migrants stated that they never stayed overnight (see Table 5–3).

Reasons for out-migrants to return to Mogura included the following: individual and household matters, such as maintenance of graves and houses, farm work and fishing; community group matters, such as extended family and temple events[6]; and community matters, such as community collaborative labors.

Table 5–4 shows the results of a multiple-choice questionnaire based on interviews and observations.

As seen in the table, many people returned to Mogura for *Obon* to pay their respects to their ancestors by tending graves or to attend temple events. Among the children of out-migrants, some said that though they could not use their house in Mogura now, they would always tend to the graves at *Obon*. One out-migrant (a male in his 60s) stated that the only thing he really wanted his son (in his 20s) to do was to tend to the graves. This shows that this tradition of visiting graves play an important role in connecting second-generation out-migrants with Mogura.

The state of house and field management varied in different households. In fields where cultivation has become difficult, different techniques are used; for example, wild vegetables

---

5) The reason that many people have relatives who are from Mogura is that approximately until the 1960s, marrying within the community was still popular.
6) Mogura has two Buddhist temples.

Table 5–1. Out-migrant Age

Age	Number (%) (n=29)
~10's	0 (0.0)
20's	0 (0.0)
30's	0 (0.0)
40's	5 (17.2)
50's	13 (44.8)
60's	10 (34.5)
70's	1 (3.4)
80's	0 (0.0)
90's~	0 (0.0)

Note: Parentheses in the table indicate percentage (Rounded to 2 decimal places).
Source: Questionnaire.

Table 5–2. Out-migrants' parent(s) residential status in home community

Parents residental status	Number (%) (n=27)
Both Parents are Resident in Mogura	1 (3.7)
One Parent is Resident in Mogura	5 (18.5)
Parent isn't Resident in Mogura	21 (77.8)

Note: Parentheses in the table indicate percentage (Rounded to 2 decimal places).
Source: Questionnaire.

Table 5–3. Frequency of Visits by Out-migrant

	Frequency of visit Number (%) (n=25)	Frequency of overnight stay Number (%) (n=24)
Over 3 days in a week	1 (4.0)	1 (4.2)
2 days in a week	1 (4.0)	0 (0.0)
One day in a week	1 (4.0)	1 (4.2)
Once in 2 week	1 (4.0)	1 (4.2)
Once in a month	10 (40.0)	2 (8.3)
Once in 2 month	4 (16.0)	2 (8.3)
Once in 3–4 month	7 (28.0)	4 (16.7)
Once or twice in a year	0 (0.0)	11 (45.8)
Never in a year	0 (0.0)	2 (8.3)

Note: Parentheses in the table indicate percentage (Rounded to 2 decimal places).
Source: Questionnaire.

Table 5–4. Reasons/motivations for out-migrants returning to Mogura

	Reasons/motivations for out-migrants returning to Mogura	Number(%) (n=26) multiple answers allowed
Individual/ Household	Visit family grave	22 (84.6)
	Keep family house	16 (61.5)
	Participate buddhist memorial service for the family	10 (38.5)
	Visit parent or parents	6 (23.1)
	Work in the farm or garden	6 (23.1)
	Cut the glass	6 (23.1)
	Collect mountain vegetables	3 (11.5)
	Collect mushroom	3 (11.5)
	Meet friens or relatives	2 (7.7)
	Manege the forest	2 (7.7)
	Fishing	2 (7.7)
	Hunting	1 (3.8)
Community group	Participate in events of the temple	17 (65.4)
	Participate in buddhist memorial service for the relatives or *Dozoku*	15 (57.7)
	Paticipate in events of *Dozoku*	6 (23.1)
Community	Participate in community collaborative labor	17 (65.4)
	Participate in events for *Dosojin* (community guardian)	7 (26.9)
	Participate in events for *Ujigami* (tutelary god)	4 (15.4)
Season	Because of *Obon*	18 (69.2)
	Because of New year vacation season	13 (50.0)
	Because of estivation	0 (0.0)
	No reason	0 (0.0)

such as bracken and Japanese butterbur were grown in some places as they are comparatively easy to manage, whereas the grass is grown in others; herbicides are used in some fields. Thus, people do what they can to prevent their property from looking bad and causing ill will among neighbors.

It is difficult to determine the boundaries of private forest, and both residents and out-migrants considered it impossible to look after. In fact, some people hoped that public bodies would buy the forest out or private enterprise would manage them.

4. Out-migrants and Community Association

The out-migrants who are the subject of this research are those who are still involved with Mogura, and, more specifically, who are members of out-migrant households who belong to Mogura's community association called *Ku*. In 2009, 63 households paid an annual fee of 5,000 JPY each to be considered members of the community association, and 40 of these households were out-migrants. Since 1970, 44 households have migrated from Mogura, of which 40 households (approximately 90%) are still paying the annual fees to be members of the community association. In those out-migrant households wherein no household members live in Mogura, heirs, such as the eldest son, play a central role as association members.

The Mogura community association is the local self-governing body in Mogura. The first chief of the association was appointed when the municipal government system was introduced in 1889 (Mochizuki, 1972: 125). Mochizuki, who was the chief in 1972, provided details regarding how each chief from the 1950s to 1970s petitioned the mayor of Hayakawa town for things such as road construction, water system improvement, childcare center construction, and construction of an extension school.

The association is mainly managed by its board members. As shown in Table 5–5, there were 29 positions in association in 2008 fiscal year; however, only 19 people were involved as some people held more than one position. The most important of these positions were those of chief, sub-chief, and proxy-chief, who were also responsible for managing the water system and community center. Council members play an ancillary role to these three positions and are asked by the chief to attend gatherings or provide assistance. At present, out-migrants are considered to be able to serve as council members, which some of them do[7].

Even if they do not serve as council members, each household is a member of the association and must participate in the community collaborative labor and attend the annual general meeting (AGM) held each year on April 29 as well as the community collaborative labor and the council member election held each year on the first Sunday of December. Households must pay a fee of 3,000 JPY each time they do not participate in the community collaborative labor, and households that are unable to participate in the AGM are requested to submit a letter of proxy.

---

7) With regard to the council members, one out-migrant served for six consecutive years from 2004 fiscal year and another out-migrant assumed office in 2010 fiscal year. Both are males in their 60s.

Table 5–5. 2008 fiscal year positions of Mogura's Community Association

Post	(another post)
Chief	
Sub-chief	
Proxy-chief	
*Adviser	
council member	
council member	
*council member	
council member	
*Leader of Shimo-group	
Leader of Shimonaka-group	
Leader of Naka-group	
Leader of Oki-group	(Community congress member)
Property management commissioner	(Community mayor)
Property management commissioner	(Community sub-mayor)
*shrine parishioner representative	(Community congress member)
*shrine parishioner representative	
*shrine parishioner representative	
Festival leader	
Water association commissioner	(Community mayor)
Sanitary association commissioner	(Community mayor)
Local welfare commissioner	
Community center commissioner	(Community sub-mayor)
Rearing commissioner	(Community mayor)
Fire company commisioner	(Community proxy-mayor)
Leader of Women's association	
Leader of elderly people association	
Forestry cooperative commissioner	
Community congress clerical	(Local welfare commissioner)
Special congress clerical	

Note: Asterisk indicates out-migrants.
Source: Recorded by author from Mogura Community Center roll.

There were 47 participants for the community collaborative labor held on April 29, 2009[8]. Many out-migrants from Yamanashi Prefecture and Shizuoka Prefecture participated, and one out-migrant from Saitama Prefecture, who always attend, were also present. Participants are divided into groups of six or seven people, each of whom then performs some task, such as cleaning the roads of fallen leaves and dirt and checking water sources and fire hydrants. Older participants were given less physically strenuous tasks such as cutting the grass around the community center. According to one male resident in his 70s, when

---

8) In 2008 fiscal year, there were two collaborative labors and 18 households were charged for not participating.

young people come for community collaborative labor, tasks that require many people are more easily executed, and out-migrants also come to help with road and water supply system maintenance[9]. When asked why he participated in community collaborative labor, an out-migrant said, "I have done it every year since I was a child".

Another out-migrant in his 60s said during work, "We have got to brighten up our home village." For out-migrants, community collaborative labor is an opportunity to return home, and for people who want to look after their home village, it is an opportunity to contribute to the community. Seeing each other and working together helps residents and out-migrants to maintain relationships.

The AGM is held in the afternoon on the same day as the community collaborative labor, and in 2009 fiscal year, 36 people attended, of which at least 13 were out-migrants[10], whereas 10 people sent in a letter of proxy. The assistant chair was an out-migrant (a male in his 50s); out-migrants actively participated in the meeting and there was no sign of any exclusionary attitude toward them. In addition to approving the financial report, budget, and board members for the new financial year, there was also discussion regarding the items that should be included in the agenda for the AGM. For example, in 2009 fiscal year, the problem of the location of the "deities that guard the demon gate," situated up in the mountains, was discussed as it was difficult for people to make a pilgrimage to them. Specifically, there were discussions regarding whether certain families or the entire community was responsible for looking after these deities, whether it was possible to move these deities to a more accessible place, and where such places could be located. The AGM is thus a place of sharing, where residents and out-migrants can discuss their views on the current situation in the community.

In this way, in Mogura, there is presently an ongoing awareness among many out-migrants as members of Mogura community association regarding their obligation to participate in community collaborative labor and in the AGM, which are opportunities to maintain relationships between out-migrants and residents and among out-migrants themselves.

---

9) The roads and water supply system are normally maintained by the residents in a usable condition. When rocks and branches fall from the hillsides onto the Mogura town Road that links Mogura with the Prefectural Road, residents often stop their vehicles to remove the rocks and branches. Some residents clean the roads with brooms after it rains, whereas others collect the fallen leaves in the autumn to put on their fields. In July, in addition to the community collaborative labor, community member wow the grass alongside the Mogura town Road. Further, although there is no penalty for not participating, approximately 15 residents and 6 out-migrants participated in 2009.

10) Calculation Method: The number of resident households (23) was subtracted from the number of attendees (36). There was one participant per household. As, in some cases, younger out-migrants (eldest sons, etc.) attended in place of older residents, it is possible that the total number of out-migrants was equal or higher than 13.

Table 5–6. Mogura Young People's Group Events (1973)

Date	Event
11th Jan.	Plenary meeting of young men's group
	Yanagimaki event
14th Jan.	*Dosojin* (community guardian) Festival (*Shisimai* (Lion dance))
15th Jan.	*Jugonichi-kazari* event, handover celemony
1st Feb.	*Yanagitaoshi* event
27th Mar.	Event before the Spring festival
28th Mar.	Spring Festival (carring portable shrine)
29th Mar.	Event after the Spring festival
21th Mar. (Old calender)	*Sanbaso* (Traditional performing art)
14th Jun.	*Dosojin* (guardian for the community) Festival (*Shisimai* (Lion dance))
11th Aug.	*Kiriko*-Festival
17th Aug.	Summer festival (Performing Sumo wrestling)
24th Nov.	Winter festival

Source: Oikawa (2007: 78) Underlined items were added by the author.

All three shrine parishioner representatives were out-migrants. Festival management, which is currently the main role of parishioner representatives, was formerly managed by the young people's group and, subsequently, the preservation association, which succeeded them. Out-migrants played a notable role in the community festivals[11]. The next section focuses on the festivals and out-migrants' involvement therein.

5. Changes in Festivals due to the Migration of Young People

5.1 Festivals Conducted by Young People

The young people's group has existed in Mogura since before the Pacific war (Oikawa, 2007: 72–77). In 1973, the young people's group was in charge of the events shown in Table 5–6.

Members of the young people's group, which comprised young men aged between 15 years and 25 years, were known as the *young males' group* (*wakēshu* in Japanese). Boys from the upper levels of primary school to junior high school were known as the *young boys' group* (*kowakēshu* in Japanese) and worked under the supervision of the *young males' group*.

---

11) Though the events considered as festivals are limited to the festivals involving tutelary god (*Ujigami*) and community guardian (*Dosojin*) that were formerly managed by the young people's group, in some cases, festivals such as the Ancestors Festival, which is managed by family groups, the Hachiman Festival, and the Wind God Festival are also included.

Eldest sons aged 25 years became leaders who were called *wakēshu-gashira* or *kashira*, which means "head" in Japanese, and were in charge of the festivals. If there were more than two eldest sons in the same year, the position was decided by election or nomination.

Out-migrants who left Mogura when they graduated from junior high school already possessed the experience of being members of the *young boys' group* until they graduated. Young boys learned by watching the older boys, being taught and scolded by them, helping with the festivals and learning how to manage them. It is said that when children watched sumo wrestling they were told to "stand at attention", and for boys, the *young males* were a presence to be feared.

Since the 1960s, the number of people who migrated after graduating junior high school increased; however, many continued their involvement in managing the festivals as a member of the *young males' group*. The strength of such youth was needed to carry the portable shrine for the Spring Festival and to perform sumo wrestling at the Summer Festival. According to Oikawa (2007: 91, 98, 109), 15 young males participated in the Spring Festival and 13 males in the sumo tournament at the Summer Festival in 1995. They sometimes took time off work for the events and traveled home to Mogura from the area around Kofu City.

As the number of out-migrants increased, the number of children raised in Mogura decreased, leading to a decrease in the number of members in the *young boys' group*. There have been no boys born and raised in Mogura since one boy who was born in 1979. The local people call him "the last *wakēshu*." After completing high school, in the spring during which he attended university in Tokyo, this boy held the position of *the leader of the group*. A 60-year old female resident commented that "he was by himself when he was a *kowakēshu*. As he was the last boy, we continued with the group and system until he became *kashira*." During the period until he became the last leader, the number of the member of *young males' group* decreased and consequently, those who were not an eldest son were allowed to become *leaders* (Oikawa, 2007: 91).

5.2 Maintaining and Continuing Festivals through the Preservation Association

Children were no longer raised in Mogura, and no boys were becoming members of the *young males' group*. As the people of the area say, "In Mogura, you will always be a youngster". Males born since the latter half of the 1960s were expected to continue as members of the *young males' group*, even after the age of 25. However, because of the strain of traditionally having to return home for every event, at the gathering of the *young males' group* in 1997, there were discussions regarding whether it was possible to increase the

Chapter 5   Task-sharing, to the Degree Possible   151

number of events organized by the Mogura community association and whether the festivals should be managed by a preservation association (Oikawa, 2007: 93).

After the last young male served as a leader, from around 1998, the *young males' group*, who had migrated, worked together with the community association to manage the festivals as the preservation association. Mr. A, a 40-year old out-migrant, served as the head of the preservation association. A 70-year old female resident said that "The preservation association started when Mr. A said he would do it for as long as he could." As shown in this comment, he had a strong sense of responsibility with regard to continuing the festivals. As Mr. A was a second son, he had not been able to serve as *a leader* when he was 25 years old. "There were three or four second sons and as none of them had served as *a leader*, we decided to create the preservation association (healed by Mr. A)." After Mr. A had served as a leader for three years, the five other second or third sons followed serving as a leader. The festivals and the enjoyment and honor of serving as a leader may have been the reasons for Mr. A, the head of the preservation association, and other out-migrants to continue the festivals. Another out-migrant Mr. B, a 40-year old male who served as a core member of the preservation association, said, "The reason we were able to continue with the preservation association was the support of the *Mujin* group members[12]. We wanted to help Mr. A because of his effort." The member met on a regular basis, which could have been the reason why they could continue the preservation association. Mr. B said, "We never demanded anyone to come;" thus, participation in the festivals was not a compulsory duty. Even though those who made an effort were applauded, those who were unable to come were not criticized.

However, the preservation association was eventually disbanded in 2005. Mr. A said, "I guess we decided not to continue with the preservation association (after the second and third sons had served as leaders) because there was no longer the pretext of allowing the second sons to experience the role of the leader". To serve as a leader, it is necessary to bear a considerable amount of responsibility and be recognized by other local people. The fact that there was no one who had not served as a leader, or who was able to serve as a leader, was one factor in disbanding the preservation association. Mr. B said, "There was no new young members and the existing members were getting older. It was such hard work that

---

12) *Mujin* is a type of traditional group. In Mogura's case *Mujin* group members meet for drinks once a month at a Japanese-style pub near Kofu and take a trip together once a year. Though number of members has decreased, the group is still continuing and this provides a venue for out-migrants to stay up-to-date with developments in each other's lives. Among out-migrants of the same generation, as far as can be ascertained, there were two Mujin groups—12 people in the 30–40 years age group and 11 people in the 50–60 years age group—, both of which have continued to this day.

we stopped. We were unable to go to work the day after the sumo tournament because we were covered in bruises. Physically, it was terrible. To carry a portable shrine, you need at least 12 people—three on each corner. It was very difficult to find that number of people. The portable shrine weighs anything from 120 kg to 150 kg. It's impossible to carry it for two hours". A 70-year old female resident reflected on this saying, "Even though he (Mr. A) was keen, other people did not get involved enough". Therefore, another factor that led to the disbanding of the preservation association was the aging of the *young males' group* members and the lack of participants. Moreover, as people grew older, married, and had children, there was a "loss of freedom", which gradually resulted in deterioration in attendance. Furthermore, as mentioned later, the children of out-migrants born and raised in the place they migrated to were not involved in the festivals as members of the *young boys' group*.

A 40-year old male out-migrant said, "When we announced that we were disbanding the preservation association, they (the local residents of Mogura) told us we did a great job". The fact that the preservation association continued to hold the festivals for approximately seven years even after young males no longer lived in Mogura was greatly appreciated by the residents of Mogura.

5.3 Present Festivals

After the preservation association was disbanded, the management of the festivals became the responsibility of the community association and shrine parishioner representatives. The carrying of the portable shrine in the Spring Festival and the sumo tournament in the Summer Festival were abolished and only children's versions of these activities were subsequently continued. With regard to present festivals, one 70-year old female resident said, "They're gradually being reduced. We only do what we can without too much burden." However, the involvement of out-migrants, such as former *young males' group members*, has not altogether ceased.

As of 2009, all three parishioner representatives were out-migrants who returned to Mogura to fulfill their roles in each event. These roles included gathering bamboo and pine prior to festivals and preparation, such as decorating in a particular way. When serving as a parishioner representative, it is necessary to possess knowledge and skills, such as knowing where bamboo and pine are growing, whether it is possible to harvest them, what lengths they should be cut to, and how to decorate.

At the 2009 Summer Festival, the task of putting up the hanging lanterns was performed by seven residents and one out-migrant (a parishioner representative), with most of the preparation being completed by residents. However, on the day of the sumo tournament (August 15), in addition to the parishioner representatives, 18 out-migrants participated in preparing and cleaning. This work included making a sumo ring out of straw, venue preparation such as raising banners, and cleaning up subsequently. Preparation was a lively affair, with out-migrants and residents chatting with each other in the process. One out-migrant (a male in his 50s) said, "When I smell the *konuka* (wood shavings laid in the sumo ring) I think to myself, 'This is summer. This is Mogura.'" Sumo is a symbol of Mogura to out-migrants and they look forward to working together. The work was not specifically directed by anyone, and as each person contributed as per what was required, it was difficult for anyone who was unaware of the requested tasks to be involved. Because out-migrants raised as members of the *young boys' group* in Mogura continued to be involved as *young males' group* members and preservation association members even after they migrated, they were considered to still possess the knowledge and skills required to be involved in the management of the festivals.

The children's sumo tournament saw participation of 19 out-migrant children in 2008 and 20 out-migrant children in 2009. One out-migrant male (in his 30s) said regarding his daughter who attended primary school, "It seems she saw me participating in the sumo tournament during a bout and now she is saying she wants to do it too (and she actually did)." Furthermore, of those children who participated in the sumo tournament, a daughter of an out-migrant drew a picture of a sumo wrestler and won an award. One boy (the son of an out-migrant), who was upset about losing the previous year, practiced all year with his father (an out-migrant). Unfortunately, he still lost and was very upset. Thus, for the children, the sumo tournament creates lasting memories, and for local residents who oversee the tournament, it provides an opportunity to see the children of out-migrants mature.

With regard to continuing the festival, one resident (a woman in her 70s) said, "We want to make it fun for the children so they know that there will be Sumo when they come to Mogura. We do it because we want them to come back." Thus, the children of out-migrants were the stars as they played an important role by providing a reason for both the residents and the out-migrants to continue with the festivals. Furthermore, "the festivals are for the shrine parishioners," said a 70-year old female resident, and people are of the opinion that the portable shrine should be carried only by shrine parishioners. The children of out-migrants have not experienced living in Mogura, but because "they have our blood flowing in their veins they are shrine parishioners," said a 70-year old female resident. Therefore, they were able to play a part in carrying the portable shrine—something outsiders are unable to do.

5.4 The Limits of Out-migrant Involvement

Residents have also played an important role in continuing the festivals, and done things that out-migrants are unable to do.

For example, when the *young people's group* managed the festivals, the sticky rice cakes (*mochi*) for the Spring Festival were prepared only by the young males at the home of their leader and women were not permitted to participate[13]. However, since the disbanding of the preservation association, the job has become the responsibility of the parishioner representatives, and women are now involved.

The *rice cakes* prepared on March 20, 2009, as part of the preparations for the Spring Festival were made by the wives (residents in their 50s and 60s) of the three community association executives (chief, sub chief, and proxy-chief) and the three shrine parishioner representatives (out-migrants in their 60s). Though preparing the rice cakes was the job of the shrine parishioner representatives, the wives played a more central role. One of these women (a female resident in her 70s) said, "Under normal circumstances, preparing the rice cakes is the job of the shrine parishioner representatives. This year, however, as all three of them lived outside the village, they asked the chief to arrange it". Therefore, the women assisted. Though the shrine parishioner representatives were in a difficult position to take the responsibility of preparing the rice cakes themselves, they were aware of the fact that it would be appropriate to ask the chief to manage the activity. This shows that they understood what was necessary to receive the support they required from residents.

It is noteworthy that the reason the shrine parishioner representatives (out-migrants) had to ask the wives (residents) of the three executives for assistance was that the wives of the shrine parishioner representatives were from outside the community. One resident (a 70-year old female) said, "If you go outside for high school and marry someone outside the community, then the wife does not know what to do for each festival. Someone from outside does not know how to make these [*rice cakes*]". In addition to the shape, color, volume and decoration of these rice cakes prepared as offerings, it was difficult for someone who was unaware of the ins and outs of the way things function to know the manner in which the community hall or shared things were utilized. In Mogura, until the 1960s, although people married predominantly within the community, it does not necessarily mean that no one married into Mogura from communities outside. In response to the author's comment, "Would it not be good if this generation of wives from outside would come to prepare rice cakes", a female resident in her 70s replied, "Yes, it would. If only they could learn how to do it".

However, one out-migrant stated that it was difficult for him to participate because he also had to participate in the community association where he presently lived, whereas

---

13) The reason for this is that women are not favored because the god at Kunitama Shrine in Mogura is actually a goddess.

Chapter 5 Task-sharing, to the Degree Possible 155

another stated that he would be unable to participate in Mogura festivals because the dates involved overlapped with those of festivals occurring where he presently lived.

Not all aspects of traditional events are managed by out-migrants. At present, it is mainly residents aged 70 years or older who play a role in traditional performing arts, such as *Sanbaso*, which is performed around March 20 under the old calendar, and the lion dance (*Shishimai*), which is performed in January and July at the Community Guardian Festival[14]. Acquiring the required level of skill to be able to perform properly requires intense practice, and none of the existing performers were imparting on their skills to the next generation. Though the fife is essential for both *Sanbaso* and the *Shishimai*, only two residents (males in their 70s and 80s) possess the required skill at present. In other words, there is no one younger.

5.5 Possibility of Future Festivals

At present, the children of out-migrants play a major role in the children's version of carrying the portable shrine and children's sumo, and they do not possess the experience of living in Mogura. Unlike the *young boys' group* or *young males' group*, these children are not involved in preparations and are not recognized as being able to manage festivals themselves. In other words, they are merely guests participating in the festivals; once they start feeling embarrassed about participating in sumo as they enter the higher years of primary and junior high school and become busy learning other subjects and attending club activities, their involvement in the festivals decreases. Mr. A (an out-migrant in his 40s), who had served as head of the preservation association, stated the following concerning stated situation.

"[Although it would be nice if I could pass these festivals on to my son,] this is not something that can be taught. It needs to be learned naturally. It is easy to let [events such as festivals] go, but once we let them go, then it is very hard to get them back. It's not that there is no one [to pass these things on to]. I have got children. But, I cannot teach them. It is difficult, to tell the truth. I would like my children to understand the value of these festivals. Not only that, but my children also do not have enough friends of the same generations in Mogura".

As is evident, Mr. A recognizes the difference between himself, who grew up with children of the same age who were involved in festivals as members of the *young boys'* and *young males' group*—something that he took for granted at the time—and his children, who

---

14) *Sanbaso* requires nine performers in total: three actors (each for *shiroki*, *kuroki*, *sende*), Japanese whistle, drum and wooden clappers, and three Japanese hand drums. In 2009 fiscal year, all nine of these performers were residents in their 70s or older.

do not have the same experiences it is difficult for out-migrants' children to play a leading role in managing festivals in the future. With regard to out-migrants' children supporting festivals in the future, many residents and out-migrants have emphasized that it will be difficult because they have few friends from the same generations in Mogura.

Thus far, this research has focused on festivals related to tutelary gods and community guardians; however, some other events are currently "on hold" in Mogura. When they cease hosting certain events, the villagers dislike saying that they are going to "quit". Instead, they say they are going to "*Yasumu*" (put them on hold). For example, the Ancestors' Festival, which is hosted by families, was an opportunity to get together to have a feast in the house of the relative whose turn it was to host it. However, the festival has lost its meaning as people's eating habits have changed. Among such families, they have started saying, "we're all old people" or "the turn to host was back to the earliest ancestor's house so we should take a break," and so the event has been put on hold. When they put Ancestors' Festival on hold, they moved the objects of their faith from the outside to the inside. Further, as pilgrimages were difficult in the mountains, preferences have changed from wooden shrines to stone shrines to prevent wooden decay. In some cases, after the festival was put on hold, people still went on pilgrimage individually. Therefore, the villagers consider that they are putting things on hold in ways that are acceptable.

6. Discussion and Conclusion

Several social factors have made it possible for out-migrants to continue their involvement in the local community of Mogura, while downsizing the scope of festivals and modifying the methods involved. Such factors can be summarized under the following four points.

First, residents and out-migrants share knowledge and skills related to natural resources and local history, as well as relationship etiquette. Not only residents but also out-migrants share the experience of living in Mogura, and out-migrants have also been able to maintain their involvement in the village even after migrating, which is all the more reason for them to be given responsibilities in support roles and be able to partake of the joy of being involved in this local community. For example, one of the responsibilities of shrine parishioner representatives has been to gather pine and bamboo and assemble the decorations required for festivals. They also know local etiquette and how to request assistance from the appropriate people for preparing *rice cakes* during festivals. Shrine parishioner representatives were given these jobs because they were deemed to know the detailed workings of the community and to be capable of fulfilling their responsibilities, even though they were out-migrants. Furthermore, it was also possible for out-migrants who did not have positions to be involved and enjoy festivals because they possessed the requisite knowledge and skills for various activities such as constructing the sumo ring.

Second, residents and out-migrants maintained a close relationship and shared a sense of camaraderie. Regardless of whether they were residents or out-migrants, they maintained this close relationship on the basis of friendship through relationships with extended family and groups, such as the *Mujin* groups. These relationships made visiting their home community and sharing work enjoyable. They maintained this relationship by meeting once a month through the *Mujin* groups in the areas where they had migrated, which provided a good opportunity for regularly meeting people of the same age group. This sense of camaraderie evoked a sense of responsibility toward each other as they had all been born in Mogura, which in turn enhanced the sense of enjoyment of spending time together.

Third, the residents and out-migrants discussed the possible scope of their involvement and reduced their involvement in an acceptable manner. When changing or reducing their involvement, it was necessary for them to discuss the issue at the community's AGM. Because they were able to use the AGM forum to discuss issues and make decisions, both out-migrants and residents were able to identify the potential scope of their respective involvement. The reason they were able to continue such festivals until now, while changing the format, was that they were able to downsize according to the situation and people who were able to help did so without having to place unreasonable demands on out-migrant young males. In other words, ascertaining and deciding the scope of available involvement meant clarifying the minimum expected scope of involvement. Both residents and out-migrants were aware of the minimum responsibilities they had to fulfill. They were also aware of the specific responsibilities associated with certain positions, and they fulfilled these responsibilities. By doing so, they avoided having to relinquish whatever little scope of involvement they had. They shared the opportunity to discuss these things and arrived at an acceptable consensus.

Fourth, people accepted that each member would be involved to the extent suited to their abilities. For example, in the community collaborative labor, the elderly were assigned work with little physical strain. When one member was unable to fully cope with the role assigned, other members provided support, for example, when an out-migrant became a shrine parishioner representative, he was supported by residents. In such situations, the member who was unable to fully cope with his responsibility was not criticized.

On the other hand, social factors that reduced the scope of possible involvement by out-migrants included the following health problems related to aging, which caused physical difficulties, such as carrying the portable shrine or engaging in sumo; difficulties in obtaining support from spouses because of the reduced number of people marrying within the community; problems relating to skills that had not been acquired, such as out-migrants who were unaware of how to perform *Sanbaso*; and the difficulty in simultaneously continuing involvement in both home community and the area to which someone had migrated, such as in events and executive positions in both locations, accorded them double responsibility at the same time.

Regarding festivals, it has been noted thus far how the relationships of out-migrants with the local community of Mogura (the village they migrated from) continued even as they changed the nature of their involvement, and the social factors that facilitated this continuance have been discussed.

It can probably be said that sharing (1) knowledge and skills related to natural resources and local history, as well as relationship etiquette, (2) a sense of camaraderie, and (3) opportunities to discuss and reach a satisfactory consensus are important for the success of collaborative governance. Out-migrants have a blood relationship with residents and have lived in the community and naturally share such relationships. On examining the involvement of outsiders, sharing these things can be considered as respecting the area and the people living there. Collaborative governance based on open-minded localism will be possible: if outsiders can endeavor to understand the common norms of the area, endeavor to be accepted as part of the community by the people of the area, and endeavor to, through discussions, form a consensus that is satisfactory for everyone involved.

Furthermore, focusing on resource management under the scenario of "underuse", the acceptance of people's involvement to the extent possible, is probably important. In the example of Mogura, the villages prevented the number of members from decreasing any further and prevented the degree of underutilization from becoming more serious by accepting members who could share roles to the extent that their abilities allowed. Sharing certain roles does not necessarily mean that each member needs to have an equal share of responsibility; rather, it is important to accept that each person assumes roles and responsibilities according to their capabilities, depending on their own situation. Therefore, the core members (mostly people living in the area) accept the inadequate and immature involvement of other members (mostly out-migrants or outsiders).

Collaborative governance would be effective not only in situations where it is necessary to coordinate the interests of a diverse range of stakeholders in situations of overuse but also in situations where underused local commons, residents, and other stakeholders are linked to create new commons. An attitude of mutual understanding and acceptance is necessary when seeking to link people from different backgrounds.

References

Ajisaka, M. 1994. "Toshi'iju-sha no Shugyo Kozo" [Occupation Structure of Urban Migrants]. In: M. Matsumoto and K. Maruki (eds.) *Toshi-iju no Shakaigaku* [*Sociology of Urban*

Migrants]. Kyoto: Sekaishisosha, pp. 83–101.

Ajisaka, M. 2009. *Toshi'iju-sha no Shakaigaku-teki Kenkyu: Toshi Dokyo dantai no Kenkyu Zohoban* [*Sociological Research of Urban Migrants: Research on Urban Village Associations, Augmented Edition*]. Kyoto: Houritsu Bunka Sha.

Araki, K. 1994. "'Shumatsu Nomin' no Jittai to Tenbo: Hiroshima Kencho Shokuin ni Taisuru Anketo Kekka kara" [The Actual Condition and the Review of Weekend Farmer from Questionnaires at Hiroshima Prefectural Office]. *Chiri-Kagaku* [*Geographical Sciences*], 49(2): 85–94.

Ashida, T. 2006. "Tashutsu-shitei no Furusato eno Kanyo Jittai to Chiiki Nogyo Iji ni Hatasu Yakuwari: Kitakanto Chusankanchiiki Noson wo Taisho to Shite" [One Case of How People Come to Their Rural Home Village and Assist Farming There: A Case Study of a Kitakanto Rural Village]. *Nosonkeikaku Gakkai-shi* [*Journal of Rural Planning Association*], 25: 473–478.

Hardin, G. 1968. "The Tragedy of the Commons". *Science*, 162: 1243–1248.

Hayakawa cho kyoiku iinkai 1980. *Hayakawacho shi* [*History of Hayakawa Town*]. Hayakawa cho Yakuba.

Hoshino, H. 2009. *Mura no Dento Geino ga Abunai* [*Crisis of Traditional Performing Art in Rural Villages*]. Tokyo: Iwata Shoin.

Ishizaka, T. 2002. "Setouchi Kaso Chiiki no Koreisha Seikatsu to Tashutsu Kazoku, Hiroshima ken Kaso Sanson no Chosa Jirei yori" [Lives of the Aged and Living-Apart Families in the Setouchi Underpopulated Area]. *Jinbun Ronso* (*Mie Daigaku*) [*Bullitin of the Faculty of Humanities, Law and Economics* (*Mie University*)], 19: 31–44.

Ishizaka, T. and N. Midorikawa 2005. "Kaso Chiiki no Koreisha to Tashutsu Shi: Mie ken Kii Nagashima cho no Chosa Jirei wo toshite" [The Aged and 'Out-migrant Children' in the Underpopulated Area]. *Jinbun Ronso* (*Mie Daigaku*) [*Bullitin of the Faculty of Humanities, Law and Economics* (*Mie University*)], 22: 111–128.

Kawada, Y. 2009. "Shizen Shihon no Kasho Riyou Mondai" [Problem in Underuse of Natural Capital]. In: K. Asano (ed.) *Shizen Shihon no Hozen to Hyoka* [*Conservation and Evaluation of Natural Capital*]. Kyoto: Minerva Shobo.

Maeda, M. and I. Nishimura 2002. "Tanada Owner Seido Sankasha no Jigyo ni Taisuru Ishiki to Kongo no Kadai: Toshi Nokon Koryu ni Okeru Toshi Jumin Chiiki Jumin no Seikatsu Kankyo eno Kouka to Kadai ni Kansuru Kenkyu Sono Ni" [The Consciousness in Projects of Participants and The Tasks at Maintaining Rice Terraces Projects: A Study on Effects in Life Environment of the Urban and Local Inhabitants and Tasks at Interaction between Urban and Rural Area Part 2]. *Nihon Kenchiku Gakkai Keikakukei Ronbunshu*, 556: 213–218.

Mastsudaira, M. 2008. *Matsuri no Yukue: Toshi Shukusai Shin Ron* [*Whereabouts of Festivals*]. Tokyo: Chuokoron Shinsha.

Matsumoto, M. and K. Maruki 1994. *Toshi Iju no Shakaigaku* [*Sociology of Urban Migrants*].

Kyoto: Sekaishisosha.

Matsuzaki, K. 2002. *Dokyosha Shudan no Minzokugakuteki Kenkyu* [*Ethnological Research of Home Village Associations*]. Tokyo: Iwata Shoin.

Ministry of Internal Affairs and Communication 2011. "Kaso Chiiki tou ni okeru Shuraku no Jokyo ni kansuru Genkyo Haaku Chosa".
http://www.soumu.go.jp/main_content/000113146.pdf

Mochizuki, Y. 1972. *Mogura ku Kyodo Shi*.

Nichiiki Kyoju Jinko Iinkai 2005. "Nichiiki kyoju no igi to sono senryakuteki shiensaku no koso". http://www.mlit.go.jp/kisha/kisha05/02/020329/01.pdf

Odagiri, T. 2006. "Chusankan Chiiki no Jittai to Seisaku no Tenkai: Kadai no Settei". In: Y. Yaguchi (ed.) *Chusankan chiiki no Kyosei Nogyo System: Hokai to Saisei no Frontier*. Tokyo: Norin Tokei Kyokai [Association of Aqrioulture and Forestry Statistics].

Ohno, A. 2005. *Sanson Kankyo Shakaigaku Josetsu: Gendai Sanson no Genkaishurakuka to Ryuiki Kyodo Kanri*. Tokyo: Nosan Gyoson Bunka Kyokai [Rural Culture Association].

Ohura, Y. 2008. "1990 Nendai Iko ni okeru Tosinosanson Kouryu no Seisakutekitenkai to Sono Houkousei" [Policy Development and Direction of Interaction of Cities and Countryside after 1990's]. *Ringyo Keizai Kenkyu* [*Journal of Forest Economics*], 54(1): 40–49.

Oikawa, K. 2007. *Yamano mura kara: Rekishi to Minzoku no Tenkanki* [*From a Village in the Mountain: History and Folk in Transition*]. Tokyo: Kindaibungeisha.

Okubo, M., M. Tanaka and M. Inoue 2011. "Matsuri wo Toshite Mita Tashutsusya to Shusshinson Tono Kakawari no Henyou: Yamanashi ken Hayakawacho Mogura shuraku no Baai"[Changes in relationships between out-migrants and their origin village:focusing on traditional festivals in Mogura village, Hayakawa town, Yamanashi prefecture, Japan]. *Sonraku Shakai Kenkyu* [*Journal of Rural Studies*], 17(2): 6–17.

Sakuno, H. 2006. "Chusankan Chiiki ni okeru Chiiki Mondai to Shuraku no Taiou" [The Problems and Expectations of Regional Development in Hilly-Mountainous Region and Correspondence of Rural Settlements]. *Keizaichirigakunenpo* [*Annals of the Association of Economic Geographers*], 52(4): 264–282.

Shibuya, M. 2000. "Dentogyoji no Densho to Chiiki Kasseika: Iwate-ken Kitakami-shi Shuraku no Kosyogatsu Gyoji no Jirei wo Chushin ni". *Sonraku Shakai Kenkyu* [*Journal of Rural Studies*], 6(2): 48–59.

Shibuya, M. 2006. *Minzoku Geinou no Densho Katsudou to Chiiki Seikatsu Ascendance of Folkloric Performing Arts and Life in Local Communities*. Tokyo: Nosan Gyoson Bunka Kyokai [Rural Culture Association].

Suga, Y. 2010. "Local Commons toiu Genten Kaiki: Chiiki Bunka Commons [local cultural commons] ron e Mukete". In: S. Yamada (eds.) *Komonzu to Bunka: Bunka ha Dareno Monoka* [*Commons and Culture: Who Governs Culture?*]. Tokyo: Tokyodo shuppan.

Suga, Y. 2013. "The Tragedy of the Conceptual Expansion of the Commons". In: T. Murota and K. Takeshita (eds.) *Local Commons and Democratic Environmental Governance*, Tokyo:

United Nations University Press, pp. 3–18.

Sugahara, M., H. Aizawa, T. Ihashi and S. Fuji 2006. "Risonsha no Shussinchi no Chiiki Shakai ni Taisuru Kanshin to Sankaku: Koreika Shita Nosanson Chiiki ni Okeru Chiiki Shakai no Arata na Un'ei Houhou" [Interests and Participation of Left-Residents in Community of Hometown: Community Management in Aged Rural Society]. *Noson Keikaku Gakkaishi* [*Journal of Rural Planning Association*], 25: 461–466.

Tachibana, S., M. Inoue, N. Yasumura, H. Okuda, N. Yamamoto, and H. Kuboyama 1998. "Jinteki Tsunagari kara Mita Shutoken Kinko Sanson no Genjo to Tenbo: Saitamaken Otakimura wo Jirei ni" [Current State and Prospect of Mountain Villages Near the Metropolitan Area through Their Human Networks: A Case Study of Otaki Village]. *Ringyo Keizai Kenkyu* [*Journal of Forest Economics*], 44(2): 67–72.

Takahashi, N. 1983. *Boson to Risonsha no Yukue: Nihon no Nogyo Asu eno Ayumi 149*. Nosei Chosa Iinkai.

Tamura, S. 2001. "Sanson no Kurashi kara Kangaeru Mori to Hito no Kankei: Yukiguni ni Okeru Shinrin Riyou to Sono Henyou". In: M. Inoue and T. Miyauchi (eds.) *Commons no Shakaigaku: Mori Kawa Umi no Shigen Kyoudou Kanri wo Kangaeru* [*Sociology of Commons: Considering Collaborative Resource Management of Forest, River and Ocean*]. Tokyo: Shin-yo-sha, pp. 55–72.

Tokuno, S. 2008. "Nosanson Shinko ni okeru Toshi Noson Koryu: Green Tourism no Genkai to Kanousei: Seisaku to Jittai no Hazama de" [Urban and Rural Interchanges on Rural Promotion: Feasibilies and Limitations of Green Tourism within Social Policies and Actual Circumstances]. *Sonraku Shakai Kenkyu* [*Journal of Rural Studies*], 43: 43–93.

Torigoe, H. 1997. *Kankyoshakaigaku no Riron to Jissen: Seikatsu Kankyo Shugi no Tachiba kara* [*Theory and Practice of Environmental Sociology: From the Standpoint of Life Environmentalism*]. Tokyo: Yuhikaku Publishing.

Ueda, K. 2007. "Kaso Shuraku ni okeru Minzoku Buyou no Hozon wo Meguru Ichi Kosatsu: Kumamoto-ken Itsukimura Kajiwara shuraku no Taikoodori no Jirei kara" [About a "Preservation" of Traditional Dance in a Depopulated Community: A Case Study of a "Preservation" of Taiko-Odori in Kajiwara, Kumamoto Prefecture]. *Sonraku Shakai Kenkyu* [*Journal of Rural Studies*], 14(1): 13–22.

Ueda, K. 2010. "Shoreisho Jusho no Kotoba" Speech for Encouraging Prize. *Sonraku Shakai Kenkyu* [*Journal of Rural Studies*], 16(2): 59–60.

Yamamoto, M. 1994. "Toshi no Dokyojin Kankei to Dokyo Dantai [Home Village Relationships and Associations in Urban Area]". In: M. Matsumoto and K. Maruki (eds.) *Toshi'ijusha no Shakaigaku* [*Sociology of Urban Migrants*]. Kyoto: Sekai-shiso-sha, pp. 103–135.

Yamamoto, N., M. Inoue, S. Tachibana, H. Okuda, N. Yasumura, and H. Kuboyama 1998. "Jinteki Tsunagari kara Mita Chugoku Chihou Sanson no Genjo to Tenbo; Shimaneken no Sanson Shuraku wo Jirei ni" [Current State and Prospect of Mountain Villages in the Chugoku Area through Their Human Networks: A Case Study of Shimane Prefecture].

Ringyo Keizai Kenkyu [*Journal of Forest Economy Society*], 44(2): 79–84.

Yamashiro, C. 2007. "Okinawa ni Okeru Kyoyukai no Keisei Katei to Konnichiteki Tenkai" [A Formation Process and Contemporary Development of Okinawan Kyouyukai]. *Kumamoto Daigaku Kyoikugakubu Kiyou Jinbunkagaku* [*Bulletin of the Faculty of Education, Kumamoto University*], 56: 99–110.

Yoshida, Y. 2007. "Kaso Sanson ni okeru Chiiki Kaihatsu Jigyo no Tenkai to Chiiki Kankyo no Kaihen: Ishikawaken Shiraminemura Nishiyamachiku no Jirei kara" [Public Works of Regional Development and Transformation of Regional Environment in Depopulated Area: A Case of Nishiyama in Shiramine Village, Ishikawa Prefecture]. *Sonraku Shakai Kenkyu* [*Journal of Rural Studies*], 14(1): 23–24.

Yusa T., H. Goto, D. Kurauchi, and K. Murakami 2006. "Chusankan Chiiki ni okeru Akiya Oyobi Sono Kanri no Jittai ni Kansuru Kenkyu: Yamanashiken Hayakawacho wo Jirei to Shite" [Vacant Houses and Their Management in Hilly Rural Areas: A Case Study in Hayakawa Town, Yamanashi Prefecture]. *Nihon Kenchiku Gakkai Keikakukei Ronbunshu* [*Journal of Architecture and Planning*], 601: 111–118.

Chapter 6
Collaborative Governance for Planted Forest Resources
Japanese Experiences

Noriko Sato, Takahiro Fujiwara, and Vinh Quang Nguyen[1]

1. Forest Ownership and the Characteristics of Planted Forest Resources in Japan

This chapter aims to consider the conditions for realizing collaborative governance of forest resources through various agents by examining two case studies concerning the management of planted forest resources in Japan.

1.1 Four Conditions of Planted Forest Resources in Japan

To debate the conditions for collaborative governance of forest resources in Japan, it is first necessary to discuss the following four characteristics. The first characteristic is that though forests account for a large proportion of national land (68.5%)[2], approximately 40% (10 million ha) comprises of planted forests. In terms of area of planted forests, Japan ranks fourth globally, preceded by China, the US, and Russia, and a significant proportion of Japan's total forests is planted forests. In recent years, Asian countries such as China, Vietnam, and Indonesia have been actively engaged in afforestation/reforestation (A/R forestation), and the conditions for collaborative governance of planted forests in Japan will expectedly be of some use by way of reference to these countries wherein the growth of planted forests is rapidly proceeding.

The second characteristic is the lengthy growing cycle—from planting to harvesting—in planted forests in Japan. This differs from the fast-growing tree plantations seen throughout Asian countries. From the late 1950s, when the growth of afforestations increased in Japan, to the early 1970s, there was massive demand for construction timber; thus, the price of timber dramatically increased during this period. Japanese cedar (*Cryptomeria japonica*) and Japanese cypress (*Chamaecyparis obtusa*)—that were suitable for use as construction

---

1) This chapter consists of sections 1–3 by Sato, section 4 by Fujiwara, Nguyen and Sato, and section 5 by Sato and Fujiwara.
2) From *Global Forest Resources Assessment 2005*, FAO.

materials in Japan—were planted and the usable harvest period for these species was 35–40 years. Producing high quality construction materials requires more than 80 years from planting to harvesting. On the other hand, as Japan's climate is precipitous with high rainfall, forests are strongly relied on to perform water and soil conservation and watershed protection functions, and long-term cutting cycle government policies have been suggested to strengthen such functions.

The third characteristic is the harsh economic environment surrounding forestry in Japan. During the past half century, Japan's socioeconomic structure has undergone many changes, the population has moved from rural to urban areas, and wages have risen. On the other hand, during the 1960s, when there was a shortage of timber in Japan, the import tax on logs was completely eliminated, leading to the large-scale importation of timber. With the transition to a floating exchange rate system during the 1970s and the wild fluctuations in the exchange rate following the signing of the Plaza Accord in 1985, the appreciation of the yen progressed[3] and the price war between domestic and imported timber escalated. Considering the price of timber (Japanese cedar) to be 100 in 1985, by 2010, the price for logs had dropped to 47.4%, with the price of standing timber in the mountains dropping to a mere 17.5%, leading to a timber self-sufficiency rate of 28%. This drop in the price of timber led to an increase in the number of forests where adequate thinning is not being implemented.

Furthermore, as Japan is located in the Asian monsoon region, the summer season is hot and humid, and plant growth there is vigorous. To establish planted forests, it is necessary to cut grass and vines right from the time of planting until crown closure (Fujimori, 2006: 363). Japan shares this characteristic with other Asian countries. During times when the forest was used after planting for the cultivation of agricultural crops (agroforestry) and the growing of grass as feed for livestock, grass was an important resource for agricultural sustainability. However, high economic growth led to the application of chemical fertilizers to agricultural land, the importation of food and livestock feed from overseas (Japan's grain self-sufficiency rate is 28% as per 2007 data from Ministry of Agriculture, Forestry and Fisheries), and the reduction of rural labor population. Forestry land is now mostly utilized for other purpose now. Thus, after World War II, single species of plants were planted on sloping forest land and the adoption of monoculture in these plantations progressed. In Japan, for seven to ten years after planting, it is essential to cut grass to allow trees to grow, and

---

3) Although the value of the yen was fixed at 360 JPY = 1 USD under the Bretton Woods agreement (gold-dollar standard) established in 1944, the system collapsed in 1971, leading to the transition to a fluctuating exchange rate system. Following the signing of the Plaza Accord in 1985, the appreciation of the yen continued, reaching 77 JPY = 1 USD in 2012.

thus, the silvicultural cost of planted forests in Japan is higher than that in places such as Europe, which experiences lower rainfall during summer. Therefore, in recent years, replanting after cutting has become no longer financially viable, leading to an increasing number of cases where reforestation is not being implemented. In other words, despite the fact that Japan is successful in planting forests and has rich plantation resources, grave problems exist from the perspective of economic sustainability.

The fourth characteristic in planted forests is the diverse range of relationships between forest landowners and the entities engaged in growing plantations.

1.2 Forest Land Ownership in Japan

Before debating the nature of these relationships, we will provide a simple explanation of forest land ownership in Japan. Modern land ownership in Japan was established by the transition from the feudal era to a capitalist era through the land-tax reform (1874–1880) and through the policies that divided forests and wilderness areas into state and non-state land (1878–1881) at the beginning of the Meiji Era (1868–1912). With the aim of fascilitating the levy of necessary taxes on landowners as a modern state, these land reforms determined that all rights regarding land (access, withdrawal, management, exclusion and alienation, as classified by Schlager and Ostrom, 1992) be imputed to specific individuals or groups. With agricultural land, Though it was not possible to freely assign the land, it was easy to determine ownership, because during the feudal era, farming households had already possessed the and had been cultivating the land for generations. However, forest land was often used as traditional communal forests (*Iriai*), and the use of resources necessary for the livelihoods of and agricultural production for residents was governed by rules unique to each settlement. In cases where there were species suitable for use as construction timber, depending on the area, there were multilayered rights, such as the utilization of useful timber by residents being restricted by the local feudal lord. Under such government policies that divided forests and wilderness areas into state and non-state land, traditional communal forests were forcibly divided into state and non-state ownership. In cases where there was no credible proof, such as documents, the forest became state-owned land. In cases where it was classified as private land, the settlement was unable to acknowledge a legal owner, and for the sake of convenience, the land was registered as being individually or jointly owned or owned in the name of a group, etc. Consequently many land disputes occurred[4]. Examining how the traditional communal rights were legally handled, Japan's Civil Code determined

---

4) Many books, such as Furushima (1955) and Nakao (2009), discuss the historical transformation of traditional communal forests in Japan and disputes and legal interpretations regarding disputes surrounding traditional communal rights.

traditional communal rights as customary rights separate from ownership rights, with access, resource utilization, management, etc., of public land by traditional communal groups (settlements) and acknowledgment of customary use of such areas. However, in the midst of changes in agriculture and lifestyles, the importance of utilizing traditional communal forests declined among local residents, and as the privatization of forests progressed, many traditional communal forests were later divided as per individual private right holders. Since land reform in the beginning of the Meiji Era, there have also been changes in terms of policies, especially with the enactment of the Law regarding the Modernization of Traditional Communal Forests in 1966, to promote privatization to private ownership by individuals and legal entities possessing group ownership/management. As of 2010, in terms of ownership, of the approximately 25 million ha of total forest in Japan, approximately 30% was state forest, approximately 15% was owned by local governmental bodies, and approximately 55% was privately owned (individuals, shared, corporations, etc.) (Ministry of Agriculture, Forestry and Fisheries, 2011). The main characteristic of private forests is that many are small-scale or micro holdings of less than 5 ha, and in recent years, the proportion of owners living in areas other than where their forest is located is increasing. In addition, there are approximately 800,000 ha of customary traditional communal forests.

To implement A/R forestation, landowners, managers, and people to enforce the plantations are required, and combinations of these entities can be divided into three patterns. The first is where three entities are one and the same (1), where A/R forestation is performed not only by individual owners of their own initiative (self-managed forestry), but also as a group by people of settlements in traditional communal forests managed by the settlements. The second is the case where the landowner and the manager, who employ laborers to implement A/R forestation are one and the same (2), and management is effected in a landowner capitalist pattern that relies on employed workers. In this case, the attributes of the landowner (national or private forest, owner who is resident or non-resident in the village, etc.) determine the nature of the relationship with local residents. The third pattern is the case where the landowner and the manager are different (3). In this case, with regard to income at the time of harvesting, profit is divided on a commission basis as previously contracted when plantation trees are cut. In other words, profit-sharing plantation is normal. Of the 10 million ha of plantations in Japan, approximately 16% (1.62 million ha) is in profit-sharing forests[5]. There are various patterns for local people being involved in profit-sharing forests,

---

5) From *World Census of Agriculture and Forestry in Japan, Vol. 13 Report on Forestry Areas*. Of 16.2 million ha, public corporations (state and prefectural) were the reforestation agents for 9.78 million ha (60.3%) and since the 1960s, there has been a rapid increase in expansive reforestation (from natural forest to plantations). Profit-sharing forests other than these have existed since the feudal era (1603–1867) and it is reported that a

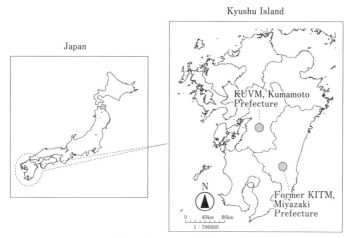

Figure 6–1. Research site.

and the pattern that is used over the greatest area is public institutions (institutions for the promotion of plantation established by the state and prefectures) investing in private land for reforestation. On the other hand, similar to collaborative forest management between state forests and local residents, which is gaining much attention throughout Asian countries[6], there are 132,000 ha of planted forests in Japan that have been planted in a profit-sharing arrangement on national land.

2. An Outline of the Case Study Areas and their Location

Both the case studies featured in this chapter are both located on the island of Kyushu. One is Kuma Village municipality (KUVM) in Kumamoto Prefecture and the other is former Kitago Town municipality (KITM) in Miyazaki Prefecture (see Figure 6-1). As shown in Table 6–1, both areas have been forestry land since feudal times with a history of feudal lords promoting A/R forestation, and the rate of planted forest is in excess of 70%. However, a closer look reveals that the type of land ownership is completely different.

The difference arose when the land was divided into state and non-state land during the transition from the feudal era to a modern nation state, as mentioned above. Further,

---

diverse range of contracts, such as afforestation/reforestation contracts, existed between feudal (many of which were incorporated into state forests) and local residents (Shioya, 1959).
6) For example, collaborative forest management in countries such as Indonesia, Laos and Bangladesh includes the sharing of profit from timber (Mahanty et al., 2009; Fujiwara et al., 2012).

Table 6–1. Overview of the survey areas

Case Study Area		Kuma Village, Kumamoto Prefecture	Former Kitago Town, Miyazaki Prefecture[1]
Ownership Type		Private forest	State forest
Present Resource Problems		Large-scale clear cutting and non-reforestation by landowners not present within the village.	Expiration of state profit-sharing forest (planted by local residents) contracts.
Forest Type		Japanese cedar plantations	Japanese cedar plantations
History of Ownership/Usage	Feudal Era	Planting of shared forest (Japanese cypress construction timber) encouraged by the Sagara Domain.	Planting of shared forest (ship-building timber) encouraged by the Obi Domain.
	Ownership Type	Non-state	State
	Prewar	Forest land lost to outsiders from the Meiji period.	Shared forest rights concentrated into the hands of outsiders.
	Postwar	There were increased sales and purchases during the 1950s/60s for asset divestment. Non-resident entities—especially from Kansai—acquired forestland. In recent years, in the midst of a drop in timber prices, there has been large-scale non-reforestation by non-resident entities.	Kitago Town office had a contract with state forests. After that, through mediation and the organization of right holders (to prevent rights from being lost outside), a reforestation association was formed with the settlement and relatives and a contract signed with the town, with individuals implementing reforestation.
Forest Ownership			
Forest Area		18,268 ha (100.0%)	15,386 ha (100.0%)
State Forest		1,775 ha (9.7%)	11,557 ha (75.1%)
State/Prefectural Corporations		2,407 ha (13.2%)	354 ha (2.3%)
Private Forest		14,086 ha (77.1%)	3,475 ha (22.6%)
Owned by Non-Resident Entities		3,604 ha (19.7%[2])	919 ha (6.0%[2])
Forest Resources			
Plantation Ratio		72.0%	74.4%
Protection Forest Ratio		18.7%	2.3%

Source: Prepared from 2000 Forestry Census and interviews with related institutions.
Note 1. Kitago Town, Miyazaki Prefecture amalgamated with Nango Town, Nichinan City to become the new Nichinan City in March 2009 and became a local body.
2. The ratio of ownership by people not living in the village is the proportion of the forestland area within the area.

77.1% of the forest in KUVM is privately owned, 75.1% of the forest in the former KITM is state owned. Furthermore, in KUVM, right from the prewar period, land was purchased by outsiders, and today, approximately 20% of all forests in the area are owned by outsiders. With regard to the relationship among the abovementioned landowners, managers, and local people enforcing the plantations, what both locations have in common is that there are large tracts of forest land that are not owned by local residents and of the aforementioned three plantation patterns, KUVM belongs to (2), while the state-owned profit sharing forests of former KITM belong to (3). As compared with (1), where owners conduct A/R forestation themselves, it could be said that both KUVM and KITM are examples of where building a collaborative governance relationship between owners and local people is difficult.

3. Governance of Large-Scale Non-reforestation Lands Owned by Non-Resident Owners
The Example of KUVM in Kumamoto Prefecture

3.1 Overview of the Area

KUVM is a mountain village located on the mid-reaches of one of the top three fastest flowing rivers in Japan—the Kuma River—and encompasses 20,773 ha, of which 87.9% is forest land. There are 18,268 ha of non-state forests, of which 14,086 ha (85.4% of non-state forests) are private forests, making it an area of predominantly private forests.

Furthermore, the area has been active in plantation forestry since feudal times (Hattori 1940; Yamamoto 2010). The Sagara Domain, which ruled the area, was basically allowed to do what they wanted as long as they did not hinder the local peasants or the Shogunate (Japan's feudal government). This type of land use was allowed, and the most common examples are that of the warriors, etc. of the domain, who planted profit-sharing forests of Japanese cedar of their own initiative (*Shichibusansashisugi*; 70% of the profit went to the state) in the forests within the territories that were given to them (*Shihaiyama*) or entrusted to them to manage (*azukariyama*) by the lord of the domain were freely sold during the Bunsei era (1818–1830)[7]. This formed the basis for determining private land during the demarcation between national and private forest ownership, leading to the land ownership structure that exists in KUVM featuring small- and micro-scale ownership by local farmers and large-scale ownership by non-resident entities (see Table 6–2)[8].

---

7) Hattori (1940) *op. cit.*, p. 110.
8) According to the 2000 census, there were 628 forestry households living in Kuma Village with holdings of 1 ha or more, of which there were 367 of less than 5 ha, 219 of 5 to 20 ha, 39 of 20 to 50 ha, and three of 50 to 100 ha, with no large-scale owners of 100 ha or more, showing that most were small- or micro-scale forest owners. Of the 3,604 ha held by non-resident owners, 1,589 ha was held by entities outside the prefecture. For further details regarding the history of ownership by non-resident entities, please see Sanson Shinkou Chosakai (1967).

Table 6-2. Present state of ownership of non-reforested land

Location	Cut Area (ha)	No. of Blocks	Location of Current Owner	Characteristics of Present Owner	No. of Times Ownership Changed	Comments
A	96.3	13 →1	Outside prefecture (Wakayama Prefecture)	Corporation (timber-related)	41	Originally jointly owned by 8 people, shares sold outside village postwar, purchased by previous-president of present corporate, used as fixed mortgage.
B	29.0	1	Outside prefecture (Kagoshima Prefecture)	Corporation (timber-related)	8	Registered in 1887, changed to outside ownership during Meiji period.
C	27.0	1	Outside village within prefecture (Kumamoto City)	Corporation (timber-related)	10	Purchased all shares of the 4 joint owners in 2004 and logged more than the permitted area of protection forest.
D	21.1	2	Outside village within prefecture (Ashikita)	Individual	26	Purchased in 1988.
E	21.0	1	Within village	Jointly owned by 5 individuals	7	Within same settlement, inherited 6 times and sold twice.
F	16.7	1	Outside prefecture (Fukuoka Prefecture)	Individual	15	Purchased in 2003 from all joint owners, purchased in 2005 by present owner.
G	14.0	2	Within village	Individual	31&27	Ownership right of all joint owners purchased in 2000.
H	13.8	2	Within village (5), outside village (1, neighboring city)	Jointly owned by 6 individuals	7	Inherited/gifted 4 cases, bought/sold 3 cases.
I	10.2	3	Outside village within city, outside village within prefecture, outside prefecture (Fukuoka Prefecture)	2 individuals/1 corporation (timber-related), jointly owned by 3 parties	25	—
J	9.4	1	Outside prefecture (Fukuoka Prefecture)	Individual	13	Registered in 1895, lost to outsiders in 1905, purchased in 2002, 2003, and 2004.
K	8.0	1	Within village	Individual	0	Registered in accordance with the provisions of Article 12 of the 1989 Traditional Commons Law.
L	6.4	1	Outside village, within prefecture (Kumamoto City)	Corporation (foodstuff-related)	3	Purchased in 2001.

Source: Prepared from data contained in copies of the land register acquired February 2006 (computerized) and March 2006 (prior to computerization).

It is thus evident that plantation forestry began in KUVM in the feudal era. However, with cutting being carried out immediately before and after World War II, most existing plantations were planted after the war and have an age class distribution of 60 years or less, which means that many are due for logging. On the other hand, as of 2010, the population of KUVM was 4,248 people, showing a drop to 63% of that in 1985, and the aging of the population (the proportion of the population aged 65 years or more) has reached 40.5%.

3.2 Large-scale clear cutting and the Characteristics of Owners of non-reforested land

Against a background of increases in usable plantation resources, since the latter half of the 1990s, logging in private forests has occurred in a random manner leading to reports of incidents of non-reforestation. Seeking to sustainably secure plantation resources and fearing a decrease in water and soil conservation functions and the degradation of the river environment, the residents of the Kuma River catchment area had researchers ascertain the actual state of non-reforested areas and analyze the socioeconomic causes (Sakai, 2003). Murakami et al. (2007) pointed out that there were large non-reforested areas in Kyushu, where timber production was active, and that there are local differences in the non-reforestation rate after logging, with the southern part of Kumamoto Prefecture—including KUVM—being one of the hotspots. According to a July 2005 Kumamoto Prefecture survey of logged areas, the total logged area was 1,410 ha (429 ha were non-reforested land where no change had been observed for three years or more after logging), with 80% (1,164 ha) of logged areas within the prefecture being concentrated in the KUVM area, and that 67% of the logged areas in the KUVM area were large-scale areas of 20 ha or more. In 2002, KUVM was particularly the focus of media attention when the occurrence of a 96 ha non-replanted area was revealed, the largest such area since 1990[9] (see Plate 6–1).

As of August 2005, 12 logged areas of 5 ha or more had been identified that had not been reforested after clear cutting ascertained by the Kuma Village Forest Owners Association. Examining copies of the land register, also enabled identification of the characteristics of the present owners, how many times the ownership rights had been transferred and in what year, as well as identify where the owners had lived (see Table 6–2). Consequently, it was discovered that eight of the 12 sites were owned by corporations or individuals outside the village, two were jointly owned by people living within and outside the village, and two were owned by people within the village. Furthermore, the ownership rights of forest land owned by people outside the village had been transferred at least eight times

---

9) *Kumamoto Nichinichi Shimbun*, July 13, 2005; *The Nishinippon Shimbun*, February 5, 2007, and other references.

Plate 6–1. Large-scale clear cutting and on-replantation in KUVM
Photos by Noriko Sato in 2005

(mainly through inheritance and sale/purchase), with the ownership right of the largest area of clear cut land being transferred 41 times, revealing that it had been bought and sold repeatedly. Sales and purchases mainly occurred during the 1890s, 1920s, and the 1950s/60s. During those times, the price of timber had been increasing and it is thought that investors from outside the village or outside the prefecture had purchased the land hoping that the price of timber would further increase in the future. In addition, according to the chairman of the forest owners association, the transfer of forestland during the 1950s/60s occurred when brokers from outside the prefecture purchased bare land after cutting or purchased forestland that had been planted several years ago by employed local people. However, during the 1990s when logging became possible, the price was much lower than expected and continued to drop. Therefore, as seen in Plate 6–1, to recover their investments before the price dropped even further, logging roads were put in, trees were cut and uprooted, and the land was left non-reforested. According to the forest owners association, the number of owners wanting to relinquish forest land is increasing, and when they do so they are dropping the price of bare land to 50,000–100,000 JPY per ha. In other words, in addition to the economic background and the weakening desire of forest owners to invest in reforestation and bear the costs of reforestation due to the stagnation in the price of timber, there seems to be a tendency for large-scale clear cutting and non-reforestation to occur among non-resident owners who have acquired forests for investment.

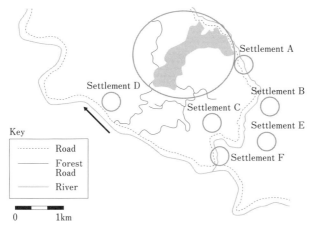

Figure 6–2. Map Showing the Location of the N Non-replanted area and Local Settlements
Source: Prepared by Akira Shigematsu based on information supplied by KUVM Office.

Under Japan's Forest Act, the clear cutting of 96 ha of forest other than protection forests is not illegal if the municipality has been notified. This is a problem in terms of the system and is a major issue that needs to be debated separately[10]. On the other hand, with regard to large-scale clear cutting and subsequent non-reforestation that would cause local residents to be concerned about landslides occurring whenever it rains, it is important to consider related local entities, historical relationships with the area, and how to cope with clear cutting before and after it occurs.

3.3 Historic Correlation between the N Non-Reforested Area and Local Settlements

Figure 6–2 shows the positional relationship of six local settlements related to the N non-reforested area. Settlement A is located alongside a tributary of the Kuma River immediately below the N non-reforested area and their farmland is being damaged by landslides. Settlements B-D are landowners adjacent to the forestry roads used for transporting logs to settlement D, which is over a ridge from the other settlements. These settlements have established a forest road association. Settlement E is located midway along the

---

10) Due to problems with the system, since 2009, the maximum logging area and guaranteed right of renewal have been debated under Forest and Forestry Revitalization Plans and in 2011 the Forest Act was amended. The law was changed to enable municipalities to issue warnings and carry out execution by proxy with regard to reforestation. However, it contained no limits regarding the cutting area in normal forests.

Table 6–3. Outline of settlements participating in discussions and a related summary of N non-replanted areas.

Settlement	A	B (Forest road association/ Region Represented)	C (Forest road association)
Surveying Entity	Ward Mayor (Male aged 70yrs +)	Previous Ward Mayor (Male aged 49yrs)	Ward Mayor (Male aged 66yrs)
Location	Clear cut area immediately below river.	Downstream	Downstream
No. of Households	18	16 (20 at peak)	12 (16 at peak)
Household Composition	6 households with elderly only, 4 households with children.	4 households with elderly only (1 to 2 in household), population 55 people, 9 people junior high school age or younger.	9 single occupant households or 2 elderly people living together, no one aged 18 or younger.
Occupation	Agriculture/forestry (elderly), carpenters or construction-related, no village office workers.	Many commute to Hitoyoshi, 1 forest association worker (processing), 1 JA worker, 4 households grow their own rice/vegetables.	Elderly households have their pensions and their own agriculture/forestry, 3 people work (carpenter, construction).
Historical Links to Non-Reforested Areas	Many residents had been engaged in reforestation at the time of initial forestation (children also earned pocket money by carrying things), or selling things to people who came from Wakayama prefecture to work, and until a few years ago some, had worked as forest wardens.	Many residents in their 60s and over had been involved in previous logging or reforestation, with some even having gone to work in Wakayama prefecture to help the owner.	One person had been a contracted foreman during the previous logging while others had gone to cut undergrowth to raise money for the PTA. They participated in the forest road association discussions.
Response/ Awareness as Logging Progressed	The turbidity of the water downstream from the settlement was terrible (heavy machinery was also in the river), and although the village was contacted, there was no reply.	As a representative area of the forest road association, discussions were held several times with the owner and logging contractor prior to clear cutting and a usage contract was drawn up. During logging, they worried about settlement A, which was immediately below the clear cutting area. Experienced forestry association workers criticized the logging methods.	They only knew that clear cutting was being carried out. They also heard people from A district and others outside the village worrying about disasters.
Influence after Non-Reforestation	They worried every time it rained, were fearful of floodwaters, and the sound of tumbling rocks (they petitioned the municipal office), and their farmland was damaged by landslides.	The river flooded due to torrential rain in August 2006; thus, they evacuated the area. Their fields were buried by soil from clear cut areas.	During the 2007 heavy rain, they thought the turbidity of the river was terrible. They informed the newspaper of the severity of the situation.

Source: From interviews with ward mayors and stakeholders (surveys carried out in August 2006 and September 2008).

D (Forest road association)	E	F
Ward Mayor (Male in 50s) Alongside forest road used for transportation.	Ward Mayor (Male aged 60yrs +) Midway along slope opposite clear cut area.	Ward Mayor Downstream
12 (20 at peak)	16	28
7 households with elderly only (1 to 2 people), 4 people junior high school age or younger.	Population 60 people, 1 single occupant household, almost all other households with eldest son and three generations, some households with wife from another town or village, festivals popular and young people actively participate.	12 households with elderly only (1 to 2 people), 10 people junior high school age or younger.
People work in an electrical shop, as carpenters or in lumber processing, with the rest living off pensions. Some women commute to work in a mountain vegetable processing plant but there are reduced opportunities for employment. Further, 4 households hold more than 1 ha (enough to grow their own rice), with serious wild animal damage to farmland; they are also close to river and prone to flooding.	10 people worked as forest association workers, while others worked for JA or as forest association staff or commuted to work in Hitoyoshi (nursing, construction). 5 households had their own agriculture/ forestry businesses, although not full-time (production/sale of tea/chestnuts). An average of 10 ha of forestland was owned and they managed these small blocks themselves.	Forestry was no more than an occasional daily job, young people commuted to work in Hitoyoshi and 3 were municipal office staff.
9 residents had been employed at the time of forestation. They had also been involved in transporting food. Wages were good (1.5 to 2 times the average local wage) and people had benefitted.	Although few in number, some had been employed at the time of forestation (the main source of income in the settlement at that time had been charcoal production). It took approximately 1 hour on foot from the settlement to the non-reforested area.	Many residents in their 60s and over had been involved in the previous logging.
When imposing conditions through the Forest Association, although elderly residents expressed opinions about logging methods, the matter was settled with money. They were shocked when the work road was formed in an unexpected way. They should have listened to what the elderly people were saying.	Observing from the opposite mountain, they thought that the access roads would become a big problem. They thought that it would be reforested, but given the present price of timber, they thought that there was nothing to do but to leave it non-reforested.	They heard that large-scale logging was being carried out.
There was damage such as turbidity in the river (Kuma River tributary) below the clear cutting area (although there was no damage to the settlement itself).	Although they received no real damage, one person farming a field in an adjacent area of the valley may have experienced some damage.	Although they heard of the matter from the municipal office, they did not understand the real scope until they saw it in the newspaper and were shocked.

non-reforested area in an area from which logging activities were clearly visible. The number of households is decreasing in the five settlements except settlement E, and the number of elderly people living alone or with one other person is increasing. In addition to forestry, a comparatively large number of the successor generation still live in settlement E who commute to work for occupations other than agriculture or forestry. They are also actively involved in the activities of the settlement, such as the communal traditional festivals.

According to the land register, the logged forestland was registered in 1894 as land jointly owned by eight people in local settlements. In 1911, an equity stake was purchased by people from outside the village and by 1947 all equity was owned by people outside Kumamoto Prefecture. Subsequently, the forest land was on-sold a number of times until it was purchased in 1953 by the father of the present owner (Mr. N, manager of a timber trading company). After purchasing it, Mr. N logged the natural forest for pulp and planted Yoshino cedar (a kind of subspecies of Japanese cedar) in the latter half of the 1950s. Many of the local residents who are now aged 60 or more worked on that land as laborers planting trees and cutting undergrowth. Some residents also carried food as children to earn pocket money or sold items to Mr. N's workers who had come to the area from Mr. N's home area of W prefecture. At that time, the price of timber was high, and they were able to earn 1.5–2 times the average local wage by working there, which generated benefits. Furthermore, until a few years ago, Mr. N had employed local people as forest wardens to look after the mountains in settlement A (see Figure 6–2).

According to an interview survey on ward mayors and stakeholders of six settlements related to N-clear harvested (see Table 6–3), many local people saw Mr. N's land as a good place to earn wages. However, approximately 50 years after the forest was planted, when it was time for logging, a large area was harvested all at once—even on slopes of 30 degrees or more—and though roads were formed traversing the forest for tracked forestry machinery and removing logs, the area was left non-reforested. Because of this, many people in local settlements raised their voices, fearing deterioration in the river environment. Moreover, the farmland in settlements A and B were damaged by landslides, causing forced evacuations during times of torrential rain. People were "scared of floodwaters and the sound of tumbling rocks"[11] and their safety and peace of mind is being threatened. Why could this situation not be controlled and how was it handled after it occurred?

---

11) From an interviewee of Settlement A.

3.4　History of N Land Logging and the Response of Local Organizations[12]

The logging of N land occurred when K logging contractor, which had purchased standing trees from Mr. N, sounded out local settlements during 2001 regarding the use of forest roads for transporting logs. As Mr. N had not invested anything when the forest roads were formed, settlements involved in the forest road association adopted a policy of not giving permission to use the forest roads. However, the issue of giving permission to use the roads was debated a number of times, and Mr. N visited the settlements involved in the forest road association together with the logging contractor in an attempt to persuade them to give permission. At that time, K Corporation requested for permission to transport logs and said that they would use cable-based logging in half of the logging area to ensure soil conservation. Finally, in 2002, with Mr. N in attendance, the forest road association and K Corporation finalized a contract with the following four clauses: (1) A payment of JPY6 million for using forest roads; (2) Forest roads would be repaired after use; (3) In terms of logging methods, vehicle-based logging would be used for only 10,000 m³ on roads, cable-based logging should be used for remaining logging so that the soil is not disturbed; (4) Efforts would be made to reforest the area after logging (indicated by word of mouth to use Japanese zelkova). Thus, a process exists whereby landowners—even non-resident landowners—can hold discussions with local residents through the forest road association to reach a consensus regarding matters such as logging methods, and local residents have the right to speak regarding the private management of forest resources within the area.

In November 2001, K Corporation submitted the logging notification to KUVM, and logging and transportation of logs was carried out on N land between November 2001 and April 2003. However, the logging and transportation was not executed by K Corporation; rather, it was done by M logging company, which had bought the standing timber from K Corporation. Of the agreed-upon clauses 1–4, only part of the usage fee of clause 1 was paid, and regarding clause 3, approximately 25,000 m³ of timber was removed using vehicle-based logging. Moreover, the feared landslides and degradation of the river environment became a reality. Clauses 2 and 4 were completely unperformed. As clear cutting progressed, settlements A, B, D, and E criticized the logging method and raised concerns, and although residents contacted the municipal office, logging did not cease. Furthermore, K Corporation, which had signed the contract, went bankrupt immediately after logging began and M corporation, which handled the logging, is also faces bankruptcy. Thus, the business failure of two corporate entities is partly responsible for N block becoming non-

---

12) With regard to the history of logging, in addition the Kuma Village Forest Owners Association, the KUVM office, the village council, and interviews in the local settlements, articles in the *Kumamoto Nichinichi Shimbun* were also referenced, July 13–16, 2005.

reforested land. This made it impossible for measures to be implemented to give consideration to the environment. The fact that there was a breach of contract with local settlements regarding negotiated matters is also noteworthy.

Regarding the bankruptcies, it has been pointed out that K Corporation had problems with the forest evaluation quote (volume and quality). They purchased trees from the owner at an unaffordable high price and after purchasing the trees, the price of timber rapidly dropped. Though the example of the non-reforested N block could be considered the result of unrealistic risk management by a single corporation, the impact on the natural environment and on the living environment (safety and wellbeing) of residents is beyond comprehension.

3.5 Government Response to non-reforested land and the Tree Planting Volunteer Program

Faced with the uncertainties voiced by residents and the fact that reforestation would not be implemented, in April 2003, KUVM, which received the logging notification in 2001, instructed the owner to implement replanting on the logged site. However, the owner, Mr. N, did not comply and stated that he was unable to afford the cost of reforestation. In 2004, the head of the Forest Owners Association, who had sounded out the purchase of the forestland, already had a mortgage of several billion yen against the forestland and was unable to come to the agreement between the forest road association organized in settlement D and K Corporation introduced by Mr. N, which resulted in a deadlock.

What greatly influenced the situation were the appeals of researchers concerned about unregulated clear cutting in such areas and the extensive articles in regional newspapers during 2005. The same articles featured not only the opinions of the government and researchers but also the voices of local forest association staffs and neighboring residents, which had a major impact on government policy. Even within the Kumamoto Prefecture Forest Bureau, the spread of unregulated logging was a problem, and in 2003, the Bureau prepared the *Manual for Dealing with Forest Logging, etc.*, to control unregulated logging. However, when logging was being closely examined as a social problem, in the fall of 2005, the Bureau prepared the *Action Plan for Dealing with Non-Reforested Areas after Clear Cutting* and created a system for dealing with non-reforested areas and controlling and monitoring new occurrences. To deal with the N block, which was the largest area of clear cutting and of greatest concern in terms of environmental conservation, the prefecture subsidized the cost of reforestation by using the Clean Water and Green Forest Promotion Tax[13]

---

13) To formulate measures for forest environment, prefectures, which are local public entities, independently introduce taxes and in Kumamoto Prefecture 500 JPY is collected each year from prefectural taxpayers. Since 2000, the number of prefectures introducing this kind of tax has increased and as of 2010, 30 of the total number of 47 prefectures had introduced such taxes.

Plate 6–2. Voluntary replantation activity in 2006 (Provided by Kuma municipality office).

independently introduced by Kumamoto Prefecture in 2005. In the summer of 2005, social science researchers held a research meeting in KUVM with the prefecture, municipal administrative staff, the forest owners association, and other local residents to discuss the problems and solutions. As this was being reported in the media, local residents were once again reminded of the vast nature of the non-reforestation issues of the N block.

At the end of 2005, specific local efforts were commenced, and a volunteer committee was established with the aim of replanting the N non-reforested block. Members consisted of the municipal government, Kuma Forest Owners Association, Kuma Agricultural Cooperative, Freshwater Fisheries Cooperative, and the abovementioned six settlements. The chairman of the committee was the head of the forest owners association and the industrial promotion section of the municipal office served as the office. The scope of volunteer replanting on the 96 ha non-reforested N block extended to the 4 ha where natural revegetation was not expected to occur and it was important to endeavor to conserve downstream areas. Logging roads where it was feared the hillside would collapse and mountainous areas in danger of collapse were selected, and maintenance, replanting, and earth retaining measures were carried out. Species planted by the volunteer committee included broadleaf species such as castanopsis, nettle trees, quercus serrata, and styracaceae. Planting was carried out with assistance from urban residents (see Plate 6–2). Many natural science researchers (from the Forestry and Forest Products Research Institute Branch Office and universities) are involved in the collecting and analyzing of scientific data such as updates on the status of logged areas and soil movement.

With regard to volunteer replanting, the committee and the landowner, Mr. N reached an agreement regarding the following five points. (1) There is no cost for using the land. (2) There are limitations on the degree of utilization (forest maintenance, such as planting and cutting of undergrowth, and simple facilities for preventing collapse). (3) Logging of

Table 6-4. State of volunteer participation and opinions/reflections

Settlement	A	B	C	D	E	F
State of Volunteer Participation	4 participants (reduced due to change in dates). Requested by the municipal office. They were of the opinion that work should be carried out on slopes above the river rather than on the ridges and flats.	Unclear as the ward mayor was in hospital and the forest association representative (also involved in conducting interviews) did not participate.	Participated at the municipal office's request. Although this was under protest, they felt it was worth doing. They were surprised at participants from other areas and people outside the village — and thankful.	They merely lent their name to the committee. There is little interest among the people of the settlement regarding the mountains and nobody participated. Although they feel guilty in this respect as compared with other areas, they do not want to participate or contribute.	The dates changed and only about 2 people from the settlement participated. Many more were scheduled to come, and there is a high level of interest as there are many people engaged in forestry. Moreover, there is much interest among residents but there is also resistance about planting on someone else's mountain.	Although they only lent their name to the project, they said they would do what they could and 2 people participated.
Opinions regarding Replanting	Just makes people feel better. Questions about selection of species planted.	Unclear as ward mayor was in hospital.	Nothing in particular.	Most people within the area replant, but they are not happy about the fact that they have to replant even though they do not own the land.	Doing what they can as volunteers. It is basically not a solution to the problem. There are many cranes in the area, but will they be OK if undergrowth is not cut (due to lack of funds)?	They are insured, but participants from the settlement are scared of possible injuries.

Source: Interviews with ward mayors and stakeholders (August 2005, September 2008).

trees and bamboo associated with such activities will be at no cost. (4) Plants and logged trees generated from such activities shall belong to the owner. (5) The landowner bears no responsibility for accidents that occur during such activities.

Various entities involved in the volunteer committee were interviewed regarding their motivations for participating in replanting in both 2006 and 2008 (see Table 6–4).

Kuma Agricultural Cooperative
Recently, the agricultural cooperative has been aware of a sense of crisis related to the increase in large-scale clear cutting and the aging of the population, and thus, after staff went to see the 96 ha clear cut block, they asked the staffs with children to participate in replanting activities. The cooperative deems it necessary to finance the living costs of the members considering logging premised on non-reforestation.

Kuma Branch of the Kuma River Freshwater Fisheries Cooperative
The cooperative considers that caring for the river begins with caring for the mountains and that there is a correlation between the long-term turbidity of the Kuma River and the increase in non-reforestation. As they have observed the fisheries cooperatives in other prefectures, they sense the importance of participation of the fishing cooperative in forest conservation measures.

Kuma Village Forest Owners Association
For quite some time, the forest association had seen the clear cutting of N block as a problem and had asked Mr. N to sell the block to them. Purchasing the block would lead to securing the business for the forest association and would enable them to use the forest promotion tax. It would also provide a chance to appeal importance of forest conservation to people living in the cities.

Local Settlements (Table 6–4)
The motives and the number of people participating differed according to settlement. Initially, the local people had despaired or were timid saying, "The municipal office asked for help" or "We have to do it to keep up with other settlements". However, when residents who had actually participated in replanting activities knew that city residents were also cooperating, they realized the significance of participating. On the other hand, residents of the village could be heard protesting against the landowner Mr. N, saying "Almost everyone living in the village does reforestation. Why do we have to do it when we don't even own it?" Furthermore, they (settlements A and E) questioned the reasons that the abandoned forest land should be replanted by volunteers.

3.6 Challenges of privately owned forest governance in KUVM

As seen so far, KUVM cannot be considered as an example of successful formulation of collaborative governance of forest resources. The following three issues can be pointed out as fuel for discussion when considering conditions for success. The first is the existence of severe economic conditions during the early 2000s when the price of timber rapidly dropped and which could lead logging companies to bankruptcy by one miscalculation of the log volume and the quality. Just because owners do not reside in the village does not mean that they can manage their businesses while maintaining absolutely no relationship with local settlements. They employed local people at the time of A/R forestation, paying them high wages, and appointed administrators. Discussions were held regarding the use of forest roads and logging methods between the owner, the logging contractor, and the local settlements. However, the break in relationship was caused by an economic dead end. The second issue is that local residents, with fewer children and an increased number of elderly people, have weakened their capabilities to resist and address landowners and appeal to the government and society. The motivation behind the formulation of response measures did not originate from the inside, but rather from researchers and journalists, who worked on the issue from the outside. The awareness of the issue as a social problem was caused by the sense of crisis of the local people and the way in which it was featured prominently through journalism. Furthermore, the reason that measures were formulated for potential disaster areas using forest environment taxes was that people who felt endangered from disasters were living there. The third issue relates to the evaluation of volunteer planting. Although it is possible to see this as simply some type of performance by the government, because of the great cost involved, it could be said that it at least turned into an opportunity for new actors such as the fishing cooperative, the agriculture cooperative, and city residents, who formerly had no interest in forest resources, to become involved.

4. Participation in State-Owned Profit-Sharing Forests[14]
Former Kitago Town Municipality (KITM), Miyazaki Prefecture

4.1 Area Overview

Former KITM is located in the center of the Obi forestry area in the southern part of Miyazaki Prefecture where forestation has been active since the feudal period. Shioya (1959) said, "If this Hyuga Obi Domain is known to future generations, then it would be on account of the Obi forestry area" (Shioya, 1959: 253) and pointed out that the Obi forestry area had the following three characteristics. (1) The growth of the forest is exceptionally good (final cutting

---

14) This section contains part of the report found in Fujiwara, Nyugen, and Sato (2010) and is the result of a joint survey.

volume in 2005 JPY was an average of 478 m³ per ha)[15]. (2) There are specialized timber usage applications, with some being used in building wooden ships called *benkou*[16]. (3) A profit-sharing forest system exists as a special production relationship. When the demarcation between state and non-state forest ownership was carried out, many forests were deemed to be state forests.

At present, there are 15,273 ha of forest in former KITM (KITM amalgamated with Nango Town to become part of the new Nichinan City in March 2009), of which 74.9% (11,444 ha) are state forests. Of these state forests, 3,304 ha[17] (464 locations) are profit-sharing forests with local residents or local bodies. In recent years, processing plants in the area around Miyakonojo have grown larger, and Miyakonojo is becoming one of the major timber production areas in Japan. These profit-sharing forest resources play an important role as a resource foundation for such timber production areas.

On the other hand, an examination of population trends reveals that during high economic growth period, there was a rapid decline between 1950 (9,465 people) and 1975 (5,638 people), followed by 2005 during which the population was stable at about 5,000 people. Prior to municipal amalgamation, the population was 5,014 people in 2008 and the aging of the population combined with a low birth rate has been continuing, with 32.4% of the population aged 65 years or more in 2008.

4.2 Changes in the Profit-Sharing Forest System in the Obi Area

With regard to the profit-sharing forest (*Bunshurin*) and the former shared forest (*Bubunrin*), in addition to the previously mentioned Shioya (1959), Washio (1979, 1980a, 1980b) gave detailed consideration to the changes in the system that occurred from feudal times to the postwar period until the 1970s and to the transition to the profit-sharing forest system. The "Shared Forest" means the forestation system that non-landowners' stakeholders (investors/workers) make a contract of sharing method of forest productions with the owners. According to Washio's findings, this area's history was that of the local people who want to continue using the forest for food production (agricultural land use and gathering of day-to-day supplies).

---

15) As the log production volume was 478 m³, the timber storage volume is calculated to be 683 m³ at an extraction rate of 70%. On the other hand, according to a 2007 Forestry Agency forest resource survey, the national average timber volume for 50-year old Japanese cedar was 396 m³, while the plantation volume per unit area for this area was more than 1.5 times the national average.
16) During the Edo period, when wooden ships were important means of transportation, Obi Japanese cedar was traded at high prices as a shipbuilding timber with a high oil content making it resistant to rot and bending. At present, as almost no wooden ships are built, Obi Japanese cedar is used as a construction material (Minamiaka Forest Association http://www.kushima-shinrin.or.jp/obisugi/ <acquired November 3, 2011>).
17) From documents of Kyushu Regional Forest Office.

It is also a history of resistance to the Domain load and the State endeavoring to expand timber production (forestry land use). It is evident that changes to the shared forest system tempered the resistance of the farmers and was a reorganization process that ensured income from timber. Listed below are points from Washio's discussions regarding changes to the system.

Conventional Shared Forests (*zairaibubunrin*) (From the feudal era to 1903)
This was the period during which the shared forest system was established, wherein one third of final harvested logs was given to the domain and two thirds was shared with the people (*sanbuichiyama*) since the Obi Domain. As the Obi Domain load was crushed by debt at the time, they encouraged the production of various special local products. The Obi Domain had good ports for transporting supplies to Osaka, which was the center of Japan's economy, and as the area's Obi Japanese cedar was highly thought of as shipbuilding timber, it was an important product for securing income for the domain; therefore planting of Japanese cedar was encouraged. Initially, while using convicts as forced labor to carry out planting, the Obi Domain severely prosecuted any logging by local residents. However, because the farmers sought their food in the forests, there was a fierce standoff between the domain and local farmers regarding forest usage. Therefore, while allowing the intercropping of agricultural crops, the domain improved the shared forest system to allow farmers to freely dispose of one third of the harvested timber at the time of logging, thus encouraging A/R forestation by the farmers. With the 1868 Meiji Reformation, the demarcation between state and non-state forest ownership, which aimed to clarify land ownership rights, deemed domains to have historically managed forests, and many of the forests of the area were classified as state forests. Around 1900, a management structure was established for state forests, and residents' management of shared forests from the feudal era was acknowledged as a conventional custom.

Shared Forests on Designated Area (*Setteiku Bubunrin*) (1904–1925)
After the Sino–Japanese and Russo–Japanese wars, as Japan quickly embraced a capitalist structure and imperialism progressed, the demand for timber increased. Thus, ways were sought to strengthen the direct management structures of state forests, even in the Obi area. However, as the farmers of the Obi area were still reliant on the forest for their food, there was strong resistance to such direct management. Therefore, the state forest management office established the following measures: (1) parameters under shared forest contracts were to be established by settlement; (2) they established contracts not with individual farmers but with settlements regarding part of the established area; (3) they changed the system to carry out planting (species and number of trees) and silviculture (including the number of times undergrowth is cut). Such areas are referred to as designated area shared forests. The sharing ratio of the final harvested income was 40% for the state and 60% for settlements; however,

Chapter 6 Collaborative Governance for Planted Forest Resources

in special cases, it was 30% and 70%, respectively. Local residents participated in planting and silviculture on a per household basis and had the right to receive a share of the income at the time of logging. However, many farmers began selling their right to receive profit share before logging and the number of such cases increased. The sale of profit-sharing rights occurred when farmers needed cash income, because farmers had to wait for required several decades until they could receive shared profits from state (40–80 years, depending on the contract). Washio (1981) described the transfer of such rights in detail and pointed out the concentration of profit-sharing rights in the neighboring merchant's capital. During the same period, there were occurrences of transfer of rights from locals to people outside the village, and illegal logging was frequently carried out by other parties to the contract.

New Shared Forests on Designated Area (*shinsetteiku Bubunrin*) (1925–1966)
To deal with the issues surrounding transfer of profit-sharing rights and frequent cases of illegal logging, the national forest management office included the local government in the form of KITM in the contract between the state and the local settlements. Furthermore, the national forest management office asked the local government (KITM) to get those local people together who were parties to the contract and provide guidance to the local people to ensure that they complied with the terms of the contract (tree planting methods and the prohibition of logging prior to the end of the contract, etc.). The profit-sharing ratio was set at 30% for the state, 7% for the local government, and 63% for local settlements. This new profit-sharing plantation framework was not implemented in systems throughout the country, but was unique to the Obi forest. However, even under the new system, the tendency among some local residents, who were poor and needed cash income, to sell their profit-sharing rights continued. Therefore, in 1954, KITM appropriated 7% of total profits from harvested logs, established a fund, and encouraged people to discuss their issues before selling their profit-sharing rights. By providing low-interest cash loans through the fund, they sought to put an end to the outflow of profit-sharing rights. While acknowledging the effectiveness of the fund in encouraging local people to retain their profit-sharing rights and in aiding appropriate management of the forests, it was also pointed out that because of a shortage of funds and the advent of high economic growth period, many local people migrated to the cities.

Latest Shared Forests on Designated Area (*shin-shin setteiku Bubunrin*) (1966–)
Historically, national forests are operated in a unique manner with a unique local emphasis in this area, where a profit-sharing forestation system unique to the area was developed. However, with the standardization of national forest management methods and the promotion of expansive A/R forestation policies through forest structure improvement measures under the Forestry Basic Law, the custom whereby parties to contracts were limited to only local residents changed and parties other than locals were able to be parties to reforestation

contracts. However, in KITM, there was no real change in giving priority to locals. The state forest signed profit-sharing contracts (with a profit-sharing ratio of 30:70 or 20:80 in special cases) with KITM. The town further promoted profit-sharing plantation through contracts with local settlements or designated groups or through direct management of the local government (by employing workers for A/R forestation). In other words, the profit-sharing forest system has strengthened the tendency of local government to coordinate relationship with local residents.

4.3 Current situation of State-Owned Forest Profit-Sharing Management

Next, we considered the state of the profit-sharing forest in recent years from an analysis of material from the Miyazaki South Regional Forest Office of the National Kyushu Forest Management Bureau[18]. A table of the forest area by sharing types on March 2008 (see Table 6–5) revealed the area of "profit-sharing forests on designated area" during 1905 and 1949. Since 1950, the area of "profit-sharing forest on designated area" was not recorded officially due to changing national policy to promote directly plantation by employees. However, in KUVM, contracts of profit-sharing forest on designated area with local communities/people had been continued, as unique local system of Obi area. The profit-sharing ratio of designated shared forest in the top row of the chart, the old customary profit sharing forest, and the memorial profit-sharing forest differs according to the customs, etc., of each area and has a rich history. This shows that in addition to the normal state 30:70 [state contractor (locals)] contract, there is profit-sharing forest with a rate of 20:80 [state contractor], which is more beneficial to the contractor. KITM has a higher proportion of profit-sharing forest than the Kyushu average, with a higher profit-sharing ratio of 20:80 for local residents in memorial profit-sharing plantation, forestry structure improvement profit-sharing plantation, and mountain village development profit-sharing plantation. Of these, memorial profit-sharing forests that were established before the war permitted A/R forestation at a special rate for local residents as a benefit from the state as an act of the Emperor or in the event that they won the war, etc. While deepening the understanding of local people regarding the enactment of the Forestry Basic Law and the enactment of the Mountain Village Development Law, forestry structure improvement profit-sharing forests that were established after the war permitted the establishment of profit-sharing forests that were especially advantageous to locals in terms of the contribution of state forests to local areas. During the 1960s, the local residents were very passionate about A/R forestation, and it is thought

---

18) The Miyazaki South Regional Forest Office manages 29,087 ha of state forest in not only former KITM but also former Nichinan City, former Nango town, and Kushima City and as of March 2008, had 7,673 ha of profit-sharing forest, of which half (3,304 ha, 464 locations) were in KITM.

Table 6-5. The area of different types of profit-sharing plantation (as of the end of March 2008).

(Unit: ha, %)

	Profit Sharing Ratio (State: Non-State)	Total Kyushu Jurisdiction		Miyazaki South area Regional Forest Office[1]		KITM	
Shared forest on designated area (1904–1949)	3 : 7–4 : 6	935	(2.6%)	856	(11.2%)	191	(5.8%)
Old customary shared forest							
School shared forest (1959–)	1 : 2–3.5 : 6.5	13	(0.0%)	13	(0.2%)	11	(0.3%)
Memorial profit-sharing forest	2 : 8, 3 : 7	1,448	(4.0%)	111	(1.4%)	53	(1.6%)
Forest structure improvement profit-sharing	2 : 8, 3 : 7, 1 : 2	3,462	(9.7%)	544	(7.1%)	357	(10.8%)
Mountain village development profit-sharing forest (1966–)	2 : 8	3,294	(9.2%)	979	(12.8%)	614	(18.6%)
Mountain village development profit-sharing forest (1968–)	2 : 8	470	(1.3%)	240	(3.1%)	177	(5.4%)
General Profit-Sharing Forest[2] — Area improvement	2 : 8	15	(0.0%)	4	(0.0%)	4	(0.1%)
Special mountain products		182	(0.5%)	0	(0.0%)	0	(0.0%)
Young People in mountainous area	2 : 8	356	(1.0%)	24	(0.3%)	11	(0.3%)
Forestry promotion		169	(0.5%)	0	(0.0%)	0	(0.0%)
Forest experience		11	(0.0%)	0	(0.0%)	0	(0.0%)
Others	3 : 7	25,389	(71.0%)	4,901	(63.9%)	1,886	(57.1%)
Subtotal		26,122	(73.1%)	4,929	(64.2%)	1,901	(57.5%)
Total		35,744	(100.0%)	7,673	(100.0%)	3,304	(100.0%)

Source: Prepared from Kyushu Regional Forest Bureau information.
Note: 1) 78% of all profit-sharing forest under Miyazaki South Regional Forest Office jurisdiction has a profit-sharing ratio of 3:7.
2) General Profit-Sharing Forest means forestation area contracted without qualification of designated area.

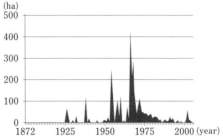

Figure 6–3. Area by Contracted Year of State-owned Profit-sharing Forest in KITM (as of the end of March 2008).
Source: Prepared from Kyushu Regional Forest Bureau information.

that strong demands to national forests by the local area for establishing profit-sharing forests that were beneficial to residents influenced the high area ratio of forest structure improvement profit-sharing forests. In addition, when considering the relationship between the national forest and the local area, it is interesting to note that contracts were signed with designated groups according to the challenges faced by those areas, such as Profit-Sharing Forests for Young People in the Mountainous area to foster successors, school profit-sharing forests for primary schools to help release funds for construction, even though such forests were small in terms of area.

According to information provided by the Miyazaki South Regional Forest Office, approximately 70% (2,288 ha, 296 locations) of the area of profit-sharing forest in former KITM by contractor is with the mayor of KITM. On the other hand, according to information from the former KITM office from the same year, the area of profit-sharing plantation is 2,465 ha. Moreover, an examination of the breakdown reveals that 465 ha is directly managed forest directly administered by the town, including the school forest. The other 2,000 ha is in contract between the national forest and the town office, which is a profit-sharing forest in which the local residents are involved in the tending and management of after plantation through the town. In addition, a little under approximately 600 ha comprises profit-sharing forest based on contracts between the national forest and local resident organization.

In terms of the area of presently-owned profit-sharing forest by contract year in former KITM (see Figure 6–3), a lot of it was placed under contract during the early 1950s and from the late 1960s to 1970, with contracts for more than 1,000 ha of profit-sharing forest located in state-owned land signed in KITM alone during the 4-year period between 1967 and 1970. The contract period is generally 50–60 years, with the longest being approximately 80 years, and now, the contracts from 1950s are reaching maturity. The areas reaching maturity and the logging area trends over the past 10 years have increased since 2000, and both the

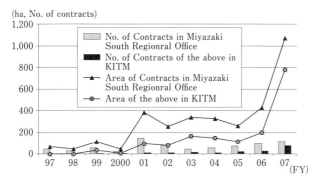

Figure 6–4. Number of Profit-sharing Contracts Reaching Maturity and Logging Area
Source: Prepared from Miyazaki South Regional Forest Office information.
Note: In terms of the annual logging volume, just because a contract reaches maturity does not mean that the area is being logged.

number of logging locations and the area involved increased rapidly during 2007 fiscal year (see Figure 6–4). The volume of timber produced from the Miyazaki South Regional Forest Office area has increased in recent years; of the approximately 70,000 m³ involved, 40,000 m³ was from profit-sharing forests that had reached maturity. While the focus has mainly been on thinning in directly managed forests, clear cutting has been carried out in profit-sharing forests. Bids are submitted for standing trees, and the purchaser pays the state and the non-state entity (reforester) separately. An examination of logging carried out in 2007 fiscal year showed that a 2.74 ha logging area with 1,730 m³ of trees was sold at a bid of JPY 6,989,000, or an average of 4,042 JPY /m³. In addition to logging companies, purchasers of trees included a large-scale processing plant in neighboring Miyakonojo, which shows the important role that national profit-sharing forests play as a foundation for raw materials for timber producing areas. One way of dealing with environmental problems is to offset clear cutting. Efforts are being made to persuade contractors to extend their contract periods for profit-sharing forests where contracts are due to reach maturity.

Can national forest that is logged after reaching maturity be managed in the future using the profit-sharing forest format? According to information from the Miyazaki South Regional Forest Office, of the number of contracts that reached maturity in the 8-year period between 1998 and 2006 and the 936 ha that were logged, only 104.7 ha (11.2%) were recontracted with the same contractor and reforested. Therefore, in recent years, the Forest Office has been calling for contractors not only from among local people but also from neighboring Miyazaki City. However, the newly contracted area is only 50.6 ha (5.4%), which means that profit-sharing forests represent only 16.6% of the logged area. Those logged areas that

Table 6-6. Trends on No. of contracts of profit-sharing forest and contracted area. (unit: ha)

Contract Period	No. of Contracts	Average Area / Contract	Minimum Area	Maximum Area	Total Area
1954–1960	28	29.20	2.62	61.57	817.57
1961–1965	19	5.97	0.66	32.96	113.39
1966–1970	69	6.99	0.31	34.48	482.27
1971–1975	58	4.64	0.12	23.90	269.22
1976–1980	29	3.71	0.14	10.10	107.53
1981–1985	26	2.67	0.09	5.69	69.54
1986–1987	5	2.06	0.86	5.62	10.30
Total					

Source: Prepared from information from former Kitago Town (2006).

are not recontracted are placed under the direct management of the national forest and the Forest Office carries out replantation or natural regeneration.

4.4 Local Plantation Entities and their Intent regarding Profit-sharing Contracts

A survey of 84 profit-sharing plantation cooperatives (234 contracts covering 1,870 ha) in 2006, prior to the amalgamation of former KITM[19], revealed that the average number of members per plantation cooperative was 20.1 (ranging from 2 to 81) and the average area per contract was 7.99 ha (ranging from 0.09 ha to 61.57 ha). Profit-sharing plantation cooperatives are voluntary organizations comprising the head of each household in each of the 29 settlements or volunteers within the settlements desiring to participate in profit-sharing plantation. A head shall be appointed for each cooperative. Of the 234 contracts, the oldest was from November 1954, while the newest was from February 1987.

An examination of the number of contracts and the area involved by contract period (Table 6-6) revealed that while the number of contracts during 1966–1970 was high, the average area per contract was largest at 29.2 ha during 1954–1960, after which it dropped to 10 ha or less, and since 1980, it has grown even smaller to 3 ha or less. Initially, many members of the settlements participated in profit-sharing plantation, but the decreasing number of people engaged in agriculture and forestry led to a decrease in the number of people participating in profit-sharing plantation.

As for the present location of cooperative members, of the total number of 2,349 members, 176 have moved outside the town, 86 have moved to another area within the

---

19) According to a KITM (2006) surveyed the state of movement of KITM profit-sharing plantation cooperative rights.

Chapter 6   Collaborative Governance for Planted Forest Resources   191

town, and 611 (26%) have died. If a cooperative member dies at the time when the contract matures, the person in charge in the town discusses the matter with someone in the family of the person who has died to decide who will receive the profit-share; however, in some cases, the matter cannot be settled as the person in charge cannot contact with any other family members of the deceased person.

The heads of two profit-sharing cooperatives were requested to distribute and collect questionnaires and 15 members of S profit-sharing cooperative and the 20 members of T profit-sharing cooperative about the relationship between their household structure and the profit-sharing forest land and their future intentions. The average age of the 35 respondents was 72.6 years, of which 24 were aged 70 years or more, with the oldest being 92 years old. There were three in the 60–70 year age group and 8 in the under-60 age group, with the youngest being 46 years old. At the time, the average number of people in a household was 3.0 (ranging from 1 to a maximum of 7).

An examination of the number of people in the household who participated in reforestation and silviculture for the profit-sharing forestland revealed that, of the 35 households, two people participated from 16 households, one participated from 13 households, and six households acquired employed labor without being involved personally. In other words, it was found that, in most cases, reforestation and silviculture was carried out by people from within the families of plantation cooperative members.

Next, in terms of the most important source of income for households, for 20 of the 35 members, it was the pension; for eight, it was permanent employment; for four, it was income from self-employment other than agriculture and forestry; and three others relied on other sources. With the aging of the population, more than half were now pensioners.

As for the extension of profit-sharing reforestation contracts after they reached maturity and the intent concerning recontracts (see Table 6–7), 24 people replied that they had no plans to recontract, 19 said they had no plans to extend existing contracts, and five said they had never thought about it. Furthermore, only six of the 26 valid responses indicated that they hoped to increase the price of standing trees by forming roads, which was thought to be an active response to the problem of the lower price of timber, while 14 replied that they had never considered it.

In terms of demands to the government (see Table 6–8), most respondents selected, "I strongly think so" for "timber price countermeasures" and "the expansion of economic support". In addition, two replied, "I strongly think so" and 23 replied "I think so" with regard to demands for "a revision of the profit sharing ratio" concerning the state profit-sharing forest system.

Table 6–7. Contractor's future plans.

	Replies	I have not thought about it.	I have not decided.	I do not have any such plans.	I intend to do so.
Aim to increase the price of trees by forming roads.	26	14	14	6	2
Recontract and carry out reforestation after logging.	27	2	1	24	0
Extend the present contract and have family members carry out thinning.	26	5	1	19	1
Extend the present contract without doing anything.	27	9	3	6	9
Want to dissolve the contract because of the low price of timber.	27	6	10	7	4
Finish after receiving income from logging at the end of the contract.	28	5	1	1	21

Source: Prepared from the results of the survey in March 2008.

Table 6–8. Demands regarding state profit-sharing forests by profit-sharing plantation cooperative members.

	No. of replies.	Strongly think so.	Think so.	No opinion.	Don't think so.	Absolutely do not think so.
Technical support	2	4	1	2	1	
Economic support	6	17	1	1	1	
Revision of profit-sharing ratio	2	23	0	1	2	
Promotion of timber sales	4	17	0	1	1	
Timber price countermeasures	7	17	0	1	1	

Source: Prepared from the results of the survey in March 2008.

4.5 Challenges of State-owned Profit-sharing

The authors considered the establishment of state profit-sharing plantation (former shared forests *Bubunrin*) in former KITM, the changes that were made to the system, and the present state of the local residents. To summarize, in this area, until the 1970s, the state and local residents were at odds and A/R forestation was carried out by applying a profit-sharing rate that was favorable to the local residents. However, even in an age when forestry could be economically independent, in the midst of the economic poverty, local residents continued to sell their profit-sharing rights to people outside the town. Therefore, through discretion at the forest office level, profit-sharing contracts were introduced between the state and local residents, and KITM became involved in ascertaining locals and providing economic support.

In this way, the system was changed in the area to make it easier for local residents to participate on an ongoing basis. Such changes made A/R forestation and management of national land successful. However, at this stage, reforestation after logging is not taking place, and the profit-sharing system itself is coming to the stage where it can no longer exist. The main reason for this is the drop in the price of timber, which make it no longer economically viable, coupled with the aging of the local residents and the dropping birth rate, which promotes the hollowing out of local resident organizations. Much of the forest land, which is not being recontracted after logging, is undergoing transition to become directly-managed national forest. From the standpoint of the management of national forests, this means that forests can be managed under a standardized policy without considering the relationship with local residents. However, on a national level, directly-managed national forests are running in the red, and it is becoming impossible to locate the personnel necessary for forest management, especially because of economic restraints (Kasahara, 2008). Therefore, what kind of management will be employed in the future is expected to be a problem related to the future resource foundation of the Obi forestry area. From the perspective of collaborative governance, local residents were aware of the forests because they were profit-sharing forests and community members of the contractors had to manage these forests. Thus, the only change would be that some residents would be employed only as workers. Local government staffs have noticed that the ability of residents to monitor the forest has weakened, and they are worried that it may lead to illegal dumping of industrial waste, general household garbage, etc.

Furthermore, from our observations of KITM we would like to point out problems not only in national profit-sharing forests but also in the profit-sharing system itself, which has promoted plantation forestry in Japan. In cases where landowners are unable to carry out plantation of their own initiative, profit-sharing forests are a way of investing A/R forestation costs from other entities and sharing the profits from logging when the contract expires in a ratio that was determined when the contract was signed. The period of the contract is determined by the number of years in the plantation cycle, which is generally 40–50 years in Japan, although in KITM, one profit-sharing forest had a cycle of 80 years. During that period, if there is a major drop in the actual income from the sale of timber from what was initially estimated because of major changes in socioeconomic conditions, then the economic sustainability of the management stakeholders (investors and reforesters) can be incredibly damaged. On the other hand, the landowner who provides the land will experience a drop in income from the initial quoted value, but has no management risk and can surely obtain income if the timber can be sold.

Therefore, in profit-sharing forest system, from the perspective of collaborative governance, a long-term framework for forestry must be designed that disperses the risk (e.g., the revision of the profit-sharing ratio according to changes in the price of timber, etc.).

5. Conditions for the Collaborative Governance of Forests from Two Examples in Japan

In this chapter, conditions for the collaborative governance of planted forests are considered by analyzing the two case studies in this chapter. In both areas, although various strategies have been tried in a process of trial-and-error process, it has still not resulted in sustainable forest management and thus cannot be considered successful examples of the collaborative governance of forests. However, the following lessons result from this failure.

The first learning is that with some variation in conditions depending on differences in land ownership, after reforestation, during the period of 40 years or more until logging, there were dramatic changes in various economic conditions, such as the movement of the population to cities for better employment conditions and the drop in the price of timber, which impacted the conditions for collaborative governance in both areas. To realize sustainable forest collaborative governance that transcends generations, a framework that assures economic sustainability by responding on the national level to the timber trade through initiatives such as border protection measures and timber price policies—in other words, the development of data for local residents—is essential. On the local level, rather than monoculture, which features a single species, multiple forest uses are important to be able to respond in a flexible manner to changes in demand according to the times. In the 1960s, demand for construction timber was high in Japan, and because the price of timber rapidly increased, Japanese cedar and Japanese cypress, which are species with a long rotation period, were planted concurrently. This unified the forest environment and was not only a minus in terms of the preservation of biological diversity but also meant that local production was forced into a set pattern. Specifically, planting only long-rotation period species meant that there was no opportunity for income until logging and that crown closure a set number of years after planting reduced the amount of forest management, which meant that the number of times people went to the forest decreased. Combining species with different cutting periods is therefore important to reduce management risk and maintain diverse relationships with the forest.

The second learning relates to the role and limitations of municipalities in forest management. The common thread that runs through the plantation of non-reforested areas in KUVM and the prevention of the transfer of profit-sharing rights to people outside the town in former KITM is the major role that local governments play. Specifically, the contribution that the local government, which is closest to the lives of local residents, makes for improving the safety of residents and improving their lives, and local governmental powers in terms of forest management are both important to collaborative governance. In recent years,

Chapter 6 Collaborative Governance for Planted Forest Resources

in Japan, in the midst of decentralization, the transfer of the authority to municipalities for managing forests is increasing. However, they need to be aware of the fact that there are many problems at catchment level, national level, and global level that are impossible to cope with on a municipal scale. As we saw in the example of KUVM, it was prefectures with forestry technical staffs that introduced appropriate management. While locating specialists in municipalities, it is also necessary to allow prefectures to share administration such as controlling logging, which cannot be handled on a municipal level.

The third learning relates to the necessity of usage restrictions for private forests. As mentioned above, during the rapid modernization process in Japan that came from the demarcation between national state and non-state private forest ownership during the Meiji period, land ownership was established, and the absolute right to use and dispose of the land was given to the landowner. Therefore, the intent of the landowner greatly influenced land use and management, which has caused conflict with traditional users in traditional communal forests (Furushima 1955: 274). With the non-resident landowners in KUVM, the local people were concerned about the logging of a large-scale area and the extraction methods employed, etc. although public organizations, such as the local government and the forest owners association, affected the owner in various ways even after non-reforestation. In the midst of deteriorating economic conditions, the owner's profit was given priority. The volunteer reforestation, which was commenced as a new effort, was not a fundamental solution for conserving the forest environment. In terms of providing an economic incentive to residents, the establishment of land ownership rights could be considered a necessary step for the collaborative governance of the forest; conversely, the restriction of private rights may also be a necessary step. In Japan, it is the time to support those who intent to appropriately manage the forests and to seek restrictions on private ownership that will restrict free use in order to conserve the forest environment. Under the Forest Act, which was amended in April 2011, regulations have been strengthened, such as allowing the government and private entities engaged in contracted management to enter forests with non-resident owners or where the owner is unclear, to intervene and conduct surveys.

The fourth learning relates to the presence of local people, without whom the collaborative governance of forests is impossible. In Asian countries, although the restriction of excessive resource utilization by local residents is a forest resource problem, the opposite problem is hindering the collaborative governance of forest resources in modern Japan. That is, in the midst of the out-migration of local residents, the falling birth rate and aging population, and the promotion of employment in other industries, while forest utilization is dropping, there is concern that settlements in the mountains will become uninhabited. There are also many cases where the boundaries with adjoining land, which is private land, are unclear, making forest management increasingly difficult. Uninhabited places lead to a hollowing out of land use, which in turn leads to a hollowing out of local society[20], which means that there is nobody to look out for forest resources. Do such things relieve the pressure of forest underutilization and contribute to conservation? In the example of KUVM, the basis for Kumamoto Prefecture's use of the forest environment tax for replanting activities was the fact that there was something to preserve, such as houses in which people lived. Furthermore, it was the local people who sensed a crisis and expressed objections regarding logging methods. Even in former KITM, underutilization and the aging of the population decreased the recontracted ratio, and the national profit-sharing forest, which is unique to the Obi area, is breaking down. For forest collaborative governance, the significance of local residents living in such areas must be reconsidered.

In recent years, in Asian countries other than Japan, industrialization is rapidly progressing and problems such as the population concentrating in cities and a falling birth rate are being pointed out (Odagiri, 2011). In conclusion, we would like to point out the significance of discussing the perspective of forest challenges in Asia, considering the forest problems caused by the decrease in population in mountain villages in Japan.

References

FAO 2005. *Global Forest Resources Assessment 2005*. Rome: FAO.

Fujimori, T. 2006 (Japanese edition). *Shinrin Seitaigaku: Zizokukanou na Shinrin Kanri no Kiso [Forest Ecology: A Basis for Sustainable Forest Management]*. Tokyo: National Forestry Improvement and Development Association.

Fujiwara, T., Septiana, R. M., Awang, S. A., Widayanti, W. T., Bariatul, H., Hyakumura, K., and N. Sato 2012. "Changes in local social economy and forest management through the introduction of collaborative forest management (PHBM), and the challenges it poses on equitable partnership: A case study of KPH Pemalang, Central Java, Indonesia". *Tropics*, 20(4): 115–134.

Fujiwara, T., V. Q. Nyugen, and N. Sato 2010. "A Lesson for Profit-sharing Forest Management between Government and Local People Based on Japanese Experiences". XXIII IUFRO World Congress, September 25, 2010.

Furushima, T. 1955. *Nihon Rinya Seido no Kenkyu: Kyodotaiteki Rinya Shoyu wo Chushin ni [Research into Japanese Forest Systems: Focusing on Communal Forest Ownership]*. Tokyo: University of Tokyo Press.

Hattori, M. 1940. "Hitoyoshihan ni okeru Ikuseiteki Ringyo: Tokuni Bubunrin no Ringyoshiteki

---

20) Odagiri (2011: 63). "Limited settlements" are settlements where 50% or more of the settlement residents are aged 65 years or over, where it is becoming difficult to maintain settlement activities, such as ceremonial occasions.

Igi" [Plantation Forestry in the Hitoyoshi Domain: The Historical Significance of Shared Forests in Particular]. Tokyo: Forestry Economics Research, 1967 Reprint, Chikyu Shuppansha.

Kasahara, Y., Shioya, H., and T. Koda 2008. *Dosuru Kokuyu Rin [What to do with National Forests?]*. Tokyo: Liberta Books.

Kitago Town Municipality (KITM) 2006.

Mahanty, S., Guernier, J., and Y. Yasmi 2009. "A Fair Share? Sharing the Benefits and Costs of Collaborative Forest Management". *International Forestry Review*, 11(2): 268–280.

Ministry of Agriculture, Forestry and Fisheries 2007.

Ministry of Agriculture, Forestry and Fisheries 2011.

Murakami, T. et al. 2007. "Distribution of Logged Areas and Non-reforested Areas Detected using Time-series LANDSAT/TM data". *Kyushu Journal of Forest Research*, 60: 173–175.

Nakao, H. 2009. *Iriaiken: Sono Honshitsu to Gendaiteki Kadai [Traditional Communal Rights: Their Real Nature and Contemporary Challenges]*. Tokyo: Keiso Shobo.

Odagiri, T. 2011. *Nosanson Saisei: Genkai Shuraku Mondai wo Koete [The Revitalization of Rural Villages: Overcoming the Problem of Limited Settlements]*. Tokyo: (Iwanami Booklet No. 768) Iwanami Shoten, Publisher.

Sakai, M (ed.) 2003. *Shinrin Shigen Kanri no Shakaika [The Socialization of Forest Resource Management]*. Fukuoka: Kyushu University Press.

Sanson Shinkou Chosakai 1967. "Kumagawa churyu Sanson no Sugata to Shinro – Kumamoto Ken Kuma Gun Kuma Mura" [The Present State and the Future for Precipitous Mountain Villages Located in the Mid-Reaches of the Kuma River: Kuma Village, Kuma-gun, Kumamoto Prefecture, FY1966]. Mountain Village Promotion Special Survey Report.

Schlager, E., and E.Ostrom 1992. "Property-rights Regimes and Natural Resources: A Conceptual Analysis". *Land Economics*, 68(3): 249–262.

Shioya, T. 1959. *Bubunrin Seido no Shiteki Kenkyu: Bubunrin yori Bunshurin heno Tenkai [Research regarding the History of Conventional Shared Forest: the Transition from Conventional Shared Forest to Profit-Sharing Forest]*. Tokyo: Rinya Kyosaikai.

Yamamoto, M. 2010. "Shinrin Shigen no Jizokuteki Kanri heno Wakugumi" [A Framework for the Sustainable Management of Forest Resources]. In: Kumamoto Gakuen University Institute of Economics and Business (ed.) "Global ka suru Kyushu Kumamoto no Sangyokeizai no Jiritsu to Renkei" [*Autonomy and Collaboration in the Globalizing Industry and Economy of Kyushu and Kumamoto*]. Tokyo: Nippon Hyoron Sha, pp. 20–32.

Washio, R. 1979. "Ringyou Keitai no Chiikisei ni kansuru Kenkyu: Obi Ringyo Hattatsu Shiron" [Research regarding the Regionality of Forest Development Morphology — A Historical Treatise on Obi Forestry Development]. *Utsunomiya University College of Agriculture Departmental Bulletin Paper*, 34: 1–128.

Washio, R. 1980a. "Senzenki no Obichiho ni okeru Bubunrin no Settei to Bubunrinken no Ido" [The Establishment of Shared Forests in the Prewar Obi region and the Transition of

Shared Forest Rights]. *Utsunomiya University College of Agriculture, Experimental Forest Report*, 16: 9–37.

Washio, R. 1980b. "Sengo Kodo Keizai Seichoki Iko ni okeru Obichiiki Ringyo no Tenkai: Kokuyu Ringyo to Bubunrin wo Chushin ni" [The Development of Forestry in the Obi Area during the Postwar Period of High Economic Growth: Focusing on State Forests and Shared Forest]. *Utsunomiya University College of Agriculture Experimental Forest Report*, 17: 1–58.

Chapter 7
Forest Resources and Actor Relationships
A Study of Changes Caused by Plantations in Lao PDR

Kimihiko Hyakumura

1. Introduction

1.1 Forest Resources: Changes in Their Perceived Value and Roles

Forest resources in tropical countries have played an important role for governments in the production of timber resources and conservation of biological diversity, endangered plants, and animals. Forests are also a source of non-timber forest products (NTFPs) for human livelihood as well as for temporary plots used by local people engaged in shifting agriculture.

Globalization of the economy and society in recent years has brought major changes in the perceived roles of forest resources. Forest areas are now being utilized to grow biofuels and for industrial plantations such as rubber and eucalyptus. Forests have also taken on new roles for dealing with global environmental issues. The ability of tropical forests to act as a carbon sink to mitigate global warming, as discussed under the United Nation Framework Convention on Climate Change (UNFCCC), is also being considered as valuable. The value of forests in supplying ecosystem services, as discussed under the Convention on Biological Diversity (CBD), is also gaining recognition. These are some of the ways by which tropical forests are being considered to have new economic and environmental value and new perceived roles.

The same changes in perceived value and roles have been evident in Laos, which until recent years experienced slower economic development than other countries in Southeast Asia. Drivers of these changes include the government's poverty reduction measures and economic deregulation policies as well as resource demand from neighboring countries and international environmental conservation initiatives. External actors from the global society, and foreign governments and companies that would normally not have any direct involvement with the forests of Laos, are also starting to recognize the new value of these forest resources because of their scarcity and the potential economic profits.

1.2 Forest Resources and Their Characteristics: New Awareness

Resources are perceived to have value if they have some application or if it is recognized at some point that they need to be managed for some purpose. Local people in day-to-day contact with forests have used forest resources mainly as a source of livelihood. Moreover, external actors such as governments and companies have obtained profits from the market through commodification of forest resources such as timber and NTFPs.

The globalization of economies and society in recent years has revealed new value for these forest resources. Various actors have become aware of the new uses and significance of forests and forest resources, and re-evaluation of their uses and the profits that they offer has given them new value. Actors who recognize the value of forest resources are entities that directly or indirectly benefit or profit from forests. These actors can be classified into four groups: (1) local people who benefit directly from living in or near forests, (2) central and local governments that administer and manage forests, (3) companies that profit from forests and land, and (4) the global society, which perceives environmental value in forests.

The value of forest resources, as recognized by the actors, can be classified into three categories: (1) as a source of forest products including timber and NTFPs, (2) as places to be conserved for livelihoods and the social environment, and (3) as land for its own value rather than as a forest. In Laos, these three categories can be used to explain the ways in which each of the actors use and manage the forest.

First, with respect to forests as a source of forest products, local people gather NTFPs such as mushrooms, bamboo shoots, and other edible forest plants, which they use as food in daily life, as well as resin and spices, which they sell for cash income. They also harvest timber from secondary forests to use as firewood and timber to construct their houses. According to Foppes (2004), NTFPs account for 40–50% of cash income in villages in Laos. As can be seen, forests function as a source of livelihood for local people. On the other hand, companies and governments have been utilizing forests for timber and NTFPs for commercial purposes. Government estimates present total annual exports of NTFPs at USD 878 million in 2005 (DOF, 2009). The government and companies consider forests as a source of profit derived from forest products. A continual supply of forest products is needed; therefore, the leading actors are motivated to ensure that the forests do not disappear, even if the forest area declines to some extent.

Second, forests are recognized as places to be conserved. Local people conserve forests for many reasons, such as for preventing soil runoff along river banks, protecting water catchments, and protecting sacred places such as spirit forests. The global society and governments have come to recognize forests as places for the protection of biological diversity and rare plants and animals. Since the late 1980s, the government of Laos has endeavored to introduce such policies on the basis of calls from the global community for

Chapter 7   Forest Resources and Actor Relationships

the establishment of protected areas. The government has designated approximately 3.3 million ha of forest as protected areas since the enactment of a prime ministerial ordinance regarding protected areas in 1993. These efforts aim to remove outsiders from the forest to ensure forest conservation. Such efforts can also be seen in government-mandated production forests. To prevent illegal logging and promote sustainable forest management[1], the boundaries of production forests are clearly defined and outsiders are prevented from entering. In such cases, the leading actors aim to ensure public benefits and prevent the loss of forests.

Third, forests are recognized as a land resource. Local people utilize some areas as fallowed forests for shifting cultivation and to grow future secondary forests. They also set aside some areas as community forests to serve as a source of forest products for the village. Further, under the current reality of economic globalization, governments, companies, and individual farmers view the land rather than the existing forests as a resource. They consider the land as nothing more than a place for cultivating fast-growing species such as eucalyptus, acacia, and rubber for the market. In such cases, actors such as companies or investors frequently acquire concessions (the right to develop forest or land) and accelerate the conversion of land use. Even forests that were seen as having low economic value until now, such as fallowed and communal village forests, have been the focus of external actors as a land resource. However, external and internal actors view the situation from different perspectives. Local people (internal actors) tend to recognize land as a resource for shifting cultivation on the basis of the nutrients in the soil of forest land. With shifting cultivation, the land eventually reverts back to a forest as part of the transition process called "fallowing." In contrast, some external actors do not recognize the value of the forest resources themselves, and their utilization of land tends to assume that deforestation will occur because of land use conversion.

This chapter discusses the newly emerging perceived value of forests as a land resource in Laos, particulaly focusing on plantations that have been expanding across the nation. An overview of plantations and classification of plantation projects will be presented. Next, an overview of rubber plantations in Laos and examples of eucalyptus plantation projects supported by aid agencies in southern Laos will be provided. Finally, how the forests and the lives of the people of Laos are being affected in the midst of this plantation boom will be discussed and the potential for collaborative governance with external actors, including companies, the government, and aid agencies will be considered.

---

1) The Decree on Sustainable Management of Production Forest Areas was promulgated on May 22, 2002, to prevent illegal logging by outside actors in production forests at the state level.

2. Merits and Demerits of Plantation Projects

2.1 Legitimization and Promotion of Plantation Projects

Recently, the area of forests has greatly diminished worldwide. The Global Forest Resources Assessment, 2010 (FAO, 2010) states that every year, 13 million ha of forests are lost, resulting in many problems. Desertification is increasing in arid areas, thus leading to deterioration of the human living environment. The loss of forests renders areas prone to flooding, which occurs when large volumes of rainwater flow from land that has lost its water-retaining capacity. The loss of forests also means that biodiversity is continuing to decline, and many species of plants and animals are now in danger of extinction. Forest loss also causes approximately 20% of global emissions of greenhouse gases (GHGs), which are a major cause of global warming. To deal with such problems, efforts are perceived made to restore the lost forests. Plantations are being seen as one of the ways of rehabilitating forest land that has been degraded.

In the case of Japan, tree-planting ceremonies are held in every part of the country to celebrate efforts to save and restore forests. Companies and non-governmental organizations (NGOs) are active players in reforestation activities. Posters and pamphlets promoting such green activities feature images of people planting seedlings. Such reforestation activities, if aimed at developing countries, would also be considered important. In developing countries where reforestation has been conducted, advertisements display areas that have once again become green, forests that have flourished, and people who are happy. People who have seen such advertisements receive the impression that companies and NGOs are working through such activities. Meanwhile, companies can boast about being active in environmental conservation and contributing to society. NGOs hope to increase the number of people participating in their green study tours. For those promoting reforestation—and for those hosting it—such activities appear to be something good.

Having studied the forests of Laos for approximately 20 years, the author has noticed the forest landscape change rapidly in recent years. Earlier, driving on National Road Route 13 toward the south of the country provided a view of forests with a diverse range of species. Today, many of those areas have been converted to monoculture plantations of rubber or eucalyptus.

In Laos, there have been three major plantation booms. The first was during the mid-1990s, when teak forests were planted, mainly in the northern part of the country, by small farmers. The second occured around 2000, when eucalyptus plantations were established in the south of the country with the support of aid agencies. Neither of these booms was occurred on a national level; rather, they occurred on a limited local scale. The third plantation boom in Laos is currently in full swing. This boom it is being conducted on a nationwide scale and is so large that it bears no resemblance to the previous booms. Rubber is the most common species being planted, followed by eucalyptus. The main characteristic of this current boom is that it is fueled by companies from both within and outside the country for the purpose of investment. This boom is also driven by the policies of the Laos government, which is actively seeking private investment.

2.2 Plantations and Forest Restoration

Though people generally tend to think of plantations when it comes to forest restoration, there exist other methods that can be employed to the same end. Natural regeneration is used to rehabilitate forests and manage open and secondary forests in an appropriate manner, and it is a typical forest regeneration technology on par with artificial forestation. Formerly, in the Japanese countryside, forest restoration was widely seen in the form of natural regeneration of fuelwood forests. When logging fuelwood forests, the bottom of the tree is left in place. New shoots grow from the stump and seeds in the ground to regenerate the forest. Under Vietnam's 5 Million Hectare Reforestation Program, natural regeneration is one of the methods used to rehabilitate forests using the potential of secondary forests and open forest (de Jong et al., 2006). Forestry Strategy 2020, Laos's mid-term reforestation plan, aims to rehabilitate 6 million ha of degraded and secondary forests using natural regeneration (MAF, 2005). Shifting agriculture also rehabilitates the forest using natural regeneration.

Though natural regeneration is an effective method of reforestation, it is almost never used in commercial forestry. It relies on the regenerative power of nature, and diverse species of trees grow, making it difficult to nurture only those species that are desired for commercial use[2].

2.3 Merits of Plantations

What exactly are the merits of plantations? On the global level, plantations contribute to the expansion of the forest area, which is otherwise on a steady decline. The planted trees also become an important source of timber. In addition, carbon is stored in tree plantations; thus, they can be considered as a means of mitigating global warming. Schemes such as

---

2) However, in commercial forests with species that have many buds, such as eucalyptus, after the planted trees have been logged, coppicing is used to grow them again in the same forest.

the Clean Development Mechanism (CDM)[3] promote the sequestration of carbon in forests, and REDD-plus[4] is used to control deforestation and forest degradation. Plantations can be seen as achieving twin goals of environmental conservation and economic development at the local level. They can promote greening while preventing desertification and can create forest cover for soil and water conservation to prevent flooding and landslides. Furthermore, by converting from native to plantation trees for timber resources, it is possible to eliminate the pressure to log native forests elsewhere. Economically, plantations make it possible to meet the demand for timber and wood fiber for pulp while also acquiring foreign currency. Plantation projects also increase employment for local people.

Therefore, from these perspectives, plantations can be considered to have merits not only for the global environment but also for the government, the local region, and the people living nearby.

2.4 Purpose of Plantations

Plantations can be broadly classified as either environmental or industrial. Environmental plantations are created for the purpose of environmental conservation, such as the creation of water catchment forests and forests for water and soil conservation, and for the restoration of devastated areas. It is necessary for such activities to be continued over a long period of time. Many people foster environmental plantations as their perception have of forest restoration through tree planting. On the other hand, industrial plantations are created to produce products such as timber and paper. After mature trees are logged, the next use of forestland is determined on a case-by-case basis. In some cases, reforestation is conducted, while in others, the land is left as it is. With industrial plantations, where the planted trees are destined to be logged, the forest cover is not maintained over a long period of time; thus, the plantation fulfills any type of environmental conservation function only for a limited period of time.

At present, industrial plantations are the main type of plantation expanding throughout Laos. Though neighboring countries such as China and Vietnam are actively implementing

---

3) The CDM is one of the flexibility mechanisms established under the Kyoto Protocol at the United Nations Framework Convention on Climate Change. With CDM projects, emission reduction projects implemented jointly in developing countries by developed and developing countries are designed so that developed countries can use the resulting carbon credits to offset their own GHG emissions. Forests used in this manner are referred to as CDM carbon sinks.

4) In addition to reducing emissions of sequestered carbon from forests through appropriate forest management, the REDD-plus (Reducing Emissions from Deforestation and Forest Degradation in Developing Countries) scheme aims to maintain and increase forest carbon sequestration and is being considered as a measure for alleviating global warming.

plantation activities, their background is different from that of Laos. In China and Vietnam, massive damage occurred through flooding and soil loss due to the loss of forests. In both countries, rapid forest recovery is essential; thus, environmental plantations aiming at environmental conservation are being created under the direction of the governments. In Laos, the situation has not yet reached the stage where it is necessary to justify the necessity of environmental plantations. Although the necessity of reforestation and the importance of environmental conservation are becoming important issues, reduction of poverty and efforts to leave the ranks of the world's least-developed countries are more pressing challenges in Laos. By amending the Foreign Investment Promotion Act in 2004 to facilitate foreign investment, the government of Laos promoted private participation in plantation activities. Consequently, many foreign companies, not only from China and Vietnam but also from developed countries such as Japan and European countries, started to plant rubber and eucalyptus plantations, leading to the third plantation boom mentioned above. As of 2005, the amounts expected to be invested by foreign companies were in excess of 400 million USD for eucalyptus plantations and 100 million USD for large rubber plantations alone (World Bank Vientiane Office, 2006). Investments have continued in the agricultural sector, such as rubber plantations, and this sector was estimated to have attracted nearly 600 million USD in investments in 2011 (Campbell et al., 2012).

2.5 Land for Plantation and their Right

It is generally assumed that both environmental and industrial plantations are created on degraded forest land that was previously devoid of trees. This impression may exist because people have a notion of reforestation as turning degraded land into a rich sea of verdant green.

The actors involved in a plantation may have determined that "degraded forests" containing nothing but scrub are suitable for plantation. However, in some cases, the people who live in villages can still gather firewood and NTFPs from "degraded forests". For the local people, these "degraded forests" are "secondary forests" that produce forest resources and are considered to be "community forest". Clear differences arise in the way in which the same forest may be perceived by different actors. This gap in perception can lead external actors to label forests as unneeded and then implement plantation activities despite the local people considering the forests to be of value.

Who do forests actually belong to? According to Food and Agriculture Organization of the United Nations (FAO) survey of forest ownership types in East and Southeast Asia[5],

---

5) Countries include Bangladesh, Bhutan, Brunei, Burma (Myanmar), Cambodia, China (Yunnan Province), India, Indonesia, Japan, Laos, Malaysia (Sabah), Nepal, the Philippines, South Korea, Thailand, and Vietnam.

approximately 67% of forests in the region are owned by central governments (Reeb and Romano, 2006). When local governments and similar organizations are included, the total proportion under government ownership increases to approximately 92%. Forest ownership by village groups and local people is virtually non-existent, and legal ownership rights have not been given to people who utilize forests on a daily basis. In Laos, all forests are owned by the central government, although land use rights are given to local people for customary use and daily livelihood.

2.6 Types of Plantation Activities

Plantation activities around the world can be broadly classified into three categories: (1) "environmental plantations" by governments, (2) "industrial plantations" by companies, and (3) plantations by people to support their own livelihood. This classification is shown in Figure 7–1.

The vertical axis in the figure shows the degree of government support, such as subsidies and tax measures. In practice, the higher the position on the vertical axis, the greater is the incidence of "environmental plantations" due to higher levels of government support; the lower the position on this axis, the greater is the incidence of "industrial plantations" with less government support. The horizontal axis shows the degree of self-motivated involvement of local people in plantation activities. Positions that are further to the right of this axis indicate, greater degrees of self-motivated local participation, and those further to the left indicate greater degrees of involvement of external actors. Hyakumura et al. (2007) described five types of plantations: company-managed, contract, direct public management, mobilization, and people-centered. Of these, three can be identified in Laos: company-managed, contract, and people-centered plantations.

Company-managed plantations undertaken for commercial motives are executed directly by companies through concessions on state-owned land. In many cases, these become large-scale plantations to maximize economic efficiency. Though the rights of local people to use the forest are restricted, in some cases, companies provide them some type of compensation. Local participation typically consists merely of employment as labor.

Contract plantations are, similarly, undertaken for commercial motives. Local people and community groups enter contracts with companies to implement plantation activities on land for which they have land use rights. Local people retain these rights. The role of companies is to purchase the timber produced, provide and sell good quality nursery stock, and provide technical guidance and the necessary funds and support in accordance with the contract.

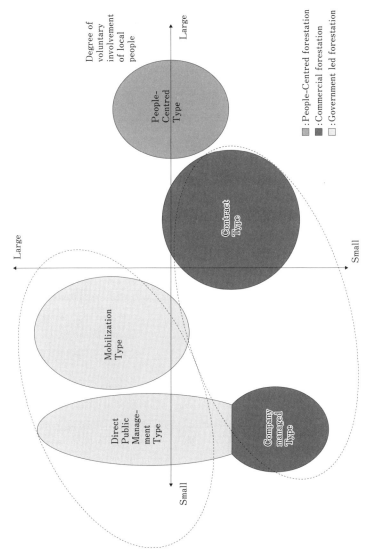

Figure 7-1. General Classification of Plantation Activities.
Source: Hyakumura et al. (2007: 3).

Table 7–1. Area of Rubber Plantations in Laos (ha)

Region	Planted in 2007	Planted in 2008
Central	2,950	25,650
South	8,700	39,000
North	16,555	75,900
Total	28,205	140,550

Source: Vongkamar S. et al. (2007) and NAFRI (2009).

People-centered plantations are implemented by people on their own land. Local people have forest utilization rights; they establish and manage the plantation in groups or by themselves and use the income themselves.

The next section introduces specific examples of plantations that have been established in Laos.

3. The Rubber Plantation Boom in Laos

3.1 Arrival of Rubber Plantations

In Laos, the number of rubber plantations has increased rapidly in merely a few years (see Table 7–1). Until 2003, rubber plantations were limited to approximately 5,000 ha across the nation (Ketphanh et al., 2006); however, this area increased to approximately 28,000 ha by 2007 (Vongkhamor et al., 2007: 5–7) and approximately 140,000 ha by 2008 (NAFRI, 2009) and is expected to reach 300,000 ha in the near future (Douangsavanh, 2009).

A number of factors are responsible for this increase in the area covered by rubber plantations. First, investment flooded into the country from overseas when the amendment of the Foreign Investment Promotion Act facilitated easier participation by foreign companies. Second, the government encouraged rubber plantations to reduce poverty and replace shifting cultivation. Through this, many people who had been engaging in shifting cultivation hoped to obtain public support from the government. Third, success stories spread widely throughout northern Laos regarding how people had received large amounts of money through rubber plantations in villages in Luang Namtha Province. These success stories are believed to have motivated many people to establish rubber plantations. It appears that the synergistic effects of these three factors were responsible for the expansion of rubber plantations.

Though rubber plantations are now present throughout Laos, there are local differences in the total area planted. Table 7-1 shows that as of 2008, the largest area under rubber plantations was in the north, with approximately 76,000 ha, which accounted for 50% of the national total area under rubber plantations. Rubber plantations in the north are mostly being established by Chinese companies, and those in the south, by Thai and Vietnamese companies. The expansion of plantation acreage in northern Laos is because of Chinese influence. Chinese consumption of natural rubber is growing, mainly due to the increased demand for automobile tires (Ma et al., 2012). The number of vehicles in China was 1.78 million in 1980, which rapidly increased to 11 million in 1996, 31.6 million in 2005 (Kowata, 2007), and 62 million by the end of 2009 (Arita, 2011). Chinese demand for natural rubber increased apace, from 1.49 million tons (2003) to 3.63 million tons (2010). Though China is the world's fifth largest producer of natural rubber, it is also the world's largest consumer, consuming one-third of all natural rubber produced[6]. When Chinese manufacturers are unable to procure enough rubber domestically, they rely on imports. In recent years, natural conditions have been conducive to rubber production, and many rubber plantations have been planted in northern Laos, which is geographically close to China. Thus, some raw materials for Chinese automobiles are being produced in the forests of Laos.

3.2 Types of Rubber Plantations

There are three types of rubber plantations in Laos: people-centered, contract, and company-managed. In people-centered plantations, the local people themselves establish and manage the plantation. They procure seedlings themselves and plant them on land for which they hold utilization rights. With contracts, local people establish and manage the plantation on the basis of contracts with companies; this is mainly seen in northern Laos. Local people acquire seedlings from companies and plant them on their own land. Income is shared according to the contract. With company-managed rubber plantations, foreign companies or entities lease land on the basis of concessions and establish and manage the plantation. Company-managed rubber plantations are found throughout the country, though they are particularly common in the south. Here, local people merely participate as laborers, and the only income they receive for their involvement is in the form of wages for their work.

Concerning the above mentioned success stories from northern Laos, though everything appears good at first glance, there are, in fact, many concerns regarding rubber plantations. First, because rubber is a relatively new species to Laos, plantation management techniques have not yet been adequately established. In spite of this, plantations have spread extensively over a short period. Local people in people-centered rubber plantations are unable to learn from neighboring farmers and are struggling to learn how to grow rubber trees. Furthermore, though they are led to expect adequate demand for their rubber, future prices are not

---

6) Source: Tokyo Commodity Exchange, Chapter 3 Rubber Demand, Section 1 World Rubber Demand (http://www.tocom.or.jp/jp/nyumon/textbook/rubber/rubber3.html#rubber03_01, as of August 29, 2012).

Plate 7–1. Land that was Secondary Forest, Logged by an Afforestation Corporation (Southern Laos).
Source: Author.

guaranteed. There is an element of speculation in natural rubber commodity futures, and sudden price fluctuations are entirely possible. In addition, contract-type rubber plantations have been multiplying; however, in some cases, companies have taken advantage of local people who lacked an adequate understanding of the contracts. Some companies have, in fact, usurped the local people's land use rights (Shi, 2008).

3.3 Changes in Perceived Value of Degraded Forests

When establishing a plantation, the area for planting must be clearly defined. In Laos, each village categorizes the forest and land according to use, with management regulations being determined according to land and forest designation. The land targeted for plantation had poor productivity, required greening, and was determined by the government to be secondary or arid forest, referred to as "degraded forest."

In many cases, though considered "degraded," the forest had still been used by local people as a source of NTFPs. Until now, the government deemed secondary and arid forests to have negligible value and showed little interest in their use, although local people used these as their own community forests. The wave of economic development brought about a change; these forests have now become prime targets for plantations (see Plate 7–1).

Company-managed rubber plantations are established by companies that have obtained concessions from the government. Though companies perform such activities after completing official procedures, in many cases, consultation with local people or consideration of land-use rights is inadequate, leading to land disputes between companies and local people (NLMA and Chieng Mai University, 2009). Companies seek concessions for land with fertile soil and good access. When such fertile soil was deemed to be a "degraded forest," competition arose with regard to the land. In Village Y in Champasak Province in southern Laos,

35 of 83 households lost farmland due to large-scale concessions being given to foreign companies for rubber plantations. Local people had received no prior notification regarding this transfer of farmland and did not receive subsequent compensation[7]. Approximately 90% of villagers in another village reportedly lost their land when another large-scale concession was given to a corporation for a rubber plantation in Champasak Province (NLMA and Cheing Mai University, 2009). Though the legislation does not allow local people to become owners of the forests of Laos, land use rights do indeed exist. If plantations are promoted by giving concessions to companies, the forest area remaining for customary use by local people will decrease.

3.4 Government Responses to Problems with Company-Managed Plantations

The government of Laos has gradually developed a sense of caution with regard to land conversion associated with company-managed plantations. At the First National Land Use Meeting (May 7–8, 2007), Prime Minister Bouasone Bouphavanh expressed alarm at the unregulated land conversion by companies and announced a temporary moratorium on the issuing of large-scale land concessions (100 ha or more)[8]. The Prime Minister pointed out that coconut cultivation was destroying the forests a fact that had been criticized by NGOs and researchers. Dense forests with an abundance of trees were being deemed "degraded forest," and after obtaining concession development permits, companies were logging these forests to harvest the timber (Hunt, 2006). No longer able to give tacit consent for such actions, the government attempted to stop company-managed plantation concessions that had been rising rapidly.

The government encouraged the introduction of a "2+3 policy" for plantation activities using private investment. This policy combines land and labor from local people with funds, technology, and access to markets provided by companies. Through the introduction of this policy, the government attempted to control company-managed plantations associated with large-scale concessions. However, they continued to multiply, and forest loss and land disputes resulting from land use conversion showed no signs of stopping. Therefore, in June 2012, the government issued a new Prime Ministerial ordinance declaring a moratorium on all new concessions, including rubber and eucalyptus plantations[9]. Therefore, it appears, that it will still take some time to solve the problem of land disputes caused by company-managed plantations that are driven by market demand.

---

7) Source: Land Issue Working Group website, "Cases from the field: Vietnamese rubber company in Thateng District" (last viewed on August 30, 2012).
   (http://www.laolandissues.org/wp-content/uploads/2012/01/Case-Rubber-in-Sekong-NEW.pdf)
8) Source: Prime Minister announces moratorium on land concessions, *Vientiane Times*, May 9, 2007.
9) Source: Government halts new mining projects, land concessions for tree farms, *Vientiane Times*, June 26, 2012.

4. Contract Plantations for Eucalyptus

This section provides an outline of contract plantations in Laos. The main advantage of contract plantations is that land disputes with local people are less likely to arise because their land use rights are maintained. In addition, examples are presented of industrial plantations established with the involvement of aid agencies.

4.1 Plantation Projects Involving Development Assistance from Aid Agencies

The Asian Development Bank (ADB) implemented the Industrial Tree Plantation Program (ITPP) in seven provinces in southern Laos between 1994 and 2003. With a total investment of USD 11.2 million, the program was conducted using a two-step loan system[10]. The aim of the program was to restore degraded forests to production forests and produce pulp (as raw material for paper manufacturing). The program comprised four parts: creation of 96,000 ha of plantations, creation of 560 ha of model forests, securing of roads for tree planting and log transportation, and capacity development of forest administrative staff (NAFES, 2003).

The following is a brief explanation of the typical process for establishing a plantation under the ITPP program. These activities were managed primarily by two local government institutions: the District Agricultural and Forestry Office (DAFO) of the Provincial Agricultural and Forestry Office (PAFO), which mainly managed technical support, and the Agricultural Promotion Bank (APB), which mainly looked after finance. The main participants in plantation activities were individual villagers who signed a loan contract with the APB, which then provided them with funds. Participants provided their land and labor and performed plantation work. After the trees matured, they were to pay back their loans and obtain income from the sale of timber. Funds for seedlings, fertilizer, and barbed wire to protect the plantation from animals were covered by the loans in the form of goods, and wages for land development were covered by the loans in the form of cash. The loans had to be repaid within seven years, but interest at a rate of 7% per annum (for local people groups and sole proprietors) was charged from the second year.

Aid agency loans were available to companies, sole proprietors, and local people's groups originally with an original emphasis on companies. However, they showed little

---

10) A two-step loan is provided through the development finance institution of the country to which the loan is extended. In the examples featured in this chapter, the Asian Development Bank provided finance to the Laos Agricultural Promotion Bank, which then financed afforestation entities, such as local residents.

interest in the program; thus, the emphasis had to be shifted to local people's groups and sole proprietors. Consequently, the number of eucalyptus plantations established by local people in the program during the latter part of the 1990s started to increase, with approximately 2,500 local households in total participating by the end of this program in 2003.

4.2 Plantations Supported by Aid Agencies: Present Status

To study plantation activities, local groups in Savannakhet Province were interviewed. One was located in Champhone District, approximately 30 km from the provincial capital, whereas the other was located in the nearby Kaysone Phomvihane District. The findings are described below.

The following is an introduction to Mr. K's group in Village N in Champhone District. In 1988, staff from the DAFO of the PAFO and the APB visited the village to promote eucalyptus plantations and explain loans available for plantation projects. The staff informed the villagers that the trees could be logged eight years after planting, that buyers would come, and that participants would earn a profits. When the head of the village asked for people to participate, many households initially expressed interest. However, only eight households actually participated, because some households were unable to pay the interest or obtain a plantation area, whereas others felt uneasy about the cultivation technology.

These eight households formed a group, and in 1999, they planted eucalyptus seedlings. Each household planned to plant 1 ha and received 1,400 eucalyptus seedlings, fertilizer, and barbed wire for plantation as well as cash for developing the land. However, the seedlings arrived without instructions or accompanying staff from the PAFO; therefore, the villagers had to think for themselves and plant the seedlings as best as they could. The area where they planted the seedlings featured comparatively good soil next to paddy fields, comprising open forest on land used for shifting cultivation. Though they were told that the appropriate size of seedlings is 30 cm, the actual seedlings that were distributed to them were smaller (approximately 15–20 cm) (see Plate 7–2). The seedlings were delivered at the beginning of the rainy season in June and were planted immediately. Around 500 seedlings died immediately because of damage caused by white ants. By 2006, the trees had grown to approximately 20 cm in diameter (see Plate 7–3). The trees were due to be logged in 2007, i.e., the eighth year after planting. Each household in the group was paying LAK 125,000 (approximately USD 13) per annum in interest. In 2005, one of the households in the group sold 11 tons of timber (approximately half of the trees they had grown) for approximately LAK 500,000 (approximately USD 52). However, compared with the price for eucalyptus in central Savannakhet Province, this figure appears to be very low even after adjusting for transportation costs.

Plate 7–2. Workers Handle Eucalyptus Seedlings Delivered by the Provincial Agricultural and Forestry Office (PAFO) (Savannakhet Province, Laos).
Source: Author.

Plate 7–3. A Mark is Placed on Each of the Trees in this Eucalyptus Plantation Planted with Funding from the ADB (Savannakhet Province, Laos).
Source: Author

Plate 7–4. Seedlings Planted on an Impermeable Layer Do Not Grow Well. The Photo Depicts Eucalyptus Plantation Planted by Local People with Funding from the ADB (Savannakhet Province, Laos).
Source: Author

Next, the details of Mr. B's group in Village P in Champhone District are as follows. Around 1999, staff of the provincial and district branches of the APB and PAFO came to the village to promote the concept of eucalyptus plantations. As in Village N, they explained that people would definitely profit in future. Similar to the situation in in Village N, though many people were interested, only five households actually participated. Personnel from both agencies visited the village three times to promote eucalyptus plantations, sign contracts with those who had decided to participate in the plantation program, and deliver seedlings. Staff from the PAFO reportedly explained what kind of areas would be suitable for planting the seedlings. Mr. B did not heed the provided explanation and selected approximately 1 ha of open forest area next to paddy fields, with comparatively good soil. Approximately 1,800 seedlings were distributed to Mr. B, though with much variation in size, and the smaller ones died before planting. The surviving seedlings were planted; however, most seedlings died within the first two years due to damage from white ants. Approximately 100 seedlings still survived as of 2008, but were not growing well. Most seedlings of the other four households were also lost due to damage from white ants, and though they were able to pay the loan interest for the first two to three years, they later lost the desire to pay due to the unsatisfactory growth of trees. Staff from the APB visit the village each year to collect interest payments, but Mr. B and others have still not paid.

Further, there was a group from Village X in Kaysone Phomvihane District. No staff from the PAFO or APB visited this group to promote plantations. The group leader, Mr. P, heard about the plantation program from other villages, and subsequently, he met with personnel from the PAFO and APB and signed a plantation contract. This information was not disclosed to the village as a whole, but six households in close relationship with Mr. P formed a group and have been participating in the plantation program since 1998. The eucalyptus trees in Village X have been growing well, and five households effected sale in 2005, seven years after planting. Mr. P, who planted 3 ha, sold 140 tons of timber for LAK 21 million (approximately USD 219); thus, he was able to repay the loan of LAK 1.5 million (approximately USD 156) plus interest and still retain a significant amount. Four other households received approximately LAK 0.75–1.5 million (approximately USD 78–156), whereas the remaining household had yet to sell the timbers, because the trees had not grown to an adequate size. There appear to be major differences in tree growth depending on soil conditions. After the timber sale, Mr. P's coppiced plantation is now growing again.

4.3 Plantation Failure and Local Ramifications

These examples indicate that not everyone can participate and succeed in eucalyptus plantation programs implemented by aid agencies and that it is important to consider the condition of the land.

Next, it was clear that the selection of the plantation site greatly influenced the growth of trees. Some people planted seedlings on land with an impermeable layer (see Plate 7–4), unaware that good growth could not be expected in such land. Local people who believed that they could plant seedlings anywhere often discovered that the trees failed to grow. Furthermore, there were problems with the seedlings. According to interviews with PAFO staff, seedlings were distributed first to settlements close to the provincial capital and then progressively further away. Consequently, it is believed that in some cases, the rainy season was ending when the seedlings eventually reached participating villages far from the provincial capital. Moreover, the poor condition of the distributed seedlings was also indicated as one of the reasons for their subsequent withering and dying.

Under this program, groups of local people and sole proprietors planted a total of approximately 9,900 ha, and companies planted a total of 3,000 ha (NAFES, 2003). These figures appear to suggest that the stated goal of establishing 9,600 ha of plantations had been achieved. However, a 2006 survey of plantation growth indicates that the area planted was somewhat smaller than officially stated, and even where planted, seedlings withered and died on 40% of the land (LTS International Ltd., 2007).

These interviews suggest that the reasons for the partial failure of this plantation program include flawed scheme design and the failure of each responsible government office to fulfill its responsibilities. Inadequate technical guidance regarding growing trees provided to the local people by the PAFO. Almost all participants who were interviewed pointed out the lack of technical guidance. On the other hand, PAFO staff opined that the program ultimately failed because of inadequate consideration of the overall plantation scheme, including aspects such as interest payments under the loan system[11].

Given the fact that the program did not achieve the target plantation area, the aid agencies that implemented the program also acknowledge the program's failure (ADB, 2005). What does the failure of a plantation program run by an overseas aid agency actually mean? This question is worth considering. Aid agencies conduct an evaluation, and the program ends with the publication of a report. For participants, however, what happens if the eucalyptus trees do not grow well and they are unable to obtain the income that they had been expecting? The ramifications of program failure ultimately affect the lives of the people of these rural villages.

---

11) Information from interviews conducted in central Laos in 2006.

5. Collaborative Governance under Plantation Programs

5.1 Plantation-Induced Poverty

Local people who participated in the aid agency contract plantations described above probably believed they could improve their lives through the program. Looking below the surface at what is actually going on, such improvement was not the outcome. Many participants were left with nothing but outstanding debt, and they regretted participating in any plantation program. Though the APB is not taking severe actions against the local people it financed, there is reason for concern about what is going to happen to these people in future.

Not all local people can participate in contract plantation programs. The first condition is that they must have excess land to use for a plantation. This means that only comparatively well-off villagers can participate. Many participants in the above-mentioned examples of contract plantation programs were already well-off, such as village chiefs and former chiefs. In other words, such a plantation program is not aimed at the destitute populations; even if such programs do succeed, the economic gap between participants and non-participants within a village can grow even larger. Moreover, if such programs fail, participants have to bear large financial burdens. Therefore, it can be said, that a major gap existed between the ideals and reality of poverty reduction through the plantation programs supported by governments and aid agencies.

In company-managed plantations for rubber and other trees, companies fence off the land. There have been examples of companies consequently usurping the land use rights of people although, in many cases, companies gain their concessions by following the proper procedures. The land subject to plantation is considered as degraded forest by the local forest administration; therefore, if the government and companies decide that the land is required for plantation, it is deemed to be useless even if that forest was being effectively utilized by local people. In one case, local people whose secondary forest was developed by companies gave this account: "This forest originally belonged to everyone in the village. The village chief said he was going to provide it to a corporation as land for a plantation. But, if all of the village forest is taken, we will not be able to use it. We discussed it in the village, and in the end, the chief decided to leave half the forest as the village forest"[12].

Apart from plantations by companies, land is also fenced off for contract eucalyptus, contract rubber, and people-centered plantations. In many cases, the key people participat-

---

12) Information from interviews conducted in central Laos in 2006.

ing in such plantation programs are already influential and belong to a relatively well-off class in the village. In some instances, the land used by people living in poverty was fenced off for a plantation. Even without the interference of companies, there is still the possibility of the affluent usurping the land use rights of the poor, who ultimately merely become laborers. Through these processes, plantation programs in Laos could, in some cases, increase the gap between the affluent and the poor.

5.2 Changes in the Livelihoods of Local People Due to Plantations

During rice shortages, secondary forests function as a safety net for people. This fact was reflected through interviews with people from hill country forests in Savannakhet Province. Poor households that experience rice shortages consume a type of yam until the next rice harvest (Hyakumura, 2002). They do not gather the yam from dense forests but from land that has been fallowed from shifting cultivation or arid dipterocarp forests near their village. If this forest is converted to plantations, these people could suffer severely during future rice shortage. The growth of plantations may mean not only the loss of forest but also the loss of a safety net.

With the accelerating pace of transition to a market economy, Laos will undoubtedly face further promotion of plantation programs. If a program ceases to be financially viable for external actors such as governments, companies, and aid agencies, they may declare the program a failure. This may mean economic losses and implications for the reputation of those external actors. However, the decision to withdraw the program would simply end the actors' association with such a program. However, the local people, would have to continue to cope with the implications on the land on which they have no choice but to stay and continue living.

5.3 Potential for Collaborative Governance in Plantation Programs

As mentioned above, in Laos, companies are among the main actors promoting plantations in rural areas in the name of poverty reduction and economic development. Presently, there is a power relationship between the haves and the have-nots. If a programs fails or is poorly managed, it is the destitute population who have to bear the brunt of it.

Company-managed plantations, prone to disputes, are still continuing. Many companies that operate plantations in developing countries are vigorously pursuing profits and have no concept of corporate social responsibility (CSR). The central government in Laos is alarmed by this situation and is seeking to introduce policies to curb such activities. Though local people have limited power against pressure from external factors, a legal framework is being developed in Laos to protect local people and help them avoid unreasonable losses[13]. For example, a Prime Ministerial Decree determines compensation for land lost through concessions. The government has also acknowledged land use rights for the customary use of forests by local people. However, locals often are unaware of their rights, and in many cases, they are tricked into ceding their land to companies. Local people should be informed of the existence of these schemes and recognize that they have such rights. In addition, they sometimes need support to proactively articulate their position in negotiations with companies. Some NGOs are already engaged in such information campaigns[14].

Such information campaigns are normally not the domain of NGOs but are conducted by regional forestry administration institutions that act as facilitators at the local level. The government is also concerned about the state of concessions[15] and has pointed out that it is appropriate for regional forest administration institutions to play such a role. This would entail standing between companies and people and mediating discussions regarding matters such as giving consideration to land and forest rights and providing assurance regarding the payment of compensation. In addition, local people need to understand the relative advantages and disadvantages of plantation programs and establish a forum where discussions can be held regarding how best to advance such programs.

Even with contract plantations, there are examples of companies usurping land-use rights through plantation programs. In such cases, local people were at a disadvantage, because the contract system did not function appropriately and they did not understand the system. In such situations, regional forest administration institutions can play an important role as an interface between companies and local people to ensure that contracts are implemented in accordance with the terms of the scheme.

Furthermore, in some regions, the land to which the poor have usage rights is fenced off by influential or more affluent individuals in contract or people-centered plantations. To prevent such problems in people-centered plantation programs, regional forest administration institutions should probably be involved in additional monitoring of these programs.

For plantation programs to continue more smoothly and equitably in Laos, it is essential for regional forest administration institutions to play an appropriate role as a buffer in the unbalanced power relationship between the government and local people, companies and individuals, investors and beneficiaries, and the affluent and the destitute. For this to

---

13) Prime Ministerial Decree on the Compensation and Resettlement of the Development Project, July 7, 2005, Laos.
14) From interviews in February 2012 at the Japan International Volunteer Center, an NGO active in Savannakhet Province, Laos.
15) "Government urged to review land compensation". *Vientiane Times*, August 18, 2012.

happen, it will be important to consider measures for increasing the capacity of regional forest administration institutions and raise awareness of the issues covered in this report.

References

Arita, K. 2011. "Shinkokoku Shijo ni okeru Jisedai Jidosha Fukyu no Mitoshi to Kanseisha Meka no Senryaku Hokosei"[The Outlook of Next-generation Vehicles Permeation and Strategic Direction of the OEM Sand in Emerging Markets]. *Mizuho Industry Focus* 103, p. 30.

Asian Development Bank. 2005. *Completion Report on Lao: Industrial Tree Plantation Project*. Vientiane: ADB.

Campbell, R., T. Knowles, and A. Sayasenh. 2012. *Business Models for Foreign Investment in Agriculture in Laos*. Geneva: The Trade Knowledge Network.

de Jong W., S. Do Dinh, and H. Trieu Van. 2006. *Forest Rehabilitation in Vietnam: Histories, Realities, and Future*. Bogor: CIFOR.

Department of Forestry, Laos (DOF). 2009. *Indicators for Monitoring of Sector Performance (Indicators 2009). Consultant for Forestry Sector Monitoring System*. Vientiane: DOF.

Douangsavanh, L. 2009. "Investment Potential and Smallholding Development in Natural Rubber Industry in Laos". Paper prepared for ASEAN Rubber Conference, 18–20 June 2009. Vientiane, Laos.

FAO 2010. *Global Forest Resources Assessments*. Rome: FAO.

Foppes, J. 2004. "The Role of Non-Timber Forest Products in Community Based Natural Resources Management in Lao PDR". Paper prepared for the Regional Workshop on Community-Based Natural Resources Management (CBNRM). 4–7 November 2003. Lobeysa, Butan.

Hunt, G. 2006. "Large Scale Plantations in Pakkading District". Report from a Visit to Pakkading Integrated Rural Development Project, 23 March 2006. Unpublished Document, Vientiane.

Hyakumura, K. 2002. "Raosu Nanbu deno Mori no Riyo: Kyuko Shokubutsu to Mori ni matsuwaru Kinki" [Forest Resource Use in Southern Laos: Taboo on Famine Plants and Forest]. *Shinrin Kagaku* [*Forest Science*] 36: 76–78.

Hyakumura, K., Seki, Y., and Lopez-Casero, F. 2007. "Designing Forestation Models Suited to Rural Asia: Avoiding Land Conflict as a Key to Success". *IGES Policy Brief* 6, Hayama: Institute for Global Environmental Strategies, Japan. pp. 8.

Ketphanh, S., Mounlamai, K., and Siksidao, P. 2006. "Rubber Planting Status in Lao PDR". Paper presented at the Workshop on Rubber Development in Laos: Exploring Improved Systems for Smallholder Production, May 2006. Vientiane.

Kowata, S. 2007. "Chugoku Jidosha Sangyo no Atarashii Ugoki: Jidosha Fukyu no Genjo to Kadai" [A Basic Study on Motorization in China: Increasing Numbers of Civil Motor Vehicles]. *Fukuoka Daigaku Kenkyu-bu Ronshu B Shakaikagaku Hen* [*The Bulletin of Central Research Institute, Fukuoka University Serious B: Social Sciences*] 2: 27–53.

LTS International Ltd. 2007. *Analysis of the 2006 Forest Inventory for the Laos Industrial Tree Plantation Project*. Unpublished Document, Vientiane: Ministry of Agriculture and Forestry, Laos (MAF).

Ma, H., Hyakumura, K., and Sato, N. 2012. "Chugoku no Tennen Gomu Yunyu Kakudai no Haikei to Kaigai Shokurin Toshi no Jittai" [Factors of Import Expansion of Natural Rubber of China and Actuality of Overseas Investment for the Plantation: A Case of Rubber Plantation Investment in Laos]. *Kyushu Shinrin Kenkyu* [*Journal of Kyushu Forestry Research*] 65: 1–4.

Ministry of Agriculture and Forestry, Laos (MAF). 2005. *Forestry Strategy to the Year 2020 of the Lao PDR*. Ministry of Agriculture and Forestry. Lao PDR, Vientiane: MAF.

National Agriculture and Forestry Extension Service (NAFES), Laos. 2003. *Project Completion Report of Industrial Tree Plantation Project ADB Loan No.1295/Lao/SF*. Vientiane: Ministry of Agriculture and Forestry, Laos (MAF).

National Agriculture and Forestry Research Institute (NAFRI), Laos. 2009. "Rubber Development in Lao PDR". *Rubber Development in the Lao PDR: Ensuring Sustainability*. July 2009.Vientiane: NAFRI.

National Land Management Authority (NLMA) Laos and Chiang Mai University. 2009. *Summary Report on Research Evaluation of Economic, Social, and Ecological Implications of the Programme for Commercial Tree Plantations: Case Study of Rubber in the South of Laos*. NLMA and Chiang Mai University.

Reeb, D. and Romano, F. 2006. "Overview". In: *Understanding Forest Tenure in South and Southeast Asia*. Rome: FAO, pp. 1–26.

Shi, W. 2008. *Rubber Boom in Luang Namtha: A Transnational Perspective*. Vientiane: Rural Development in Mountainous Areas of Northern Lao PDR (RDMA) and Deutsche Gesellschaftfür Technische Zusammenarbeit (GTZ).

Vongkhamor, S., Phimmasen, K., Silapeth, B., Xayxomphou, B., and Petterson, E. 2007. *Key Issues in Smallholder Rubber Planting in Oudomxay and Luang Pranrb Province, Lao PDR*. Vientiane: Upland Research and Capacity Development Programme, NAFRI.

World Bank Vientiane Office. 2006. *Lao PDR: Economic Monitor, November 2006*. Vientiane: World Bank.

Chapter 8
Whom to Share With?
Dynamics of the Food Sharing System of the Shipibo in the Peruvian Amazon

Mariko Ohashi

1. Various Stakeholders Surround Tropical Forest in Amazon

Though, even now in the 21st century, ethnic groups who have had no contact with "modern society" are being found in the depths of the Amazon and are attracting much global attention, it is evident that many other local societies existing within the Amazon forests have been influenced over several decades by outside societies.

1.1 Indigenous People and Global Environmental Problems

Today, almost everyone is aware about global environmental problems. As issues such as "climate change" and the "preservation of biodiversity" are being presented as problems that should be dealt with on a global level, the Amazon, as the world's largest tropical forest, is being looked to as a "carbon sink" and "genetic resource" and is attracting the attention of outside actors. On the other hand, it could be said that the interference of many actors who will never actually live in these forests is being legitimized. In other words, the forest riches of the Amazon are seen and desired as an attractive resource by outsiders.

In the face of such global environmental problems, changes in local society caused by contact with outsiders became referred to as local environmental problems (Inoue, 2010: 3). In local commons arguments, there has been much debate about the relationship between global and local environmental problems, and these two issues are said to be closely related. Thus, ascertaining how local society is changing due to the involvement of outsiders is important (Inoue 2010: 3). However, even prior to the appearance of these global/local environmental problems in recent years, several local societies existing in the Amazon forests have been subject to outside influences, such as that of Christianity and the market economy for many years. As mentioned by Tanaka (Introduction: 6), "Collaborative governance is not merely an attempt to discover how resource management should be executed or how to make it a success. Rather, collaborative governance is a new vision of society, one in which all stakeholders can interact in a collaborative manner to govern certain resources". This chapter discusses the types of social changes that have occurred in the local society of the upper Amazon basin of Peru resulting from outsiders' involvement.

1.2 Interaction with Outsiders in the Peruvian Amazon

First, it is necessary to examine the situation in the Amazon forests to note what type of outsiders have been involved, for what purposes, and where. This chapter specifically focuses on Ucayali River basin in the Peruvian Amazon.

In this area, local people had initial contact with outside world when missionaries arrived there in the mid-1600s (Eakin et al., 1986). Such contact rapidly spread during the 1800s because of increased rubber cultivation in the latter half of the 1800s, and from the 1970s, logging by international corporations for timber exports began in earnest. As people migrated from the highlands and cities seeking cash income, increasing numbers of such migrant laborers encroached for logging in the forest where locals lived. Subsequently, during the 2000s, the indigenous people who had lived there and used the natural resources were designated by governments, national, and overseas NGOs as living in "poverty" because they had no cash. Projects producing and selling cash crops and timber were introduced to increase the villagers' income. During the mid-2000s, along with increasing the income of local residents, guidance was provided by the government regarding forest management under the banner of "sustainable forest management". Now, in the 2010s, guidance is being provided to local societies based on discourse regarding global environmental problems, such as "climate change" and the "preservation of biodiversity", on the national and international level, which forms the basis for implementing sustainable forest management. This intrusion into local society in the form of projects operated mainly by European countries and the US has been perceived by some as the advent of a second age of conquest. Nevertheless, since the mid-20th century, we can see that this interference by various actors surrounding forest resources for different reasons—especially in Central Peruvian Amazon— has continued.

As the range of actors involved in local societies becomes more diverse in response to global movements, as seen with the emergence of discourse regarding global environmental problems, this chapter focuses on the customary sharing system of the Shipibo people, who make their living based on natural resources, as an example and elucidate the type of changes that have occurred—especially throughout the 20th century—as outsiders got involved.

Chapter 8 Whom to Share With?   225

2. How did the Sharing System Change?

2.1 The Shipibo in Earlier Research and Research Objectives

From 2008, the another stayed intermittently in a village of Shipibo belonging to the Pano linguistic family. The Shipibo live in the watershed of the Ucayali River, which snakes its way north for approximately 600 km as it crosses Peru. The climate of this area features a rainy season from October to April and a dry season from May to September. The repeated flooding in the rainy season followed by drops in water level during the dry season creates oxbow lakes that are separated from the main channel of the Ucayali River. These oxbow lakes are connected to the main channel by multiple narrow tributaries.

The Shipibo are the largest ethnic group in the central Peruvian Amazon basin and mostly make their living on the floodplain alongside the river (Lathrap, 1970; Bergman, 1980; Tournon, 1988). They cultivate bananas, plantain, and cassava in a pattern of shifting cultivation as well as engage in fishing, hunting, and gathering (Bergman, 1980; Tournon, 1988). It is customary for them to share the game meat and fish that they obtain on a daily basis among their kin (Behrens, 1992). Previous researchers have focused on this sharing system. Prior to the 1960s, the Shipibo had a highly mobile lifestyle; they changed their place of residence every four to five years and did not stay in any one place for a long time (Hern, 2003). Subsequently, during the 1960s, they became more settled as the government improved school education and built hospitals (Hern, 1992). During the 1970s, the establishment of administrative areas as "native communities" progressed; this will be discussed in detail in the further sections of this chapter. In addition to being the first to be incorporated into the market economy during the 1960s, the Shipibo are also said to have been an ethnic group that was influenced by *mestizos* who are of both Spanish and Native American descent (Hern, 1992).

Behrens (1981, 1992) reported that with the introduction of the market economy, the Shipibo sharing system underwent changes with regard to the sharing of game meat; changes were also reported concerning the sharing of bananas (Ohashi et al., 2011)[1]. However, in these previous studies, there have been no attempts to observe changes in the sharing system in the midst of the increasing involvement of outsiders.

Therefore, this chapter analyzes how the customary sharing system changed as outsiders became more deeply involved in a Shipibo settlement administrative established in 1984. Specifically, after investigating the historical changes that have occurred in the sharing

---

1) The description of bananas in this chapter is based mainly on Ohashi et al. (2011).

system since the village was established, the present state of the system and the changes that have occurred will be clarified based on the results of a quantitative food survey. Finally, the influence that "collaborative governance" will have on customary activities in the future will be discussed.

The author spent approximately 15 months intermittently from December 2008 to September 2013 performing fieldwork around the native community of Dos de Mayo. Data collection in the survey village was performed while living with the villagers, participating in their daily activities, and conversing with them in Shipibo. The methods used were informal interviews and participatory observation.

2.2 Research Site and Basic Information

[**Research Site Overview**] The native community of Dos de Mayo (hereafter, "D village"), which is the site of this research, is located approximately 120 km upstream from Pucallpa City (also a developmental base of this region) on the shores of an oxbow lake on a branch of the Ucayali River (see Figure 8–1). The population of D village is 102 people, comprising 15 households (as of September 2013). It takes approximately one day by canoe and ferry to reach Pucallpa city. The villagers make their living based on natural resources by growing multiple crops, fishing, hunting, gathering, and cultivating plantain, bananas (*Musa* spp., hereafter, "bananas"), and sweet cassava (*Manihot esculenta*). Furthermore, they grow and sell corn (*Zea mays*), rice (*Oryza sativa* L.), and log timber to generate an income. There are also some villagers who have, in recent years, left to work outside the area. Dos de Mayo was registered as a native community on May 2, 1984[2], after which it saw the commencement of government food assistance programs during the latter half of the 1980s and the establishment of a kiosk managed by the village using logging royalties during 2006. Since 2006, "poverty" reduction projects run by the government, IIAP (Instituto de Investigaciones de la Amazoía Peruana), and NGOs have been introduced to help them earn cash income by producing timber as a community.

[**History of the Village**] The village was developed in 1977 by a shaman who is believed to be able to contact spirits. In 1974, during Juan Velasco Alvarado's administration, who was known as Indigenismo (an advocate of the rights of indigenous people), a system was introduced whereby the state granted the indigenous people the right to occupy and use land[3] around *colonia*, which were the bases from which Christianity spread, and around natural villages. Such recognized land was referred to as native communities (*comunidades nativas*).

---

2) Dos de Mayo means "May 2" in Spanish.
3) As ownership rights are vested in the government, the sale of land within indigenous community groups is prohibited by law.

Figure 8–1. Map of the research site

Under this policy, it was publicly stated that ethnic indigenous groups could directly receive royalties from logging and land use by accepting business from domestic or international corporations; this policy aimed at enabling the villagers to independently acquire cash income.

In light of these political moves, with the aim of creating a new settlement, the shaman who developed D village moved upstream to a settlement in a native community, Amaquería, which already existed at the time. Initially, he built a new house near the settlement and then gathered his kin together to form a separate settlement. These actions were not thought of favorably by the people of settlement of Amaquería; in 1977, he left to develop new land in an area far from the main channel of the Ucayali River, where people had never lived and which was considered at the time to be a virgin forest. This is said to be the origin of the village of Dos de Mayo. Subsequently, within a few years, the family and friends of the pioneer moved there, cleared the forest, and then dispersed and lived in their kinship groups.

3. Changes in the Sharing System

3.1 The Shipibo Sharing System

The Shipibo customary sharing system is unique to this ethnic group. Previous research stated that the Shipibo cultivate their land individually or jointly and that it was customary for them to share food that had been hunted or gathered among kin and friends (Behrens, 1992). Furthermore, if someone were passing by during mealtime, even if that person were not Shipibo, they would be invited to share the meal as long as that person was an acquaintance (Behrens, 1992). In other words, the Shipibo sharing system can be divided into two parts: sharing at the ingredient stage and sharing the meals.

Even today, in D village, when a large amount of certain food resources is available, people still feel the need to share, which some of them do. In addition to such sharing of meals on a daily basis, the clearing of trees and vegetation during shifting cultivation is jointly executed. Whenever such work is carried out, the host always prepares the meals, and those involved share a meal together. According to village elders, if a Shipibo meets a person from another settlement and even if they do not know each other, the person would still be invited to have a meal together if they got along well with each other. In some ways, this custom has an aspect of obligation toward other members. On the other hand, even today, in situations where the person is not really that close, "Come and have a meal together (*Fumanpine*)" is simply said as a polite greeting.

Thus, Shipibo society places emphasis on sharing food. However, this "sharing of food" is not something that was firmly entrenched in tradition; rather, it is something that developed among people in the midst of their daily lives. This chapter traces the state of this "development" and the changes that have occurred.

3.2 Foods and Economic Activities

Today, though villagers' main foods include bananas, fish, and game meat, they also purchase other items including pasta, rice, beans, and canned food provided by the government. Apart from meals, villagers also gather and eat fruit and nuts on a daily basis.

The following section mainly focuses on bananas, fish, and game meat when considering the customary sharing system. Further it sheds some light on the acquisition process as their economic activities are based on three activities: farming, fishing, and hunting.

Bananas are considered a major part of the Shipibo's diet in terms of nutritional value (Bergman, 1980), and are found most frequently in D village. In addition, the word for meal in Shipibo (*piti*) is the same as that for fish. It could be said that the standard by which food is judged to be meal or a snack is whether fish is part of it; "it is not a meal unless there is fish". Thus, fish also form an important part of the villagers' daily diet. Moreover, game meat is a daily food resource, but over the past few years, its volume is decreasing so it can no longer be consumed on a daily basis. Though it is not as frequently consumed in meals as bananas and fish, it has still been a culturally significant food.

The following sections examine the changes in the overall sharing system by focusing on three customary foods: bananas, fish, and game meat from farming fishing, and hunting, respectively. Specifically, after tracing the historical changes, the state and changes in sharing today are reported based on the results of a quantitative food survey. Historical changes will be divided into the following three periods: (1) the period between when the pioneer first settled in 1977 and the village was recognized as a native community in 1984 (1977–1983); (2) the period between 1984, when the village was recognized as a native community, and 1999; and (3) the period between 2000 and 2008, when people changed their behavior as a result of resource decrease.

4. Historical Changes in the Sharing System

4.1 Before Recognition as Native Community: 1977–1983

[**Bananas**]   When the pioneer first settled in the area, there was nothing resembling a settlement, and houses were located on plots of land cleared from the forest. According to villagers, during this pioneering time, after migrating to the area, Shipibo males first had to build a house, and they cultivated land. In 1977, the pioneer settled by himself and broke in the land, followed by two more people who settled there and worked hard to develop land for agriculture. They performed everything from planting to harvesting in six months, initially planting crops that can be harvested early such as cassava and corn. During the time it takes to become self-sufficient through agriculture, though they brought in crops from their previous settlement, they also asked neighboring settlements to share agricultural crops with them. Older people who were present at the time said, "There were no lazy people" because "lazy people do not get to eat".

[**Fish and Game Meat**]   It is said that the men provoke before daylight every day to catch fish from the river. Though they mostly went fishing by themselves, they sometimes went with close kin such as parents, children, siblings, and cousins. Bows and spears were used for both fishing and hunting, and there was no clear distinction between these two activities at the time. Furthermore, they would catch whatever they encountered (fish or animals) on a daily basis as they walked through the forest or ventured out on the river in a canoe.

They caught fish such as pirarucu (*Arapaima gigas*), one of the largest freshwater fish in the world, or catfish. Someone who could catch a pirarucu was considered especially skillful. Moreover, they considered that one fish would suffice for an entire household[4], and though if a large number of fishes were caught, they would share with other families, fishes were not generally caught excessively. Normally, villagers only caught the amount that would be consumed on the same day. In addition, if animals such as the bald uakari (*Cacajao calvus*) or monk saki (*Pithecia monachus*) that regularly ventured near the settlements were hunted, then people would eat game meat almost every day. Game meat is a type of food that everyone wants to eat, and if asked for some meat, the villagers did not refuse. On the other hand, those skilled at catching such food were sometimes envied by others, and they feared that jealous people would use magic and perform rituals to prevent them from catching fish.

Even though magic was prohibited under Christianity, it is said that if the villagers were asked for something by someone, it was difficult for them to refuse without a good reason, because they feared becoming the target of magic if they were stingy. There was also the uneasiness associated with not wanting to be considered stingy. If larger animals were caught, such as tapirs (*Tapirus terrestris*) or capybara (*Hydrochoerus hydrochaeris*), men from other households also helped as it was not possible for one person to skin and cut them. In such situations, the meat was either shared with the men who helped or the villagers ate it together. Thus, even the families of those who were not good at hunting could eat meat.

When the meal was ready, they would shout "Let us eat together (*Pie bukane*)", and everyone would bring their food; sit on the ground around their food in separate groups of men, women, and children; and eat together.

4.2 After Recognition as Native Community: 1984–1999

[**Bananas**]   In addition to the fields around their houses, the villagers started searching for land where they could grow bananas on a semi-permanent basis on the floodplain alongside the river near the houses and started to break it in[5]. If there were no bananas on their own land, they would ask other households to share their product and would harvest them directly from other household's land. Unfortunately, during the 1990s floods, all the

---

4) Shipibo had polygyny system. A skillful man could have multiple wives, but when D village was settled, there was only one household where the man had more than one wife. It is said that he gave a catch to each of his two wives.

5) Generally, in the Amazon, there are three different geoscientific categories of land, the floodplain, the lowlands (which are not part of the floodplain but receive nutrients), and the highlands. The lowlands are the area on the boundary of the floodplain where semi-permanent cultivation is possible. See Hiraoka (1982) for further details.

land that they had broken in alongside the river was destroyed, causing everyone in the settlement to use the land near their houses.

[**Fish**]   In 1988, commercial fishing boats started to sail up the Ucayali River near D village. After a few years, it was no longer possible to catch large number of fish with bows, fish spears, and harpoons, and there was a drastic drop in the fish catch. Villagers who were present during that period stated that they had to spend three to four hours fishing each day to procure the volume necessary to sustain their own household. In 1992, some Shipibo friends from another settlement showed them how to catch a large number of fish in one attempt using a gill net. Inspired by this idea, two villagers grew rice together and bought a gill net with the profits. Though they could catch a large number of fish in a short period of time with a gill net, since they could not control the volume of fish caught, the excess fish were shared with households who did not own a net. Eventually, more households purchased nets and fishing flourished to the extent that sometimes each household would have a rest from fishing because the excess fish were frequently shared with other households. Thus, because of the introduction of fishing nets, bows, fish spears, and harpoons were no longer the main tools used for fishing. When meals were ready, the cry of "Let us eat together" would ring out from the houses, and inviting people for meals was a daily occurrence.

[**Game Meat**]   In the first half of the 1980s, a group of three to four people seeking to log trees to sell as timber came close to the settlement. In exchange for directing them to the location of the particular species of trees that they wanted, the villagers received goods such as pan and sugar from them.

In 1984, the settlement was recognized as a native community, and the first village mayor called in a logging company. With the logging came the *mestizos* and they stayed in the village for some weeks. They brought their own guns and would go hunting on their own. The villagers would occasionally help the *mestizos* with their work, and on those occasions, the villagers and the *mestizos* would share the game meat that they had caught. The *mestizo* workers would also ask the villagers to sell them bananas, cassava, fish, or game meat.

Over a period of time, people became aware that they could catch only a small number of the animals for game meat. One reason for this was that the animals did not like the noise of the machinery used by the logging companies and therefore escaped deeper into the forest. Besides, the animals became highly wary of humans and quickly fled from people when they went hunting. Prior to this, if a villager wanted game meat that someone else had caught, they would simply ask for it. During this time, however, some villagers started to imitate the people from the logging companies, saying that they wanted to buy the meat. Other than in cases where the cash was immediately produced, people replied by saying "Take it". Sharing meat among villagers was deemed the right thing to do, and the sale of game meat among villagers was almost unheard of.

4.3 Recent Changes in the Sharing System: 2000–2008

[**Bananas**] The lowland surrounding the settlement was already used for cultivating bananas and as secondary forest by the same villagers, making it difficult for others to break in new land. Therefore, in 2000, some villagers began developing land alongside tributaries far from the settlement; they traveled by canoe to reach this land. Other villagers also followed and developed land in the vicinity. Around 2006, a cow was purchased as communal property with the profits obtained from commercial logging, but when it was allowed to graze within the village, it damaged the banana seedlings. Therefore, the fields surrounding the settlement were abandoned and converted into a secondary forest thereafter, there was no one left to develop land in this area. In 2006, trees were being grown for cash under a timber production project, and land was developed in areas where such trees were growing, with bananas being grown amongst them.

Thus, while the number of villagers developing land far from their homes increased, even if they ran out of bananas and were unable to harvest them immediately, they could cater for their meals by requesting villagers with land in the lowlands surrounding the village to share bananas[6]. In 2007, the owner of a shop in the neighboring village started to frequent D village approximately once a month to purchase bananas. The villagers bought commodities that were not available in the village on credit and paid for them in bananas. Though bananas became a source of income, if someone asked for bananas, then they would not be refused, and they continued to be a resource that people gave and exchanged.

[**Fish**] People began realizing that the amount of fish they were catching was decreasing. However, at the time, they were able to secure enough for their own family by asking those households who were fishing with nets to share with them when they were unable to fish. They began using gill nets more frequently than before, and those who did not go fishing were emphatically called "lazy people" (*chiquish*). In 2005, a young villager showed keen interest in net fishing so that he could sell fish. In addition to sharing with close kin, he also started selling fishes on a daily basis to people who had no fish, and those who did not set their nets started to buy on credit or paid cash. Around 2006, a commercial Shipibo fisherman living in the neighboring village would occasionally give the villagers fishes that were not good enough for market when he was fishing close to the village. Eventually, the size

---

6) Refer to Ohashi et al. (2011) for details regarding the exact location of the land used for growing bananas in D village.

of the mesh used on the gill nets started reducing, and in the rainy season, when it became especially difficult to catch fish with a gill net, all that the villagers could do was catch enough fish to feed their own families. In the midst of this situation, the older villagers started to say "You set your nets and set your nets, but you are so bad at fishing that you can not catch anything". Moreover, as it was a hassle for villagers to go fishing every day, some of them started to store salted, dried fish. Thus, the spontaneous sharing of excess fish, as had been done until then, decreased. On the other hand, if they had a big catch, they would voluntarily share among kin and neighbors and invite people for meals, and if someone with no fish came to that household and asked for fish, it would still be shared in the same way as before.

[**Game Meat**]　　In the 2000s, people became aware of the fact that opportunities for catching wild game were decreasing; people started feeling embarrassed to ask others to share meat at no cost. Over time, when people went to buy game meat, it would simply be given away because the person was a villager, but in other cases, they would actually take the money; such sales became increasingly common. Around mid-2000s, when large game such as a wild boar (*peccary tajacu*) was caught, some villagers started to sell game meat in neighboring villages. Though villagers would only ask to buy meat when a large animal was caught, the people who were asked would find it difficult to refuse even if they wanted to.

5. The Current Sharing System

5.1　Quantitative Food Survey Results

The following is a brief discussion of content of meals based on the results of the quantitative meal survey along with a clarification of whether each food is from a person's household or is shared food to ascertain how food was acquired. Then, the changes occurring in the sharing system currently and those that have occurred since 2009 will be clarified in terms of foods such as bananas, fish, and game meat. The meal survey during the dry season of 2009 among 13 of the 15 households[7] in the village. Generally, a household is the unit for food consumption. Each household was visited twice a day to conduct an inventory survey by observing and interviewing participants to ascertain the number of times each food featured in meals and how it was acquired.

---

7) The nuclear family was the basic type of household. Some village households are extended families, or grandparents with son(s) or daughter(s) and the respective spouse(s), and some men live by themselves eating meals in other kitchens. Therefore, in this chapter a household is defined as a group of people living together in the same house.

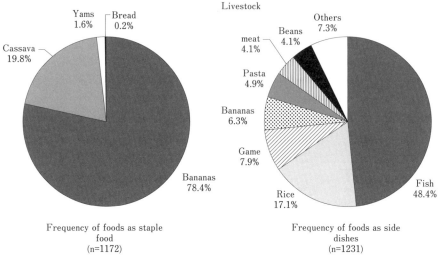

Figure 8–2. Frequencies of Various Foods in Meals
Note 1: The research was conducted from May 9 to May 28, 2009, from June 9 to June 29, 2009, and from July 6 to 18, 2009. Thirteen households were selected, and women or household members who prepare meals were principally interviewed. Observations and interviews were conducted a total of over 1213 times.
Note 2: In the right-hand figure, the category "Others" includes canning fish (1.5%), eggs (0.7%), maize (0.6%) shellfish (0.6%), and cassava (0.2%).
Note 3: Occasionally, villagers ate only a staple food or only a side dish. Also, they sometimes ate two or more side dishes together.
Source: Ohashi et al. (2011).

Results of this survey indicated the frequency of various foods as a staple food and as a side dish (see Figure 8–2). Bananas featured in meals most frequently as a staple food, and fish featured as a side dish. When fish was unavailable, then food such as rice was eaten, and in terms of nutrition, food generally classified as staple food was the side dish. Purchased goods, such as pasta, and food supplied by the government, such as rice, were also consumed. In terms of the way in which meals are eaten, even today people sit around their food in separate circles of men and women.

5.2 The State of Sharing for Various Food Resources

[**Bananas**] First, the results of the quantitative survey conducted at the place of harvest for three foods—bananas, fish, and game meat—are explained.

In many cases, the clearing of land for cultivating bananas and the enormous task of removing vegetation are performed jointly by the villagers, with the harvest being subsequently carried out by household members[8]. Figure 8–3 shows the results of my investigation regarding whether bananas, used as a staple food in the 13 households within the

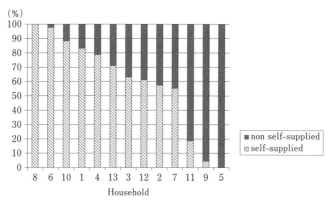

Figure 8–3. Self–sufficiency Rates for Bananas in Meals in Each Household (n=913)
Note 1: When a household cooked and mixed bananas self-supplied by its members and bananas acquired from other households, both examples were counted by 0.5. But such instances were observed only 11 times.
Note 2: Because a household member worked for a timber production operation, household 6 was researched from May 6 to 29 and from July 6 to 18, 2009, for a total of 37 days.
Source: Ohashi et al. (2011).

settlement studied in this survey, were harvested from land owned by the households themselves or by other households[9].

Of the 13 households, 10 consumed bananas harvested from their own land more than 50% of the time, but only one of the households (8) was completely self-sufficient. On the other hand, three households (11, 9, and 5) did not actively produce bananas themselves, but received more than 50% of the bananas consumed from other households. For example, household 9 used the land of those kin who did not live with them, whereas household 11 had returned to the village and did not have land for harvesting, and, therefore, actively sought bananas from other households. In addition to the sharing of bananas that occurs in person within the settlement, bananas are also shared by asking people to harvest them directly or to get them in the future. Sharing of bananas occurred as follows: on 129 occasions, people asked for bananas; on 37 occasions when bananas, simply given away; on 14 occasions, people were given bananas in gratitude or in exchange for receiving something; and on one occasion, a person purchased them. Thus, many households consume bananas acquired from other households for their meals.

---

8) Refer to Ohashi (2013) for details regarding banana cultivation.
9) In this chapter, information regarding the sharing of bananas is based on Ohashi et al. (2011). Refer to the mentioned study for further details regarding the method of sharing bananas prior to and after harvest in D village and the quantity shared.

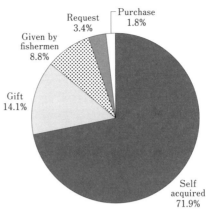

Figure 8–4. Rate of fish acquisition method (n=772)
Note 1: For details of the research period, see Note 1 in Figure 8–2.
Note 2: Multiple answers allowed.

[**Fish**] Different fishing techniques are used in the wet season and dry season. During the wet season, in addition to gill nets, hooks and hunting bows are frequently used, whereas in the dry season, during which the quantitative survey was conducted, gill nets were used more frequently. Villagers would set their nets in the evening and pull them out the following morning at dawn. At the time of the survey, all households possessed gill nets.

Regarding the use of fish in meal, it was featured 723 times. When a meal contained fish whose channels of acquisition differed, it was counted multiply according to the number of those channels. Consequently, the total number of examples rose to 772. Among them, 71.9% used fish acquired by their own household members, whereas 14.1% were fish gifted by other households (see Figure 8–4). They acquired fish either from people who voluntarily gave it to them or by lobbying for it. Within the village, whenever game is caught, there is a rule that the catch is shared with those who participated in the hunt— even with people other than kin and friends, with whom things are not usually shared on a daily basis.

This "practice" implies that they ensured that they were in the right place at the right time. The number of times that they were given fish also includes instances such "lobbying" where they did not necessarily directly ask for fish. On the other hand, 3.4% was fish that was acquired by direst request. Fish received from fisherman, which accounted for 8.8%, was shared by Shipibo fishermen living in neighboring settlements, who was one of the villager's friend, and was shared at no cost because the fish was not suitable for the market. Sales accounted for 1.8%.

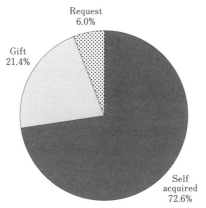

Figure 8–5. Rate of game meat acquisition method (n=84)
Note 1: For details of the research period, see Note 1 in Figure 8–2.

[**Game Meat**] Today, not all males go hunting. Methods of hunting employed as of May 2009 included guns, dogs, bows and arrows, and ready made traps in addition to machetes. Although the type of tool used depends on the type of animal being hunted, essentially anyone can hunt with a bow and arrows (or a spear). As of May 2009, only three households hunted with guns, one household hunted using dogs, and two households hunted with traps. Animals observed during the survey period included an agouti (*Dasyprocta variegata*) that was shot with a gun and a yellow-footed tortoise (*Geochelone denticulate*) that was caught using dogs (see Plate 8–1).

Game meat featured 189 times, of which 72.6% of the meat was caught by members of the same household, 21.4% was gifted by other households, and 6.0% was received on request from other households (see Figure 8–5). In addition to active voluntary sharing, which accounted for the second highest proportion after self-sufficiency, sharing included occasions where the "practice" was carried out (see Plate 8–2).

Today, villagers consider it normal to sell game meat to people other than their kin. Though many people want to eat game meat even if they have to buy it, during the meal survey period, no instances of sale of game meat were observed.

5.3 Current Changes by Food

[**Bananas**] In D village, people who clear land are considered "owners" of cultivated land. If the owner of the land is not a village resident, crops have sometimes been harvested without consent. This can be summed up in the villagers' saying "Bananas on land where the owner is absent are there to be taken by others". However, in 2009, even when the

Plate 8–1. Get pleasure from agouti.
Taken by Mariko Ohashi.

Plate 8–2. A neighbor came to watch dressing kinkajou (*Potos flavus*).
Photo by Mariko Ohashi.

owners were residing in the village, bananas were often harvested by other people. Furthermore, though people were allowed to use the land of people who were not residing in the village, this practice is being considered as theft in recent years.

[**Fish**] Sharing is common among close kin, such as parents, children, and siblings. Villagers who had not set their gill nets the night before are often heard saying "We have got no fish. I am just going to meet so and so to see if they have any", and then see them wander around the village in the morning asking for fish. When they do receive fish, they mostly give bananas in return to express their gratitude. If the other person was someone with whom they often trade food, they would often say something like "Let me have some fish, and I will give you some in return in a couple of days", or "Let me have some for now as I am on my way to go fishing, and I will give you some later". In cases where a person with whom sharing was not a daily occurrence caught a lot of fish, one would explain why they had not been able to go fishing for example, by saying "My child was sick, so I was unable

Chapter 8   Whom to Share With?

to go and set my net"—and ask that person to share. Furthermore, by asking to buy fish, anyone could negotiate regardless of whether they had a relationship of sharing food with the other person.

An example of what happened in 2009 after the village decided to accept a group of *mestizos* to implement intensive cultivation of papayas for commercial sale is given below. A young couple in the village used their dividend in Pucallpa City to buy commodities to sell in the village. Consequently, people began selling fish even to those with whom they had a relationship of giving fish at no charge when asked to sell. Furthermore, people who bought fish were unhappy about having to buy fish that should not be sold in the village, that had almost no value in the market, and that they formerly received from the fisherman at no cost.

When having their meal, people continued to call out to others, inviting them to share a meal together. They invited even those whom they did not particularly want to invite. However, in some cases, people passing by became more reticent and would walk away quickly from those who were having a meal. Even if they were called, they would refuse by saying "You go ahead without me", and continue on their way. Furthermore, some villagers became friendly with their *mestizo* employers as they had worked for them in the papaya plantation and invited them to have fish soup.

[**Game Meat**]   Game meat is generally consumed within the family. However, if one is going hunting and villagers hear about it, it becomes difficult to avoid them when they say things like "If you get something, can I have a leg?" or "If you catch something, I will buy 1 kg"; people are urged to share. Therefore, some people hide the fact that they are going hunting and say that "I am going to pick some bananas". People are now embarrassed to be there when hunters cut game meat because it seems like they are "lobbying"; they are also embarrassed to ask to buy or share meat. Some hunters dislike the thought of their own or their family's portion being reduced and thus, even if someone passes by, they do avoid inviting them for a meal.

However, if they catch a large animal such as a capybara, the owner of the meat shares a little meat with other households, and people who want a certain part, such as the ears or tail, will still ask for it. After cooking the meat, people invite others or ensure that a plate of the cooked meat is taken to other households so that even households that are unable to hunt can eat meat. In addition, some villagers take a gun with them when working at the papaya plantation, and if they catch an animal, they eat it together with their employer.

6. Toward Dynamics and Complexity in Sharing System

6.1 Changes in the Sharing System due to Involvement with Outsiders

This chapter mentions three food resources and investigates the state of the sharing system while focusing on changes in the economic activities of hunting and gathering resources.

From the establishment of D village until today, it is evident from the phenomenon of the sharing system that a truly diverse range of actors have been involved in the resources and lives of the people of D village. For example, after the villagers obtained land use rights through the establishment of the native community system by the government, within a short period of time, entities such as logging companies, commercial fishing operations, and papaya plantation employers sought to use the village resources on a large scale. The following is a brief summary of the type of changes that occurred in the sharing system in D village through such involvement with outsiders, focusing on the three foods mentioned before. Further, an explanation is provided regarding game meat, fish, and bananas (in that order), concerning which major historical changes have occurred.

After commercial logging was accepted, people realized that the volume of game meat as a resource was rapidly decreasing, with an increased consciousness of the possibility of introduction of cash income into their society. Therefore, people found the idea of having something shared with them at no cost embarrassing, which made it difficult for them to ask for game meat at no cost. In the midst of this, through contact with the outside social system in the form of their experience of selling game meat to commercial loggers, peoples' manner of asking for something changed from "Can I have some?" to "I would like to buy some". Furthermore, today, with the decreasing volume of hunting resources and the decreasing frequency with which people can catch game, there is an increasing tendency among villagers to be more reticent about communicating the fact that they would like to buy game meat.

Similarly, since commercial fishing boats commenced operations in 1988 around D village, due to the reduction in the volume of fish that can be caught in the village and the resulting decrease in opportunities for sharing, the method of fishing changed from using bows, spears, and harpoons to using gill nets, resulting in temporary active sharing. In contrast with game meat, where a limited number of people can successfully hunt, all households can go fishing if they use a gill net. Thus, because each household could catch fish, a person who has no fish could ask someone with whom they deal on a daily basis to share with them with the promise of giving fish in return later. If they did not have a sharing relationship with the person, they could ask to buy fish. This way the other person could be convinced to share and not worry about decreasing their supply of fish. It was also a way of avoiding being called "lazy" for not going fishing. Furthermore, because of their experience of obtaining income from the papaya plantation and selling commodities within the village, some villagers started selling fish that had previously been shared at no cost. This led to discord among the villagers. Thus, villagers' perception differed regarding how things should be shared or sold.

While such changes were occurring and there was some confusion regarding whether to share or sell game meat and fish, during the period since the commencement of cultivation until 2009, bananas were never sold among villagers. Ohashi et al. (2011) reported examples of how trouble in interpersonal relationships led to refusal to share. One household used to exchange bananas with another for fish. When the fish resource that the latter were able to catch decreased, the former became reluctant to give them bananas and, later, ended up giving bananas as the wives of the two households had a quarrel. In this case, prior to the occurrence of any trouble, members of that household had made no effort to break land and were lazy in their work, right from the beginning, relying on being able to use other people's land, which other households were dissatisfied with. In Shipibo, sharing is called *aquinquin*, which means not only "sharing with each other" but also "helping each other". Villagers who did not take the concept of "helping each other" seriously were punished. However, from another viewpoint, as long as no problems were caused on a personal level, it may have been possible for the villagers to continue receiving bananas in a one-sided manner. This shows that though there are resources such as game meat that people are hesitant to buy, there are also resources such as bananas that will continue to be shared. Different changes have occurred with regard to the sharing of each type of resource.

Overall, opportunities to share among villagers decreased. Nevertheless, even today, the concept of sharing food resources is firmly entrenched, as evidenced by people sending cooked food to other households and inviting them for meals. Furthermore, sharing meals is not limited to villagers, but it is also seen being implemented with outsiders, such as commercial loggers and papaya plantation employers. That is, in the midst of changes in the amount of the resource and outsiders' influence on methods used to acquire these resources, the sharing system is being affected not only by outsiders but also by locals, and the villagers have made attempts to identify the best method of sharing.

6.2 Significance of using Commons Theory to Examine the Sharing System

The correlation between people's lives and the utilization of resources by societies that are reliant on natural resources for their livelihoods has been debated using the commons theory. However, if resource sustainability is prioritized, it may be necessary to limit the scope of utilizing forest and ocean resources. However, considering the societies wherein people utilize such resources on a daily basis, in terms of resource management, it is probably not enough to merely focus social mechanisms and systems on certain resources. Steins and Edwards (1999) criticized the fact that much of conventional commons theory was polarized around the analysis of single-use common pool resources (CPRs), which pay attention to one resource use ignoring other, multiple use. In the case of D village, among the multiple resources that are utilized on a daily basis, food is considered an important resource that supports their lives and the sharing system is one of the ways in which such resources can be jointly utilized. In the examples provided in this chapter, where food resources are shared among people and sometimes leads to discord in human relationships, this endeavor to share—the sharing system—is an integral part of Shipibo society and helps define it[10].

Until now, when dealing with a single resource in commons theory, the way in which the resource is jointly managed (utilized) in a certain society and the systems that surround it are frequently in focus. Thus, little attention focused on how a resource was utilized outside the household after it was caught or acquired[11]. Furthermore, when focusing on multiple resources, the theory did not consider the interaction among each of the different resources. Considering D village as an example, it is clear that among households and other people, there is a close relationship between the way resources are shared after harvest/acquisition and the economic activities involved in acquiring such resources. This relationship can also be seen, for example, in various households that could take a break from fishing because of catching large amounts of fish owing to the introduction of nets and in villagers who are breaking in land for banana cultivation in places far from the settlement, being able to avoid the hassle of traveling far for bananas by receiving bananas from those with land near the settlement. People are aware of these things and act accordingly. From examples such as giving bananas as a gift in return for receiving fish and the fact that the fish resource volume is related to the sharing of bananas[12], as shown in the example featured in Ohashi et al. (2011), it is apparent that the sharing of each of these resources is interrelated. Examining resource sharing in this manner and considering the Shipibo as an example, it could probably be said that as long as resource management is viewed as a matter of how

---

10) This sharing of food has thus far been dealt with under cultural and ecological anthropology. However, as this debate mainly focuses on ownership theory and egalitarian theory, there is debate regarding what became whose, under what conditions, and how or how the sharing of possessions is related to the structure of society; the sharing of food has merely been one of those topics.
11) When debating commons theory, Akimichi (1995; 2009) stated that it is necessary to consider the target resource and its joint management both pre- and post-harvest. Food sharing could be considered as belonging to the post-harvest category. In this chapter, this should be considered an example of how the concept of post-harvest use influences the concept of pre-harvest use.
12) This household correlates to household 5 in Table 8–3. See Ohashi et al. (2011) for further details.

people should manage nature—that is, in terms of a relationship between nature and people—resource management will rely on relationships among people.

For the people living on the Amazon floodplain, the sharing system is an important collaborative activity for such societies. Though societies with systems, customs, and access related to resource utilization and interaction regarding how resources should be utilized and shared post-harvest/acquisition cannot be considered to exist throughout the entire Amazon basin, they are certainly not limited to the Shipibo. Today, outsiders already interfering in the Amazon forests to solve global environmental problems and are attempting to teach the local inhabitants to manage resources to avoid resource depletion. Thus, when considering the relationship between resource sustainability and the local societies whose livelihoods are based on such resources, is it not necessary to first ascertain details regarding the lives of the people living there? In such cases, it is important to focus on the multiple resources that people utilize and gain a clear understanding of the entire flow, from the production and acquisition of such resources to their consumption.

6.3 Future Trends for D Village

In 2011, based on the keywords "climate change" and "biodiversity", the GIZ (the German Federal Enterprise for International Cooperation) and the local government designated "virgin forest" as protected and introduced a project targeting the adjoining of approximately 70 native communities. The project outlines "sustainable forest management" by local residents, and D village lies within its scope. Within the protected area, though large-scale logging for timber is prohibited, hunting/gathering, collecting construction materials, and cultivating—for personal use only—is permitted. This was a government policy that took into consideration the preservation of local peoples' livelihood. On the other hand, there is an emphasis on the values dictating that "Resources within the village should be used by the people living there". In other words, "resident = the exclusive user of resources", which people were not previously aware of.

Originally, the Shipibo did not live in one particular place (Hern, 1992). They are a society exceeding the boundaries of today's government village with knowledge of utilizing resources to efficiently complement the movement of people by allowing such things as permitting an unspecified number of people to have access to bananas belonging to people who are absent from the village (Ohashi et al., 2011). In such a local society, in terms of livelihood and consumption, there is more of an emphasis on sharing with others than outsiders realize. They are open to outsiders who come with the aim of taking the resources of their village—as is evident in the way that they sustain not only their kin but also others who are present at the time, in the way that they extend their sharing system to people they know well and live with and people of the same tribal group as well as others, and also by the way they invited the *mestizos* from the papaya plantation who hired the villagers as non-permanent workers. In other words, notably, for the Shipibo, both livelihood and consumption are activities that are enabled by people working together, who are not limited to units such as "government village" and "ethnic group".

However, through the aforementioned introduction of the Western system of values by outsiders such as "resources within the village should be used by the people living there", this indigenous system of joint consumption that extends beyond the boundaries of such villages has led to discord among villagers in terms of rights related to existing forest resources and has resulted in people refusing to eat together.

This chapter has investigated the changes in the customary sharing system within D village. Future research should clarify how, as collaborative governance progresses, local residents will be able to build relationships with outsiders to develop local society, while focusing on how outsiders will be involved in such customs in the future, how the involved actors differ, and what constitutes legitimacy for their involvement. The author hopes to continue investigating the state of their society.

References

Akimichi, T. 1995. *Nawabari no Bunkashi: Umi Yama Kawa no Shigen to Minzokushakai* [*Cultural history of Territory: Resources of Sea, Mountain and River, and Folk Society*]. Tokyo: Shogakukan.

Akimichi, T. 2004. *Komonzu no Jinruigaku: Bunka, Rekish, Seitai* [*Anthropology of the Commons: Culture, History, Ecology*]. Kyoto: Jinbun Shoin.

Bergman, R. W. 1980. *Amazon Economics: The Simplicity of Shipibo Indian Wealth*. Ann Arbor, MI: University Microfilms International.

Behrens, C. A. 1981. "Time Allocation and Meat Procurement among the Shipibo Indians of Eastern Peru". *Human Ecology* 20(4): 435–460.

Behrens, C. A. 1992. "Labor Specialization and the Formation of Markets for Food in a Shipibo Subsistence Economy". *Human Ecology* 20 (4): 435–460.

Eakin, L., E. Lauriault, and H. Boonstra 1986. *People of the Ucayali: The Shipibo and Conibo of Peru*. Dallas, TX: International Museum of Cultures.

Hern, M. W. 1992. "Shipibo polygyny and patrilocality". *American Ethnologist* 119 (3): 501–521.

Hern, M. W. 2003. "Shipibo Ember". In: *Encyclopedia of Sex and Gender: Men and Women inthe World's Cultures Topics and Cultures 2* (eds. Ember, Carol R. and Melvin Ember). pp. 806–815. New York, NY: Plenum Pub Corporation.

Hiraoka, M. 1985. "Floodplain Farming in the Peruvian Amazon". *Geographical Review of Japan* 1: 1–23.

Inoue, M. 2009. "Shizen Shigen 'Kyouchi' no Sekkei Shishin: Rokaru kara Gurobaru he" [The Design Guidelines of Natural Resources 'Collaborative Governance': Connecting the

Local and the Global]. In: T. Murota (ed.) *Gurobaru Jidai no Rokaru Komonzu* [*The Local Commons in the Global Era*]. Kyoto: Minerva Shobo, pp. 3–25.

Lathrap, D. W. 1970. *The Upper Amazon: Ancient Peoples and Places*. London: Thames & Hudson.

Ohashi, M. 2013. "Amazon Hanrangen ni okeru Banana no Jikyuteki Saibai" [Subsistence Cultivation of Banana in the Floodplains of the Amazon-A case of Shipibo in Peru] *Biostory* 19: 85–94.

Ohashi, M., T. Meguro, M. Tanaka, and M. Inoue 2011, "Current Banana Distribution in the Peruvian Amazon Basin, with Attention to the Notion of 'Aquinquin' in Shipibo Society". *Tropics* 20 (1): 25–40.

Steins, N. A., and V. M. Edwards, 1999. "Collective Action in Common-pool Resource Management: The Contribution of a Social Constructivist Perspective to Existing Theory", *Society and Natural Resources* 12: 539–557.

Tournon, J. 1988. "Las inundaciones y los Patrones de ocupación de las Orillas del Ucayali por los Shipibo-Conibo". *Amazonía Peruana* 16: 43–66.

Part III
Sharing Information
Extending Collaborative Governance

Chapter 9
Providing Regional Information for Collaborative Governance
Case Study Regarding Green Tourism at Kaneyama-machi, Yamagata Prefecture

Nobuhiko Tanaka

1. Introduction

To promote collaborative governance in a certain area effectively, it is necessary for a diverse range of stakeholders to recognize and understand common information regarding the characteristics of the area in a fair, scientific, and objective manner. In other words, creating a scientific and objective knowledge base that is shared among stakeholders within the area is of paramount importance. Especially concerning basic information such as land coverage, scenery, and geography, it is quite important for all stakeholders to build a shared framework. Appropriate land use plans and local plans should be promoted based on this shared framework.

However, considering the state of affairs in Japan, provision of such information or efforts to share such information is still insufficient. The execution at the ground level is not undertaken appropriately. As each stakeholder either has different information or lacks information, there are a number of cases where collaborative governance is inefficient among stakeholders with vastly different views. In such cases, it usually takes a long time to reach a consensus.

To provide a more specific example, when people are confronted with and personally witness situations such as large-scale landscape destruction or obvious exploitation of resources, it is easy to reach a consensus even if there is a diverse range of stakeholders in the area. Because there is a slight difference in shared awareness, they will determine the direction of appropriate land use plans in many cases. However, sometimes there are cases when areas change in gradual and unexpected ways. Even if this is undesired by the majority of stakeholders, these changes will not be suspended because of problems such as (1) information that is difficult for people to notice on their own and (2) broad information that exceeds the stakeholder's scope of awareness. In such cases, neither is there shared awareness among stakeholders nor is information shared in an appropriate manner after the change is established. Consequently, goals are not achieved, appropriate consensus is not reached, and the area is administered in a manner that is unsatisfactory to all stakeholders.

Bearing such issues in mind, this chapter discusses the feasibility of various stakeholders providing information when determining future land use and the direction of local development. Further, it also focuses on green tourism in the context of the Mogami Region of Yamagata Prefecture, where Kaneyama-machi is located. Moreover, this chapter attempts to develop a method for the various stakeholders to obtain shared awareness concerning the geographical relationship between existing tourism and green tourism.

There are two main reasons for examining green tourism in this research. The first reason is that large expenditures on tourism development in hilly and mountainous villages have resulted in varying degrees of success and failure. Moreover, local people have developed various strategies for resistance, adaptation, and collaborative governance with regard to tourism development. Thus, tourism has been an ongoing challenge for the local communities in Japan.

The second reason can be plained as follows: (1) in recent years, interest in green tourism as a form of community-based tourism run by local people has been increasing in Japan; (2) it is difficult to ascertain geographical information because the promotion of local areas through tourism is conducted over a wide area, which falls outside the sphere of daily activities; and (3) adequate surveys and research have not been conducted in terms of geographical analysis of existing tourism and green tourism in Japan.

This chapter first summarizes the state of green tourism in Japan and provides an overview of the position of green tourism in tourism industries. Next, it discusses the expectations regarding green tourism and the process of creating new tourist destinations, followed by a geographical analysis of green tourism in the context of the Mogami Region of Yamagata Prefecture. Finally, this chapter compares the geographical status of green tourism with that of existing tourism. This will help to provide information to clarify, in an easy-to-understand manner, the geographical characteristics for all stakeholders.

2. Current State of Green Tourism in Japan

2.1 Establishment of Green Tourism

The lifestyle of vacationing and spending leisure time in a rural village emerged mainly in Europe around World War II. Though green tourism has different names, such as "rural tourism" in the UK and *tourisme vert* (green journey) in France, there is no significant difference in its meaning. In many manuals and handbooks in Japan, it is simply introduced as "green tourism." Now, almost 70 years after the end of the War, green tourism has been established as a type of tourism in Europe.

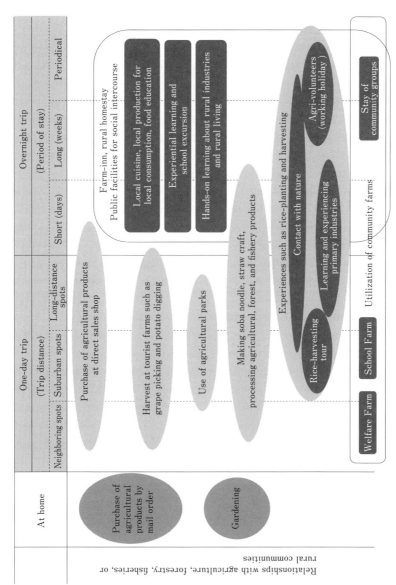

Figure 9–1. Examples of Green Tourism at Agriculture, Forestry, and Fishery Areas

Influenced by these European trends, the Ministry of Agriculture, Forestry and Fisheries of Japan advocated the promotion of green tourism as a new form of tourism in 1992, following which "the Act on Promotion of Development of Infrastructures for Leisure Stay in Rural Areas" was enacted in June 1994.

Reviewing the structure of the green tourism promoted by the Ministry of Agriculture, Forestry and Fisheries of Japan, reveals that it can be organized as shown in Figure 9–1. The Ministry classified green tourism according to differences in factors, such as the depth of connection with rural industries and village life, travel distances, and length of stay. It is clear that green tourism activities were spread over a diverse range of fields associated with rural industries (agriculture, forestry, and fisheries). These activities included purchasing goods for direct delivery from areas of production, day trips to tourist farms, rural homestays with hands-on experience, school trips, and working holidays.

Though these activities cover a diverse range of fields, they have one thing in common. All were once everyday activities that people were not required to pay for until the middle of the Showa Period (1926–1989), whereas now they have become an industry and are regarded as special tourist activities that people need to pay to experience. This evolution means that, with a little ingenuity, the average rural village can be managed as a tourist destination without the need for large-scale development of tourist facilities.

2.2 Green Tourism as New Tourism

As green tourism is a comparatively new type of tourism that has emerged in recent years in Japan, it is often recognized in tourism studies as a type of new tourism.

In the White Paper on Leisure in 2007, new tourism is defends as "a concept of tourism that has seen a rise in interest amid the diversification of the needs and styles of tourists as the style of tourism sought by the Japanese is changing from transit-type and group-type sightseeing and pleasure-seeking to experiential, interactive, and individual-type tourism where people can connect with the nature, culture, and people of the area that they visited." The following types of tourism are included in this new tourism:

- a. Extended stay tourism (so-called "vacations")
- b. Ecotourism (tourism where people can experience the natural environment along with the associated traditions and customs)
- c. Green tourism
- d. Cultural tourism (tourism that satisfies intellectual inquisitiveness regarding local history and culture)
- e. Industrial tourism (visits to factories and ruins of historical and cultural value, hands-on experiences and visits to traditional industries)

f. Health tourism (tourism that allows people to experience healing in nature and related meals and recreation)
g. Other new tourism (such as flower tourism and film tourism)

The White Paper on Leisure also identifies the following general characteristics of new tourism:

a. Themes (tourists select themes that they are interested in)
b. Contribution to the local area or local characteristics (the characteristics of the local area, such as its unique attractiveness. Destination-based tour plans are also developed that will contribute to the local area economy)
c. Participation and experience (emphasis is placed on participation in experience-based tourism or programs, rather than simply sightseeing)
d. Interaction with the local community (enjoying interaction and communication with local people)

Applying these characteristics to green tourism shows that green tourism has the following characteristics in common with new tourism.

a. The theme has a close relationship with agriculture, forestry, and fisheries.
b. The setting is rural villages that have unique activities related to agriculture, forestry, and fisheries.
c. It basically features experience-based tourism.
d. It deepens interaction with local rural people engaged in agriculture, forestry, and fisheries.

Comparing the characteristics of green tourism in Japan with those of rural tourism in Europe, we can see a few differences emerge. In addition to the former having a short history as new tourism, it comprises a comparatively short period of stay and consists of few Lifestyles of Health and Sustainability (LOHAS) activities where people relax and do nothing. Furthermore, during times of recession, green tourism attracts much attention as a new industry in the hilly and mountainous areas of Japan where it is difficult to make a living solely through rural primary industries.

Further, Considering the position of green tourism among the seven types of new tourism classified in the White Paper on Leisure, green tourism is not being developed completely independent of other types of new tourism. For example, it is quite natural that nature observation in ecotourism, forest bathing and forest therapy in health tourism, and experiencing traditional industries in industrial tourism have a close relationship with green tourism. It is no exaggeration to say that depending on the situation, these activities are being conducted as green tourism. Furthermore, it is untrue to state that new tourism has nothing to do with existing tourism and recreation activities; in fact, in many cases, green tourism is developed in close relationship with attractions such as hot spring bathing and mountain climbing. Therefore, green tourism and existing tourism cannot be clearly separated.

2.3 Expectations for Green Tourism

Despite the fact that green tourism somewhat overlaps with other diverse types of tourism, in many cases green tourism is an independent form of tourism. The reasons that support this idea are as follows.

As mentioned earlier, green tourism is expected to be a great form of local promotion strategy wherein people are invited to visit regular rural villages and areas in the hilly and mountainous regions of Japan, which they would probably never visit under the existing conventional forms of tourism. Such expectations may form the basis for treating green tourism as separate from other forms of tourism. To be more precise, many activities conducted under the banner of green tourism were formerly everyday activities in rural primary industries. However, because these activities are no longer recognized as being a part of everyday life, it has led to the idea that people should pay money to experience them as tourist activities, leading to an industry in and of itself. Therefore, there is a possibility that these rural villages, which are otherwise lacking in existing tourism resources, may be revitalized through tourism through conversion them into new tourist destinations.

Another perceived advantage of green tourism is that it makes it easy for local stakeholders, such as local residents, authorities, and organizations, to establish revitalization initiatives. Green tourism allows destination-based tours to be developed by the local community, unlike origin-based tours that are planned and proposed at the initiative of the tourism industry in the cities—which has thus far been the norm in the Japanese tourism industry. In other words, the reason that green tourism is gaining such attention is that it is possible to easily frame plans for the tourism industry that match the realities of the local society in terms of scale, budget, and time schedules.

Of course, certain challenges persist such as reaching a consensus between stakeholders who are keen on green tourism and those who are not, and how best to handle the burden placed on local authorities regarding waste disposal, public safety, and fire services in conducting green tourism.

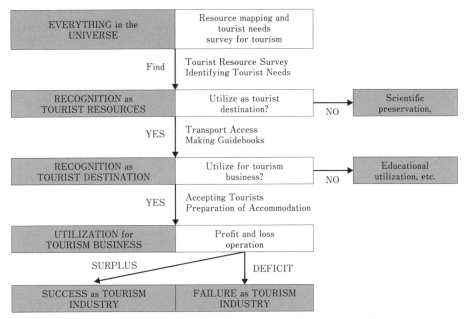

Figure 9–2. Evolutionary Process of Industrialization of a Tourist Destination

2.4 Process of Creating a New Tourist Destination

As mentioned previously, green tourism is expected to create a tourist destination that can be managed at a community's own appropriate pace. Generally, however, with the exception of major theme parks such as Disneyland, as shown in Figure 9–2, for a place to succeed as a new tourist destination, it is necessary to identify resources from among everything that exists or can be introduced in an area that can be converted into tourism resources; from those resources, certain assets that can be objectified for tourism must then be industrialized to generate income.

To convert everything that exists in an area into a resource for tourism, it is first necessary to compile an inventory of the resources available and then sort through them to determine what can be developed locally to meet the needs of tourism. Next, to objectify the resources that have been identified as usable by people, it is necessary to secure access to enable tourists to reach the concerned area, and then create information that can be used to communicate the value of the tourism resource to tourists. It is also necessary to consider the financial viability of developing accommodation and bringing in tourists to progress to the industrialization stage where economic activities are conducted using the resource that has been objectified. After industrialization, it can only be considered as an economically successful tourist destination if it proves to be financially viable.

Incidentally, during the abovementioned process, those assets that do not reach the industrialization stage (assets that were identified as resources and objectified) need not necessarily be considered a complete waste. From the perspective of nature, history, and folklore, it is entirely conceivable that things, which were identified as resources, may have sufficient value to the community to be preserved academically. Furthermore, assets that were objectified may be utilized as local teaching materials for environmental education and study of the local area, which can turn out to be activities that may not be economically viable. These assets may not be of direct advantage to the tourism industry, which has to be economically viable. However, for tourist destinations to reflect the catchphrase of Japan's tourism policy, which is "a good place to live, a good place to visit," it is vital to manage not only tourism resources that are directly financially profitable but also other resources and facilities that are not up to the standard. This may help to enhance the attractiveness of the entire area and improve its habitability.

Furthermore, even if these things in themselves cannot directly bring in customers, they often play an adequate role as secondary resources. These resources can be used by tourists who visit the area for other purposes. It is critical to link the places that these tourists visit with such secondary resources and to engage in community development that can contribute to revitalization.

Given the creation of new tourist destinations, there are two perspectives to consider. One is a narrow definition of tourism where direct profitability as an industry is considered important and the other is a broad definition of tourism that also includes resources that are not necessarily financially viable on their own. To realize the revitalization of hilly and mountainous regions, it is necessary to have a broad planning perspective that considers the broad definition of tourism.

In green tourism, it is undeniable that the ultimate goal is to ensure the success of industrialization to achieve the revitalization of a local area. In reality, however, in the settlements that are the bases for green tourism, the main rural primary industries—agriculture, forestry, and fisheries—must first be in place before engaging in tourism based on such rural industries. Therefore, the idea that stands in the closed world of narrowly defined tourism, which requires all resources in the local economy to be financially viable, is not required. On the contrary, it is vital to consider green tourism from the perspective of the broad definition of tourism to place a greater emphasis on the "good place to live" part of the policy catchphrase and to contribute to the revitalization of the community while continuing to identify and objectify resources that are not directly economically viable as an industry.

Table 9–1. Purpose of Research and Study Area

[Purpose of Research] Analyses of numerical geographical relationship between green tourism area and existing tourist destination
[Study Area] Mogami Area, Yamagata Prefecture (11 Municipalities), JAPAN

3. Method for Providing Regional Information

The purpose and study area of the research presented in this chapter are summarized in Table 9–1.

3.1 Purpose of Study: Future Development in Green Tourism

The ultimate research goal of this chapter is to consider the direction of green tourism development in Kaneyama-machi. To achieve this, four levels of development goals have been proposed. The first level is "the development of methods for geographically and quantitatively analyzing the degree of success of green tourism as new tourism in creating a tourist destination in a new place where there were no existing tourist destinations." In other words, there is a great deal of interest in "the ability of green tourism to contribute to the creation of new tourist destinations and the revitalization of areas with no existing tourist destinations through the utilization of rural industries." However, the existing research in Japan provides almost no examples of geographical and quantitative consideration regarding "whether green tourism is actually succeeding in creating tourist destinations in new areas that are different from where they have existed in the past."

Many traditional tourist destinations exist in rural villages in Japan. For instance, these destinations appear as scenic sites in Waka poetry such as *Man'yoshu*[1], which described sightseeing and vacation spots of more than 1,000 years ago. Then, some shrines and temples in the Edo Period became popular destinations among the general public in the 17th century. Areas of wilderness with natural beauty became popular since the Meiji Period in the late 19th century, followed by the hot spring boom that occurred in the 20th century. One can easily imagine that there is no small number of areas with long traditions as existing tourist destinations in rural villages.

This research will consider whether the bases for green tourism that have appeared since the start of the Heisei Period in the 21st century are simply rehashes of existing tour-

---

1) *Man'yoshu* (literally "the anthology of a thousand leaves") is the oldest collection of Japanese poetry, compiled sometime after 759A.D.

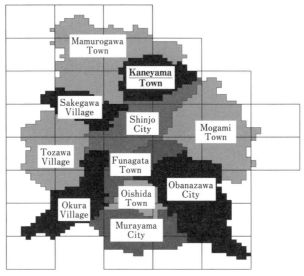

Figure 9–3. Study Area
(11 municipalities in Mogami area, Yamagata Prefecture)

ist destinations, or new tourist destinations that have been created independently with no connection to existing tourist destinations. In order to do so, it is necessary to develop appropriate methods.

The second and third levels are to apply the methods developed for the first level to the Mogami River basin in Yamagata Prefecture. The results are then used to clarify the characteristics of green tourism in Kaneyama-machi, which is located in that area.

Lastly, based on the abovementioned results, the direction of green tourism development in Kaneyama-machi will be discussed for the final research level.

3.2 Study Area

One of the common areas considered in this project is Kaneyama-machi in Yamagata Prefecture. However, because of geographical scale factors, the study area of analysis in this chapter has been expanded to 11 municipalities in the Mogami River basin surrounding Kaneyama-machi (see Figure 9–3). This is because when carrying out a geographical analysis of tourist destinations, one municipality alone is too small to analyze. Generally, it is rare for tourists to stay in one municipality alone, and they usually engage in activities over a number of different municipalities. Therefore, when considering the direction of green tourism development in Kaneyama-machi, it is necessary to analyze information regarding not only Kaneyama-machi but also the surrounding municipalities, and then to clarify the relative rank of Kaneyama-machi based on those results. Thus, 11 municipalities of the Mogami River basin have been included.

Kaneyama-machi is close to the borders of Akita Prefecture and Miyagi Prefecture, and tourism activities are expected to spill over into these prefectures. However, as until recently tourism data was collected in Japan on different bases from different prefectures, this research includes only Yamagata Prefecture in the scope of analysis.

3.3 Analysis Method

There was a need to determine whether "areas where green tourism is concentrated" within the 11 municipalities of the Mogami River basin in Yamagata Prefecture—which is the target area—are the same areas with "existing tourist destinations." The procedures were designed considering this need. Figure 9–4 shows the procedures used in the analysis mentioned in this chapter.

First, green tourism bases and existing tourist resources and facilities were identified. For this, existing databases that included the database of Yamagata Green Tourism Promotion Council[2] and the database of Japan Travel and Tourism Association[3] were used. Then, bases and resources/facilities were scoured for green tourism operations and existing tourist destinations, respectively. With regard to the weighting for green tourism, a slant distribution for each base was plotted depending on how many categories of activity could be experienced. As shown in Table 9–2, in the Yamagata Green Tourism Promotion Council's database, there are 10 categories of activities. In accordance with those categories, one point was given to each type of activity per category in each green tourism base. After assigning points, it was clear that though the maximum was 10 points, only one to four points were assigned.

For weighting for existing tourist destinations, the existing research of Tanaka et al. (2002) was followed. Specifically, as shown in Table 9–3, one to four points were assigned for each resource/facility using the point weighting for each type of resource/facility given their relationship to the surrounding forests and forestry. When assigning these points, the location of each base or resource/facility was simultaneously verified, for which the address of each was plotted by hand on a 1:25,000 scale map to determine which tertiary meshgrid (almost 1 km^2) they belonged to on the tertiary meshgrid of digital national land information system.

---

2) Data obtained from http://www.gt-yamagata.com/
3) Data obtained from http://www.nihon-kankou.or.jp/index.php

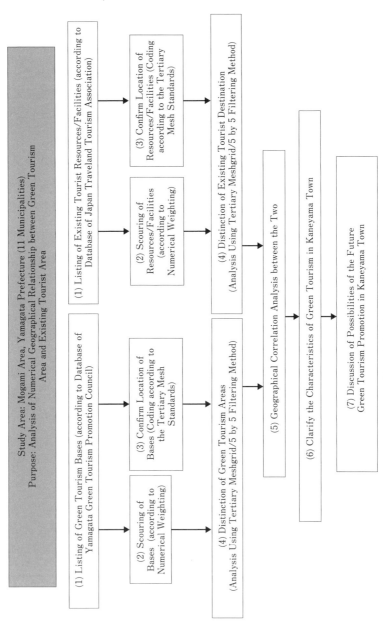

Figure 9-4. Flowchart of Survey

Table 9–2. Activities in Green Tourism

1	Farm Stay
2	Experience (Agriculture, Forestry, and Fisheries)
3	Experience (Craftwork, etc.)
4	Experience (Cooking, Processing, etc.)
5	Experience (Nature, Outdoor, or Ecotour)
6	Eating
7	Purchasing
8	Purchasing (at Marketplace)
9	Owner Farm, Rental Farm, etc.
10	Others

Source: Database of Yamagata Green Tourism Promotion Council.

Table 9–3. Score of Traditional tourist resources and facilities (by numerical weighting)

Resource Type	Score	Facility Type	Score
Mountains	4	Principal Local Products	1
Highlands	4	Visitor Information Center (including Roadside Station)	1
Lakes and Ponds	4	Museums	2
Falls	4	Zoos and Botanical Gardens	4
Valleys	4	Parks	4
Riparian Landscape	4	Industrial Facilities	1
Special Geographical Features	4	Observatories	4
Animals	4	Center Facilities	3
Plants	4	Sports Facilities	4
Big Trees	4	Cycling Courses	4
Spa	3	Hiking Courses	4
Temples and Shinto Shrines	3	Nature Trails	4
Gardens	4	Orienteering Courses	4
Streets	2	Camp Sites	4
Historic Highways	2	Skiing Areas	4
Historic Sites	4	Tourist Farms	4
Historic Architects	4	Recreational Fishing Areas	3
Stone Monuments	2	Leisure Lands	1
Cemeteries	3	Tourist Shops	2
Statue of Buddha or Deities	1	Local Cuisine Restaurants	2
		Sightseeing Boat Areas	4

C1	C2	C3	C4	C5
C16	B1	B2	B3	C6
C16	B8	A	B4	C7
C14	B7	B6	B5	C8
C13	C12	C11	C10	C9

Calculation method of scores at 'Meshgrid A' (P(A))
(1): First, P'(A) is calculated as written below.
   P'(A) = A + 0.5(B1 + B2 + B3 + B4 + B5 + B6 + B7 + B8)
(2): The calculation (1) is repeated and the final potential score of the Meshgrid A, P(A), is calculated.
The method of calculation is shown below.
   P(A) = 3A + 2(B2 + B4 + B6 + B8) + 1.5(B1 + B3 + B5 + B7)+ 0.75(C3 + C7 + C11 + C15) + 0.5(C2 + C4 + C6 + C8 + C10 + C12 + C14 + C16) + 0.25(C1 + C5 + C9 + C13)

Figure 9–5. 5-meshgrid by 5-meshgrid filtering Method

With the information obtained from scouring and locating resources/facilities, areas where green tourism bases and existing tourism resources were concentrated were then identified and quantified. Then, a tertiary meshgrid (almost 1 km^2) analysis of the results using "the 5 meshgrid × 5 meshgrid filtering method" developed by Tanaka et al. (2002) was performed (see Figure 9–5).

The two meshgrid maps—"areas where green tourism and areas with existing tourism resources are concentrated"—were obtained from the analysis results using the filtering method, and these two maps were then overlaid. By comparing the points for each meshgrid, the study analyzed the geographical correlation between the two maps was analyzed to determine whether green tourism had formed new tourism bases.

Finally, on the basis of the results obtained, the characteristics of green tourism in Kaneyama-machi in the Mogami basin and the direction of future green tourism development in Kaneyama-machi were clarified.

4. Results and Discussion

4.1 Analysis of Areas where Green Tourism Bases are Concentrated

According to the database of the Yamagata Green Tourism Promotion Council, there are bases for green tourism in 140 locations within the 11 municipalities comprising the study area, where it is possible to experience 164 different categories of activities (see Table 9–4).

In Kaneyama-machi itself, it is possible to experience 23 categories of activities at 14 different locations (see Table 9–5).

263

Table 9–4. Green Tourism Activities that can be experienced in the 11 Municipalities of Mogami Region

	Categories							
Farm Stay	Experience (Agriculture, Forestry, and Fisheries)	Experience (Craftwork, etc.)	Experience (Cooking, Processing, etc.)	Experience (Nature, Outdoor, or Ecotour)	Eating	Purchasing	Purchasing (at Marketplace)	Owner Farm, Rental Farm, etc.	Others	Total
8	17	18	19	26	24	26	13	11	2	164

Table 9–5. Green Tourism Facilities in Kaneyama-machi and Categories of Experience Programs

Name of Facilities	Categories										Total	Code Number of Tertiary Meshgrid
	Farm Stay	Experience (Agriculture, Forestry, and Fisheries)	Experience (Craftwork, etc.)	Experience (Cooking, Processing, etc.)	Experience (Nature, Outdoor, or Ecotour)	Eating	Purchasing	Purchasing (at Marketplace)	Owner Farm, Rental Farm, etc.	Others		
1 Kurashi Kobo Green Tourism Village	1	1	1							1	4	5840–3310
2 Learning Facility "Shiki no Gakko Taniguchi"	1		1			1					3	5840–2274
3 Schönes Heim Kaneyama	1										1	5840–2391
4 "Iwanaya" River Fish Center				1		1					2	5840–2382
5 "Tyoho-ya" Agrishop							1				1	5840–2257
6 Yamagata "Yugaku" Recreational Forest "Komorebi House"		1	1	1	1						4	5840–2390
7 Exploring Taniguchi Silver Mine Ruin					1						1	5840–2274
8 Nature Observation and Adventure Experience					1						1	5840–2390
9 Snow Trekking					1						1	5840–2390
10 Hobby House "Toubei"			1								1	5840–2391
11 Agri-life Museum "Kyubei"			1								1	5840–2298
12 "Yamato" Craft Center			1								1	5840–3209
13 Kumuro Farm Club			1								1	5840–2391
14 Forest Learning House (Pottery School)			1								1	5840–2391
Total	3	2	7	3	4	2	1	0	0	1	23	

Though the number of bases for green tourism in Kaneyama-machi is less than that in other municipalities, where there are 14.9 locations on an average, the number of different categories of activities at a single location was greater in Kaneyama-machi (1.64 categories for Kaneyama-machi compared with an average of 1.17 categories in the other 11 municipalities). It was clear from the categorization that there were more categories where one could experience crafts (7 of the 18 locations within the 11 municipalities were in Kaneyama-machi) as well as more accommodation categories (3 of the 8 locations within the 11 municipalities were in Kaneyama-machi).

The points awarded to each base for green tourism at each of the other 140 locations (164 points in total) were examined, after which a tertiary meshgrid (almost 1 km^2) analysis of the results was performed using the 5 meshgrid × 5 meshgrid filtering method, and tertiary meshgrid codes were assigned. The results show 33 locations where green tourism bases were concentrated within the 11 municipalities (see Figure 9–6).

4.2 Analysis of Areas where Existing Tourist Destinations are Concentrated

According to the database of Japan Tourism Association, 312 tourism resources and 88 tourism facilities—a total of 400 locations—can be found within the 11 municipalities comprising the target area. Of these, some locations were removed because their tertiary meshgrid codes could not be determined (folk entertainment, etc.) and some tourism resources were counted twice (tourism resources counted under multiple municipalities, such as the summits of mountains on the boundaries). Finally, a total of 302 locations were selected for analysis (see Table 9–6). Incidentally, there were a total of 25 locations for tourism resources and tourism facilities within Kaneyama-machi, of which four could not be given a tertiary meshgrid code and were excluded from the analysis. In addition, one location (Mt. Kamuro) was counted under both Shinjo City and Kaneyama-machi; but, it was included as one object. Ultimately, 21 locations within the town were considered for the analysis (see Table 9–7).

After completing the analysis, the results showed that compared with the average for the 11 municipalities (27.5 locations), there were slightly fewer existing tourism resources and facilities in Kaneyama-machi. Especially obvious was the low number of campgrounds (one) and traditional restaurants (two) among the existing tourism facilities.

Then, a tertiary meshgrid (almost 1 km^2) analysis of the 302 existing tourism resources and facilities with one to four points using the 5 meshgrid × 5 meshgrid filtering method. The results showed that there were 47 locations where existing tourism resources and facilities were concentrated within the 11 municipalities (see Figure 9–7).

Table 9–6. Number of Total Objects and Analysis Objects of Tourist Resources and Facilities

Resource Type	Total Objects	Analysis Objects	Facility Type	Total Objects	Analysis Objects
Mountains	20	18	Visitor Information Center (including Roadside Station)	2	2
Highlands	5	5	Public Tourism/Recreational Facilities	0	0
Lakes and Ponds	9	8	Museums	8	8
Riparian Landscape	17	17	Art Museums	0	0
Coastal Landscape	0	0	Zoos and Botanical Gardens	1	1
Special Geographical Features	2	2	Aquariums	0	0
Natural Phenomenon	0	0	Parks	9	9
Animals	4	1	Industrial Facilities	4	4
Plants	28	28	Observatories	2	2
Spa	19	19	Center Facilities	10	9
Castles	0	0	Sport Facilities	3	3
Temples and Shinto Shrines	50	50	Cycling Courses	3	1
Gardens	2	2	Hiking Courses	4	4
Streets	1	1	Nature Trails	6	5
Historic Highways	4	4	Orienteering Courses	1	0
Historic Sites	19	19	Camp Sites	9	7
Historic Architects	3	3	Golf Courses	0	0
Modern Architects	0	0	Field Athletics	0	0
Other Sites	45	43	Skiing Area	5	5
Rites and Festivals	32	0	Ice Skate Links	0	0
Folk Entertainment	24	0	Swimming Beach	0	0
Traditional Handicrafts	9	0	Marinas and Yacht Harbors	0	0
Local Customs	1	0	Tourist Farms	1	1
Local Landscape	0	0	Tourist Pastures	0	0
Local Cuisine	0	0	Recreational Fishing Area	3	3
Principal Local Products	18	1	Leisure Lands	1	1
Sub-total	**312**	**221**	Tourist Shops	1	1
			Local Cuisine Restaurants	12	12
			Sightseeing Boat Area	3	3
			Sub-total	**88**	**81**
			Total	**400**	**302**

Table 9–7. Existing Tourist Resources and Facilities in Kaneyama-machi

No.	Resource or Facility	Name of Resource/ Facility	Categories	Temporal Categories	Remarks	Analysis Object
1	Resource	Mt. Kamuro	Mountains		Boundary between Shinjo City	Yes
2	Resource	Mt. Dai	Mountains			Yes
3	Resource	Carp in the Oozeki Waterway	Animals			Yes
4	Resource	Old Growth of Kaneyama Cider	Plants			Yes
5	Resource	Fresh Green and Autumn Colors at Shunezaka Mountain Pass	Plants			Yes
6	Resource	Fresh Green and Autumn Colors at Masuzawa Dam Site	Plants			Yes
7	Resource	Autumn Colors at Mt. Rouba	Plants			Yes
8	Resource	Kamuro Spa	Spa			Yes
9	Resource	Ensyouji Temple	Temples and Shinto Shrines			Yes
10	Resource	Hachiman Shinto Shrine	Temples and Shinto Shrines			Yes
11	Resource	Houenji Temple	Temples and Shinto Shrines			Yes
12	Resource	Ushu Traditional Highway	Historic Highways			Yes
13	Resource	Tateyama Castle Ruin	Historic Sites			Yes
14	Resource	Taniguchi Silver Mine	Historic Sites			Yes
15	Resource	Stone Monument of Isabella Lucy Bird	Other Sites	Historic Sites		Yes
16	Resource	Stone Image of Iwaen Jizo Deity	Other Sites	Temples and Shinto Shrines		Yes
17	Resource	Avenue of Pines for Daimyo's Alternating Edo Residence	Other Sites	Plants		Yes
18	Resource	Stone Monument of Boshin War	Other Sites	Historic Sites		Yes
19	Facility	Kamuro Camp Site	Camp Sites			Yes
20	Facility	Sakaeya Local Cuisine Restaurant	Local Cuisine Restaurants			Yes
21	Facility	Shinogeya Local Cuisine Restaurant	Local Cuisine Restaurants			Yes
–	Resource	Osaito Festival	Rites and Festivals		Difficult to locate geographically	No
–	Resource	Kaneyama Festival	Rites and Festivals		Difficult to locate geographically	No
–	Resource	Fireworks Display	Rites and Festivals		Difficult to locate geographically	No
–	Resource	Inazawa Bangaku Ritual	Rites and Festivals		Difficult to locate geographically	No

4.3 Analysis of the Geographical Correlation between Meshgrid Points for Both Areas

Next, the abovementioned results were overlapped for each area type and analyzed the correlation between the points assigned to the meshgrid codes was analyzed. This analysis was conducted to determine whether there was a high degree of geographical overlap between areas where green tourism is concentrated and areas where existing tourism resources and facilities are concentrated. If there was low correlation between the two types of areas on the basis of the points given, then it could be concluded that green tourism is creating tourism activities in new areas. However, if there was a high correlation between the two types of areas, it could be concluded that there was a high possibility that this was because green tourism merely built on the foundation of existing tourist destinations rather than created tourism activities in new areas.

Figure 9–8 shows the results of the analysis of the correlation between these two types of areas. As can be seen, the correlation between them is a mere $R^2 = 0.17$, which means that there is low correlation between the two. Therefore, it can be concluded that bases of green tourism have been responsible for bringing tourists into areas in the 11 municipalities of the Mogami River basin where tourists have never come before, and that it has been quantitatively demonstrated through geographical analysis.

However, the state of dispersion in Figure 9-8 shows that there are a few points that can be plotted from the origin to the top right corner in a straight line. Thus, it can be deduced that bases for green tourism are not exclusively located in places that are different from areas with existing tourism resources and facilities. In some cases, bases for green tourism are also located in the same areas as existing tourism resources and facilities.

Though the research attempted to elucidate the trends regarding what types of areas doubled as existing tourist destinations and bases for green tourism, no definite answer was established. Nonetheless, the research hypothesis was that bases for green tourism related to food, such as local markets for rural products and restaurants, overlap with areas within urban areas where existing tourism is concentrated (including areas with historical and cultural assets such as shrines and facilities such as museums).

4.4 Characteristics of Green Tourism in Kaneyama-machi

As discussed above, it was possible to quantitatively demonstrate through geographical analysis that green tourism has been responsible for attracting tourists to the 11 municipalities of the Mogami River basin in Yamagata Prefecture, which have not been tourist attractions in the past. This section, however, explores the types of characteristics of Kaneyama-machi with regard to both types of areas that are investigated in this project.

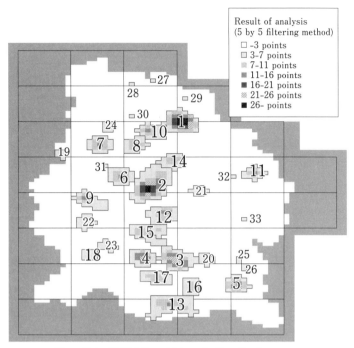

Figure 9–6. 33 High-Potential Areas for Green Tourism
Source: Database of Yamagata Green Tourism Promotion Council

As already mentioned, one of the characteristics of Kaneyama-machi in terms of bases for green tourism is that although it does not have many bases compared with other areas, each base facilitates experiencing activities from different categories, many of which are related to crafts and accommodations.

Furthermore, in comparison with other municipalities, although the number of existing tourism facilities was extremely low, that of existing tourism resources is not very low. Of the 33 locations shown in Figure 9–6 indicating concentration of green tourism bases, those located in Kaneyama-machi are indicated by triangular districts between Sugisawa, Kamuro, and Yūgaku-no-Mori Recreational Forest, which had the highest number of points, and the Taniguchi silver mine district, which had the 10th highest number of points.

In contrast, of the 47 locations shown in Figure 9–7 where existing tourism resources are concentrated, those located in Kaneyama-machi are the Kaneyama-machi Center, with the 5th highest number of points on the list; the Kamuro Spa district, 27th on the list; the Masuzawa Dam district, 36th on the list; and Mt. Kamuro, 42nd on the list.

Figure 9–9 shows the geographical correlation between the abovementioned 6 locations (two in areas where green tourism bases are concentrated and four where existing tourism resources

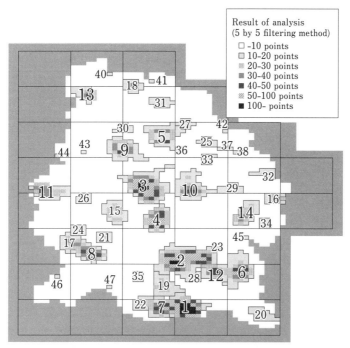

Figure 9–7. 47 High-potential Areas for Existing Tourism
Source: Database of Japan Travel and Tourism Association

and facilities are concentrated). Examining the relationship between these locations shows that the triangular district between Sugisawa, Kamuro and Yūgaku-no-Mori Recreational Forest (1st on the list), which has a high potential as a base for green tourism, and the medium-ranked Taniguchi silver mine district (10th on the list) are neither too close nor too far from the Kamuro Spa district, which is a mid-ranking existing tourist destination (27th on the list), and Kaneyama-machi Center (5th on the list), which is a mid-ranking existing tourist destination. The locations are either in the same settlement or within one or two valleys of each other. Settlements that are separated by a valley differ from areas in which existing tourism are concentrated. As these settlements were formerly ordinary rural villages, it can be said that they succeeded in developing tourism in a new place[4].

Kaneyama-machi stands out only because of its three existing tourism facilities; however, this lack of facilities is somewhat compensated for by the fact that in the area

---

4) No bases for green tourism could be found in the area surrounding the Masuzawa Dam district (36th on the list) and the Kamuro mountain area (42nd on the list).

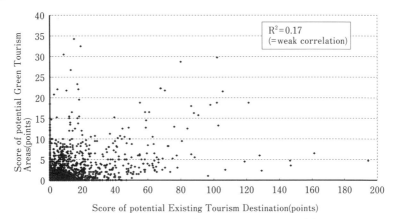

Figure 9–8. Correlation between the scores of Green Tourism Areas and Existing Tourism Areas

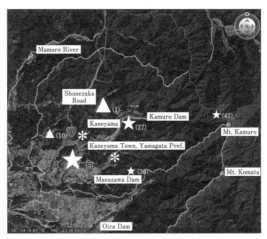

▲ : **Green Tourism Area**
▲(1) : Sugisawa, Kamuro and Yugaku-no-mori Rec-Forest Area
▲(10) : Taniguchi-Ginzan Area
★ : **Existing Tourist Destination Area**
★(5) : Central Kaneyama Area
★(27) : Kamuro Spa Area
★(36) : Masuzawa Dam Area
★(42) : Mt Kamuro Area
✻ : **Blank area where green tourism will be accommodated**
Note : Size of legends means the potential as tourist destination.

Figure 9–9. Geographical Relationships between Existing Tourist Areas and Green Tourism in Kaneyama-machi, Yamagata Prefecture, Japan
Geographic details are indicated on Google Earth®.

surrounding the Kamuro Spa district (which, of course, has hot springs), there are base facilities, such as facilities for accommodation and hands-on experience (Schönes Heim Kaneyama and Kurashi-kobo) as well as learning facilities (Yūgaku-no-Mori Recreational Forest and the Forest Academy). Furthermore, near the urban Kaneyama-machi Center, hostels which double as bases for green tourism, such as Shiki-no-Gakkō (Four Seasons School) Taniguchi, are being developed to attract groups that want to come to stay and learn, and these will evolve into places with an advantage in terms of geographical relationships.

4.5 Direction of Green Tourism Development in Kaneyama-machi

According to a supplementary poll conducted in the town along with this analysis, there are places where planning has been conducted together with existing tourist destinations in bases for green tourism in Kaneyama-machi, such as Schönes Heim Kaneyama, which was planned in conjunction with the East Japan Railway Company's Yamagata Shinkansen Line extension to Shinjo City. However, there are also places like Taniguchi where, for example, the accidental use of the abandoned school.

The results of this analysis show that regardless of whether it is planned or incidental, green tourism in Kaneyama-machi has the advantage of having a geographical relationship with existing tourist destinations in terms of location; this has helped support the new facilities and experiential activities of green tourism.

Considering the direction of green tourism development in Kaneyama-machi, the areas that stand out are those within the town where green tourism is currently not being implemented. There are also settlements within Kaneyama-machi where green tourism is not being implemented, such as areas surrounding the Masuzawa Dam and in the northeastern part of the city as indicated by asterisks (*) in Figure 9–9. Regional development could be accomplished in such areas through green tourism.

Of course, as green tourism is a destination-based industry where development can be archieved at the initiative of the local people, there is no need to desperately promote green tourism in such areas without respecting the local people's wishes. If and when there are plans to further develop green tourism within the town, these areas can be potential sites in planning concepts.

Nevertheless, by mapping and quantifying using the analysis method, it is possible to ascertain the state of green tourism development for different places and provide information for making a decision regarding future plans for green tourism in Kaneyama-machi.

5. Future Challenges

This research considered the direction of green tourism development in municipalities by analyzing the 11 municipalities of the Mogami River basin as a whole.

In existing research, there are almost no examples that have geographically and quantitatively considered whether green tourism is succeeding in creating tourist destinations in new places. However, using analysis methods, it is possible to specify an area and conduct quantitative analysis regarding the development of green tourism that tends to be promoted through image and qualitative judgment.

However, mere research into green tourism planning concepts must conclude when it provides information for determining the direction of development in the future. Through incorporating green tourism as a part of the process of revitalizing areas, it is critical to analyze the potential to attract tourists based on the desires, incentives, and marketing methods of the people living in that area. After considering these factors together with the geographical and quantitative analysis, it will be necessary to make final judgments regarding the overall direction of green tourism development. Therefore, it is necessary to develop future research along those lines.

Furthermore, when it comes to actually applying these methods in municipalities throughout Japan, it is believed that the element of consulting rather than research will become more prominent. Therefore, popularizing these research results may also be a challenge in the future.

By various stakeholders aiming at collaborative governance in an area, ascertaining local geographical tourism properties, and engaging in discussion to further scientific and objective understanding, it is believed that it will be possible in the midst of the abovementioned challenges to bring all stakeholders to the same awareness. Moreover, on the basis of this awareness, new possibilities regarding the state of collaborative governance can be developed.

References

Japan Productivity Center 2007 (Japanese edition). *Leisure Hakusyo 2007: Yoka Juyo no Henka to New Tourism* [*White Paper on Leisure: Change of the Leisure Demand and New Tourism*]. Tokyo: Japan Productivity Center, pp. 87–121.

Tanaka, N. and T. Watanabe 2002. (Japanese edition with English Summary). "The Structure of Forest Areas Important for Tourism and Recreational Opportunities in a Hilly and Mountainous Region of Japan". *Landscape Kenkyu* [*Journal of the Japanese Institute of Landscape Architecture*]. 65(5): 615–620.

Chapter 10
Simulating Future Land-cover Change
A Probabilistic Cellular Automata Model approach

Arief Darmawan and Satoshi Tsuyuki

1. Introduction

1.1. Why do we have to Predict Future LULCC?

Understanding the human- or climate-induced changes in a tropical environment requires knowledge of the current status of landscape, the extent of land-cover types vulnerable to change, and the causes and impacts of such changes. Monitoring land-use and land-cover change (LULCC) is definitely the very first step for understanding the past and current states of our environment. Time series and accurate change detection of the Earth's surface features provides historical land-cover change information as a result of interactions between human and natural phenomena.

Knowledge of historical land-cover change preserves important information to foresee future changes using a modeling approach. Modeling and simulating land-cover change can provide an outlook of what, where, why, and how to mitigate future LULCC under any condition. Moreover, modeling approach is a prerequisite under the post-Kyoto Protocol climate change mitigation regime, for example, Reducing Emissions from Deforestation and Forest Degradation Plus (REDD-plus), for forecasting business-as-usual emission or reference emission level.

On the other hand, from the perspective of local people, they cannot clearly recognize LULCC even though they are familiar with the land area. Although the change gradually proceeds over a long time, They say "Oh! I did not know the landscape was changing so much like this!". Hence, it will surely be difficult for them to imagine LULCC for a wide area of several hundred km². Furthermore, when they are requested to express their own opinion at government-organized group meetings that explain future land use policy alternatives orally or by paper, it is conceivably very difficult for local people to concretely predict by themselves the manner in which each policy alternative will affect their living environment and surrounding landscape. To make land-use plans comprehensible to (and encourage the involvement of) the local people, it is desirable that local people understand, first, the LULCC in the area from past to present; second, why such changes occurred and how it affected their life and environment; and third, how each policy alternative will affect their landscape. Finally, after considering the abovementioned criteria, they should select a policy alternative. In such situations, visual information and not just verbal or written explanation will be very effective, because it is easy for the local people to imagine a concrete situation using visual information. Although satellite remote sensing imagery currently provides such visual land-use/land-cover (LULC) information, other types of prediction models have to be used to provide future information.

1.2. How can we predict future LULCC: Cellular Automata?

Over the past decades, a range of LULCC models have been developed to meet land management requirements and address the future role of LULCC (Veldkamp and Lambin, 2001). A few examples of spatial modeling approaches are multi-agent models, artificial neural network models, statistical models, and cellular automata models (Singh, 2003). Though, by definition, all models fall short in terms of incorporating all aspects of reality, they provide valuable information about the system's behavior through scenario building.

Cellular automata (CA) derive their name from the fact that they consist of "cells", similar to the cells on a checkerboard, and that the cell states may evolve according to a simple transition rule, the "automaton". The mechanism for defining the next state of a cell based on its actual state and the actual state of the cells in the neighborhood makes CA a very simple mechanism with the following main characteristic: spatial interactivity.

CA uses a "bottom up" approach in which a final global structure emerges from purely local interactions among the cells. CA not only offers a new way of thinking for dynamic process modeling but also provides a laboratory for testing decision-making processes. CA have a natural affinity with Geographical Information System (GIS) and remotely sensed data (Torrens et al., 2001). Nevertheless, the definition of the transition rules of a CA model is the most essential element for obtaining realistic simulations of LULCC (Verburg et al., 2006). To date, standardized methods for defining the transition rules have not been provided; hence, more attention must be paid to efforts to formulate such methods.

CA were employed in almost every land-use change model for urban environments (Pontius and Malanson, 2005; Barredo et al., 2003; de Almeyda et al., 2003). Few studies focused on the land-use dynamics of rural or more natural landscapes; examples are provided by Messina and Walsh (2001), who studied LULC dynamics in the Ecuadorian Amazon. Ménard and Marceau (2006) simulated the impact of forest management scenarios in an agricultural landscape using geographical CA.

1.3. Chapter Objectives

This chapter aims to determine the possibility of elaborating CA and the probabilistic transition rule for simulating land-cover changes in the Middle Mahakam area, West Kutai District, East Kalimantan Province, Indonesia. The predictive power and reliability of several probabilistic CA-based LULCC models are assessed and future land-cover states in this area will be simulated on a scenario basis using the most reliable model.

2. Materials and Methods

2.1. Situation in West Kutai District

Middle Mahakam has an area of 10,898.5 km^2 and is located in the southeastern part of West Kutai District, East Kalimantan Province, Indonesia (Figure 10–1). Geographically, it is situated between 0°40′33″ N and 0°49′25″ S and between 115°25′24″ E and 116°23′30″ E.

Most of the study site comprises lowland areas (below 200 m asl), ranging from nearly flat areas situated downstream of the Mahakam river including its wetland complex to tropical lowland areas situated upstream. This area is a typical form of lowland tropical ecosystems and includes swampy ecosystems (e.g., peat swamp forest and freshwater or riparian forests), heath forests (*kerangas* forest), and lowland tropical rain forests.

Based on a forest land-use allocation map (*Tata Guna Hutan Kesepakatan*: TGHK), the study area comprises 3,396.8 km^2 of permanent production forest (*Hutan Produksi*: HP), 1,189.7 km^2 of limited production forest (*Hutan Produksi Terbatas*: HPT), 309.4 km^2 of protection forest (*Hutan Lindung*: HL), 58.5 km^2 of conservation forest (*Cagar Alam*: CA), and 5,944.1 km^2 of non-forest area (*Areal Penggunaan Lain*: APL).

The annual rainfall ranges from 2,150 mm to 3,580 mm. Under normal conditions, the wet season lasts from November to May followed by the dry season from June to October.

The study area is a representative of complex factors that give rise to LULCC, such as forest logging, agro-industrial estate development (e.g., palm oil and rubber estates), shifting cultivation, transmigration program, mining operation, as well as forest and land fires. In addition, as West Kutai is a newly formed district since 1999, the economic and infrastructure development issues are quite strong, which could further influence LULCC in this area (Casson, 2001).

2.2. Procedure to Predict Future LULCC

The flow of the study is depicted in Figure 10–2. Two land-cover maps for different years, i.e., 2000 (LC2000) and 2003 (LC2003), derived from satellite remote sensing data were used.

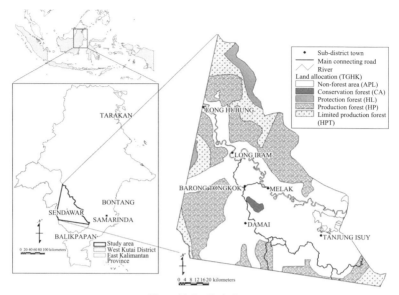

Figure 10–1. Study Area

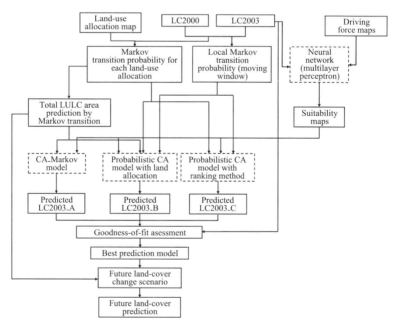

Figure 10–2. Flow of the Study

These data were employed because during 2000–2003 land-cover change rate was relatively moderate compared to that during 1992–1997 or 1997–2000.

Three CA-based LULCC models were investigated during this study namely the CA_Markov model, probabilistic CA model with ranking method, and probabilistic CA model with land allocation. The CA_Markov model considers neighborhood effect and land-cover suitability map, followed by an iterative procedure of reallocating land cover until it meets the area totals predicted by the Markov transition. Apart from incorporating neighborhood effect and suitability maps, the other two models incorporate land-cover transition probabilities that were assigned to each pixel (cell). Global and local transition probabilities were experimented within these models. The probabilistic CA model with ranking method partially adopts methods proposed by Zamyatin and Markov (2005) in which the transition rule is determined by the highest probability of potential land cover. In contrast, the probabilistic CA model with land allocation adopted an iterative procedure of reallocating land cover suggested by the CA_Markov model. For each model, six neighborhood sizes with distance decay effects were examined.

The resulting models were evaluated using two error matrices, i.e., the goodness-of-fit within the "change" area and that within the "no-change" area. Goodness-of-fit was measured using percent of correct pixels (total accuracy). Separation between goodness-of-fits of the "change" and "no-change" areas is necessary because it is common to attain a large percent of correct pixels from a null model that predicts pure persistence (i.e., no change) between time t and time $t + 1$ due to temporal autocorrelation between the reference maps of t and $t + 1$ (Pontius and Malanson, 2005). In most of the land-change modeling literature, the typical amount of change in the landscape was approximately 10%; therefore, a null model of pure persistence would be 90% correct, which is a high measure of accuracy according to most naive interpreters. Hence, separating these "change" and "no-change" areas will provide a more robust goodness-of-fit assessment for the resulting models.

Then, prediction scenarios were employed in the best prediction model. Scenarios are plausible views of the future based on "if, then" statements—if the specified conditions are satisfied, then future land use and land cover will be realized in a particular manner (Alcamo et al., 2006). Scenario analysis is the procedure by which scenarios are developed, compared, and evaluated. This analysis does not eliminate uncertainties regarding the future but provides a means to represent current knowledge in the form of consistent, conditional statements about the future.

In this study, scenarios were constructed on the basis of transition probabilities that fall under different forest land-use allocation policies (TGHK). A scenario was formulated to visualize how different policies implying different levels of human activities were implemented in the entire study area. The first scenario was if LULCC in the study area was determined by the transition probability of the entire study area, then a "business-as-usual" LULCC was implied. The second scenario was if LULCC in the study area was determined by the transition probability of the non-forest area (APL), then a "rapid deforestation" condition was implied. The third scenario was if LULCC in the study area was established by the transition probability of the production forest area (HP), then a "rapid forest degradation" condition was implied. Finally, the fourth scenario was if LULCC in the study area was determined by the transition probability of the protection forest area (HL), then a "rapid forest regrowth" condition was implied.

2.3. CA model
Cell space

The digital space in CA consists of a rectangular grid of square cells (pixels), each representing an area of 90 m × 90 m or similar to the spatial resolution of Shuttle Radar Topography Mission (SRTM) terrain image. A raster resampling procedure was performed to standardize the size and orientation of pixels used in this study. In the case of LC2000 and LC2003, these data were filtered by a 3 × 3 modal filter before resampling (using the nearest neighbor resampling method), because their initial spatial resolution was 30 m (see Figure 10–3).

Cell state and time step

The models used four land-cover categories as the cell state; natural forest (NF), fragmented/degraded forest (FF), shrub and grassland (SG), and bare land (BL). Furthermore, the models employed cell states that represent spatially distributed variables for creating a suitability map. These variables (i.e., driving force maps) include the landscape's physical properties (i.e., elevation, slope, aspect, and soil type); its climatic properties (i.e., rainfall distribution map); and proxies of human influence properties (i.e., distance from nearest roads, distance from nearest rivers, and distance from nearest settlement) (see Figure 10–4).

A three year time step was used in this study similar to the time difference between LC2000 and LC2003.

Neighborhood

The operations that are used to calculate neighborhood characteristics are called convolution, spatial filtering, or focal functions. In many studies, the size of the neighborhood was arbitrarily chosen and only the direct neighborhood of a location was considered (Verburg et al., 2004). Others have argued that human activities are influenced by a wider space resulting in a flexible definition of the neighborhood essential (White and Engelen, 2000).

Six neighborhood sizes were considered for convolutions, including 3 × 3, 5 × 5, 7 × 7, 11 × 11, 15 × 15, and 21 × 21 spatial filters. A distance decay weight neighborhood suggested by White and Engelen (2000) was adopted by taking an inverse of Euclidian distances of neighbor cells from the cell assigned to the spatial filter (see Figure 10–5). Hence, cells

0.07	0.07	0.08	0.08	0.09	0.09	0.09	0.10	0.10	0.10	**0.10**	0.10	0.10	0.10	0.09	0.09	0.09	0.08	0.08	0.07	0.07
0.07	0.08	0.08	0.09	0.09	0.10	0.10	0.11	0.11	0.11	**0.11**	0.11	0.11	0.11	0.10	0.10	0.09	0.09	0.08	0.08	0.07
0.08	0.08	0.09	0.09	0.10	0.11	0.11	0.12	0.12	0.12	**0.13**	0.12	0.12	0.12	0.11	0.11	0.10	0.09	0.09	0.08	0.08
0.08	0.09	0.09	0.10	0.11	0.12	0.12	0.13	0.14	0.14	**0.14**	0.14	0.14	0.13	0.12	0.12	0.11	0.10	0.09	0.09	0.08
0.09	0.09	0.10	0.11	0.12	0.13	0.14	0.15	0.16	0.16	**0.17**	0.16	0.16	0.15	0.14	0.13	0.12	0.11	0.10	0.09	0.09
0.09	0.10	0.11	0.12	0.13	0.14	0.16	0.17	0.19	0.20	**0.20**	0.20	0.19	0.17	0.16	0.14	0.13	0.12	0.11	0.10	0.09
0.09	0.10	0.11	0.12	0.14	0.16	0.18	0.20	0.22	0.24	**0.25**	0.24	0.22	0.20	0.18	0.16	0.14	0.12	0.11	0.10	0.09
0.10	0.11	0.12	0.13	0.15	0.17	0.20	0.24	0.28	0.32	**0.33**	0.32	0.28	0.24	0.20	0.17	0.15	0.13	0.12	0.11	0.10
0.10	0.11	0.12	0.14	0.16	0.19	0.22	0.28	0.35	0.45	**0.50**	0.45	0.35	0.28	0.22	0.19	0.16	0.14	0.12	0.11	0.10
0.10	0.11	0.12	0.14	0.16	0.20	0.24	0.32	0.45	0.71	**1.00**	0.71	0.45	0.32	0.24	0.20	0.16	0.14	0.12	0.11	0.10
0.10	**0.11**	**0.13**	**0.14**	**0.17**	**0.20**	**0.25**	**0.33**	**0.50**	**1.00**	**0.00**	**1.00**	**0.50**	**0.33**	**0.25**	**0.20**	**0.17**	**0.14**	**0.13**	**0.11**	**0.10**
0.10	0.11	0.12	0.14	0.16	0.20	0.24	0.32	0.45	0.71	**1.00**	0.71	0.45	0.32	0.24	0.20	0.16	0.14	0.12	0.11	0.10
0.10	0.11	0.12	0.14	0.16	0.19	0.22	0.28	0.35	0.45	**0.50**	0.45	0.35	0.28	0.22	0.19	0.16	0.14	0.12	0.11	0.10
0.10	0.11	0.12	0.13	0.15	0.17	0.20	0.24	0.28	0.32	**0.33**	0.32	0.28	0.24	0.20	0.17	0.15	0.13	0.12	0.11	0.10
0.09	0.10	0.11	0.12	0.14	0.16	0.18	0.20	0.22	0.24	**0.25**	0.24	0.22	0.20	0.18	0.16	0.14	0.12	0.11	0.10	0.09
0.09	0.10	0.11	0.12	0.13	0.14	0.16	0.17	0.19	0.20	**0.20**	0.20	0.19	0.17	0.16	0.14	0.13	0.12	0.11	0.10	0.09
0.09	0.09	0.10	0.11	0.12	0.13	0.14	0.15	0.16	0.16	**0.17**	0.16	0.16	0.15	0.14	0.13	0.12	0.11	0.10	0.09	0.09
0.08	0.09	0.09	0.10	0.11	0.12	0.12	0.13	0.14	0.14	**0.14**	0.14	0.14	0.13	0.12	0.12	0.11	0.10	0.09	0.09	0.08
0.08	0.08	0.09	0.09	0.10	0.11	0.11	0.12	0.12	0.12	**0.13**	0.12	0.12	0.12	0.11	0.11	0.10	0.09	0.09	0.08	0.08
0.07	0.08	0.08	0.09	0.09	0.10	0.10	0.11	0.11	0.11	**0.11**	0.11	0.11	0.11	0.10	0.10	0.09	0.09	0.08	0.08	0.07
0.07	0.07	0.08	0.08	0.09	0.09	0.09	0.10	0.10	0.10	**0.10**	0.10	0.10	0.10	0.09	0.09	0.09	0.08	0.08	0.07	0.07

Figure 10–5. Various Neighborhood Sizes (Bold Line) with Distance Decay Weight Obtained from Inverse Euclidian Distance.

that are more distant in the neighborhood will have a smaller effect on the evaluated central cell.

Neighborhood value was calculated on the basis of the contiguity rule (Pontius and Malanson, 2005). The contiguity rule generally has the effect of predicting the growth of a category near locations where that category already exists. To achieve this contiguity rule, a focal mean filter operation was performed for each binary state of land-cover category in time t. Thus, four grid layers consisting of NF, FF, SG, and BL neighborhoods were produced for each neighborhood size.

2.4. Role of Suitability Map

For an urban growth model, the so-called suitability map is widely used (Pontius and Malanson, 2005; Barredo et al., 2003). A suitability map contains information about the urbanization probability of some area given a function of different driving factors that influence the urbanization level in a specific location. Few examples of such driving factors are distance to roads, distance to markets, distance to water facilities, and slope information. To apply such suitability map perspective to the land-cover change model, a more complex and intractable task of suitability map design should be involved. This task requires formalizing several different data such as a variety of driving factors and land-cover types at the investigating area.

An effective solution, which allowed the use of such complex and non-formalized data for designing the suitability maps, is the application of artificial neural networks (ANN) (Yeh and Li, 2003). The multilayer perceptron (MLP) with a hidden layer was used for solving such tasks. In this study, the ANN was trained by the standard back-propagation algorithm to reveal the influence of the eight driving force maps on each land-cover category in LC2003. Then, the trained ANN was recalled to produce a suitability map of each land-cover category (i.e., NF, FF, SG, and BL). The suitability map has a range of values with 0 representing not suitable and 1 representing very suitable. Next, normalization was performed for each pixel to make the sum of suitability of the four land-cover categories as 1.

3. Results and Discussions

3.1. Land-cover Suitability Map

The learning process by back-propagation algorithm showed that the eight driving factors could explain as much as 47.3% of land-cover categories in 2003. This result is satisfactory, as the resulting suitability maps illustrate that the potential of one land-cover category will exist in one specific location. The above result implies that the potential of the land-cover can be explained by driving force maps with driving factors as much as 47.3%, whereas the remaining part can be explained by other factors or processes such as neighborhood effects, transition probabilities, market factors (e.g., prices, access cost, and property right) (Angelsen, 1999) as well as social accessibility, and land tenure (Mertens and Lambin, 2000). In this study, neighborhood effects and transition probabilities were examined. Figure 10–6 depicts land-cover suitability maps of the study area.

Land-cover modeling usually requires preparation of various development alternatives under different assumptions, which are then illustrated on a suitability map. Wu and Webster (1998) proposed a method for generating alternative development patterns by integrating multi-criteria evaluation (MCE) techniques with CA. Parameter values could be adjusted to correspond to various planning objectives. In addition, MCE techniques became the standard tool for deriving suitability maps in the CA_Markov model within the IDRISI software (Eastman, 2006). However, there were uncertainties in this method because determination of parameter values was quite relaxed (Yeh and Li, 2003). Furthermore, MCE techniques were highly subjective because they required a decision-making process; therefore, the results would depend on the person who made the decision (evaluation). Using ANN for creating a suitability map is more robust because data-fitting procedures were performed by strictly using back-propagation algorithm for the training. The parameter values could automatically be obtained through appropriate training of neural networks; therefore, a subjective decision was no longer included.

3.2. Evaluation of Goodness-of-fit of Future Prediction Using the LULCC Model

The goodness-of-fit of all models within the "change" area is shown in Figure 10–7, whereas that within the "no-change" area is shown in Figure 10–8. The CA_Markov model's (I) accuracy did not change when experimented with different neighborhood sizes. The total accuracies of the CA_Markov model in the "no change" and "change" areas were approximately 91.3% and 5.9%, respectively. These results were mostly obtained because the CA_Markov model only considers suitability maps and the neighborhood effects; eventually, different input of neighborhood sizes resulted in a slight difference in accuracies. Nevertheless, these results emphasize the previous argument that combining suitability maps and neighborhood operation will increase the prediction accuracy, as suitability maps alone can only explain as much as 47.3% of land-cover categories in 2003.

A comparison between the probabilistic CA models with ranking method (II–V) and the probabilistic CA models with land allocation method (VI–IX) reveals that for each neighborhood size within the "no-change" area, the probabilistic CA models with ranking method resulted in higher accuracies than those with land allocation method. In contrast, within the "change" area, the probabilistic CA models with land allocation method produced higher accuracies than those with ranking method for each neighborhood size. The above different trends were mainly influenced by different methods for allocating potential land cover in time $t + 1$. In the probabilistic CA models with land allocation, predicted land cover in time $t + 1$ was determined by the total LULC area predicted by the Markov model. Therefore, in some areas, higher ranking of one potential land cover (which normally occurred in the "no-change" area) did not automatically turn into that land cover in the time $t + 1$, which resulted in slightly lower accuracies than the probabilistic CA models with ranking method. Moreover, a similar condition occurred within the "change" area, which resulted in higher accuracies for probabilistic CA models with land allocation.

A very distinctive trend is shown by an increase in neighborhood sizes. Accuracies in the "no-change" area decreased with increasing neighborhood size. On the contrary, the accuracies in the "change" area increased with increasing neighborhood size. These results suggest that the contiguity rule also occurred within the natural landscape and is not exclusively applicable within urban models (Pontius and Malanson, 2005; Barredo et al., 2003; White and Engelen, 2000). Though generalizing the contiguity rule indicated that a slight decrease in accuracy would occur within the "no-change" area, it considerably influences the predictive power of land-cover change modeling within the "change" area. As the distance decay effect was employed in this study, the accuracies of using neighborhood sizes larger than 21 × 21 are considered to be static in nature. Therefore, the largest neighborhood size (21 × 21) was the highest accuracy in this study.

A comparison between incorporating global and local transition probability images reveals that incorporating a local transition probability image resulted in better accuracy than incorporating a global transition probability image. This is reasonable because the probability of change from one land cover in time t to another in time $t + 1$ is not similar throughout the study site, as many driving factors work at a local scale. Moreover, each locality has a special characteristic that differs from the other. This result is comparable with that of Ménard and Marceau (2006) who argued that local dynamics is an important driver of land-use change and therefore reinforces the adequacy of using CA as a modeling tool. However, these results should be considered with care; in fact, a smaller window size will provide better accuracy. Figure 10–7 shows that within the three local transition probability images, the smaller size of the moving window (11 × 11) (VII) provides better accuracy than the larger one (31 × 31) (IX). A smaller window size (e.g., 3 × 3) suggests that higher influence of spatial dependency will exist where transition probability will turn into a binary expression of "change" and "no change" for a specific land-cover category in time $t + 1$. This binary expression will lead to a static model in which the model can predict the time $t + 1$ very accurately; however, it is static for times $t + 2, t + 3$, and so forth. Therefore, a very small window size has a critical weakness for predicting future land-cover change. Hence, in this study, an 11 × 11 window size (or approximately 1 km^2) was chosen as the smallest size for calculating local transition probability. This is because within this size, extensive land use can be undertaken, for example, dry shifting agriculture requires each household to clear approximately 0.02 km^2 of land each year (Seavoy, 1973).

Figures 10–7 and 10–8 describe the probabilistic CA models incorporating local transition probability using an 11 × 11 window size and a 21 × 21 neighborhood size, which exhibit the highest accuracy in the "change" area and acceptable results (accuracy higher than 88%) in the "no-change" area. These models were significantly better than the CA_Markov model in the "change" area and slightly inferior to the CA_Markov model in the "no-change" area. In addition, the probabilistic CA models were significantly better than the null model in the "change" area and slightly inferior to the null model in the "no-change" area. As suggested by Pontius and Malanson (2005), a reasonable minimum criterion for goodness-of-fit assessment would be that the goodness-of-fit of the prediction model should be better than that of a null model.

Figure 10–9 depicts the comparison of predicted LULCC areas by Markov transition

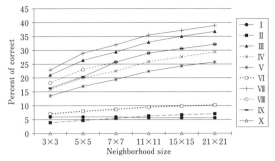

Where:
I: IDRISI CA_Markov module
II: Using Global Markov transition image with ranking method
III: Using Local Markov transition 11 × 11 image with ranking method
IV: Using Local Markov transition 21 × 21 image with ranking method
V: Using Local Markov transition 31 × 31 image with ranking method
VI: Using Global Markov transition image with land allocation (MOLA)
VII: Using Local Markov transition 11 × 11 image with land allocation (MOLA)
VIII: Using Local Markov transition 21 × 21 image with land allocation (MOLA)
IX: Using Local Markov transition 31 × 31 image with land allocation (MOLA)
X: Null Model

Figure 10–7. Goodness-of-fit of All Models within the "Change" Area

probability and the best CA model using ranking rule, and Figure 10–10 describes the comparison of predicted LULCC areas by Markov transition probability and the best CA model by land allocation. As shown in Figure 10–9, the probabilistic CA model with ranking method did not appear realistic because the predicted forest area was abruptly increasing during predicted periods. In contrast, the probabilistic CA model with land allocation seems reasonably realistic (see Figure 10–10) because it follows a common logic that forest area will decrease relative to the time step. Based on these results, the probabilistic CA model with land allocation using a 21 × 21 neighborhood size and incorporating local transition probability at an 11 × 11 window size was selected as the best model for predicting LULCC in the study area.

3.3. Comparison of scenario-based LULCC Simulation Results

Figures 10–11-14 show the predicted LULCC areas under various scenarios. In the "business-as-usual" scenario, the rate of change is rather moderate, as the land-cover change from 2000 to 2003 was not very abrupt (see Figure 10–11). The NF cover gradually decreases,

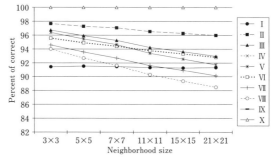

Where:
I: IDRISI CA_Markov module
II: Using Global Markov transition image with ranking method
III: Using Local Markov transition 11 × 11 image with ranking method
IV: Using Local Markov transition 21 × 21 image with ranking method
V: Using Local Markov transition 31 × 31 image with ranking method
VI: Using Global Markov transition image with land allocation (MOLA)
VII: Using Local Markov transition 11 × 11 image with land allocation (MOLA)
VIII: Using Local Markov transition 21 × 21 image with land allocation (MOLA)
IX: Using Local Markov transition 31 × 31 image with land allocation (MOLA)
X: Null Model

Figure 10–8. Goodness-of-fit of All Models within the "No-change" Area

whereas the FF cover gradually increases. Change in SG and BL covers is very minor in this scenario. This condition occurs because the study area falls under a different land-allocation policy, which consequently results in a different land-cover change pattern. Land-cover conversion is permitted only in the APL area, whereas other forest areas can only degrade or regenerate. However, as the APL area covers approximately 60% of the study area, the contribution of this land allocation is higher than the others. Therefore, gradual NF cover loss is inevitable.

In the "rapid deforestation" scenario, the NF cover decreases faster than that in the other scenarios (see Figure 10–12). Moreover, the SG cover significantly increases compared with the FF cover. This is because forest conversion is assumed to be permissible for every part of the study area. A significant increase of the SG cover indicates that many new converted or degraded areas are formed under this scenario.

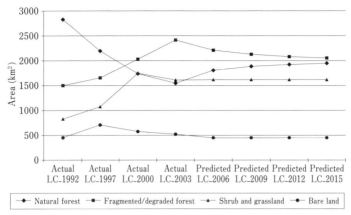

Figure 10-9. Predicted LULC area comparison I (Markov transition prediction and the best CA model by ranking rule)

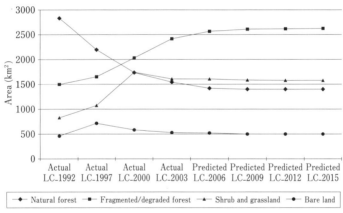

Figure 10-10. Predicted LULC area comparison II (Markov transition prediction and the best CA model by land allocation)

In the "rapid forest degradation" scenario, the FF cover increases faster than that in the other scenarios (see Figure 10–13). This trend emerges because forest conversion is assumed to be prohibited, whereas forest exploitation using the selective cutting system is implemented in the entire study area. By this assumption, the FF cover will grow faster either because of logging activity or vegetation regrowth. An interesting trend occurs where the NF cover decreases at first (in 2006) but then becomes constant and tends to increase very slightly. This trend implies that as long as good logging practice is implemented, the forest cover will be maintained in a continuous form.

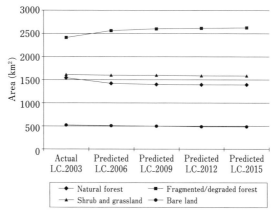

Figure 10–11. LULCC in the "Business-as-usual" Scenario

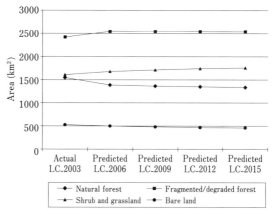

Figure 10–12. LULCC in the "Rapid deforestation" Scenario

An extremely progressive change is shown in the "rapid forest regrowth" scenario in which an unrealistic forest regrowth occurs within the study area (see Figure 10–14). As the protection forest area (HL) is relatively inaccessible, human influence is negligible. Under this condition, if there is any small opening or disturbance in the NF cover, the natural system will rapidly restore this disturbance to the NF cover. Though it seems unrealistic, this "rapid forest regrowth" provides a different perspective in which forested cover in a tropical environment may possibly be regained as long as human influence on the forest is minimal. In other words, if one considers securing the remaining tropical forest, major

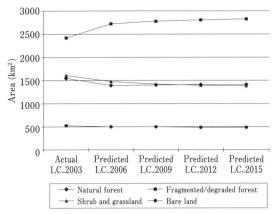

Figure 10–13. LULCC in the "Rapid forest degradation" Scenario

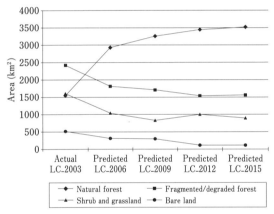

Figure 10–14. LULCC in the "Rapid forest regrowth" Scenario

human influence such as large development projects or large forest conversions into another land use (e.g., plantation estates, open mining, etc.) should be minimized.

Figure 10–15 illustrates an example of spatial results of allocating LULCC in a "business-as-usual" scenario. The results of the contiguity rule using a large neighborhood size (21 × 21) were clearly visible, particularly when one compares the actual land cover in 2003 to the predicted land cover in 2006. The "salt and pepper" areas were likely to first change the major land-cover category around their location. However, some patchy areas that were dispersedly located within the cloud mask mostly did not change because, to some extent, the neighborhood operation did not significantly affect their scattered location. In the end, the model yielded a more localized pattern of landscape where a clear boundary of land-cover categories could easily be defined.

4. Conclusion

Various methods of elaborating CA and probabilistic transition rule were undertaken for simulating land-cover changes in the Middle Mahakam area, West Kutai District, East Kalimantan Province, Indonesia. The generated probabilistic CA model with land allocation using a 21 × 21 neighborhood size and incorporating local transition probability at an 11 × 11 window size was selected as the best model for predicting LULCC. Cloud mask affects the reducing neighborhood effect for some patchy areas that were dispersedly located within the cloud mask. Consequently, these areas mostly did not change throughout time steps.

The scenario-based LULCC model yields interesting results. A "rapid forest degradation" scenario implied that as long as good logging practice was implemented, the forest cover would be maintained in a continuous form. In addition, a "rapid forest regrowth" scenario seemed unrealistic; however, it provides a different outlook in which a tropical forest may possibly be regained as long as human influence on the forest was minimal.

Future LULCC prediction was conducted using four land-cover categories with a mesh size of 90 m, and LULC maps were produced under four scenarios with different levels of human activities. The resultant maps were indeed visual information, but because of the coarse mesh size and the small number of land-cover categories, these maps may still be inadequate to convince local people when used in the group meeting. Certainly, the prediction model need to be revised to produce more realistic maps. Moreover, if simulated scenery images (photographs) can be produced based on LULC maps that can show more concrete imagery of landscapes, they can surely be utilized as more practical tools.

References

Alcamo, J., K. Kok, G. Busch, J. A. Priess, B. Eickhout, M. Rounsevell, D. S. Rothman, and M. Heistermann 2006. "Searching for the Future on Land: Scenario from the Local to Global Scale". In: E. F. Lambin and H. J. Geist (eds.) *Land-use and Land-cover Change: Local Processes and Global Impacts*. The IGBP Series. Berlin and Heidelberg: Springer-Verlag, pp. 137–155.

Angelsen, A. 1999. "Agricultural Expansion and Deforestation: Modelling the Impact of Population, Market Forces and Property Rights". *Journal of Development Economics,* 58: 185–218.

Barredo, J. I., M. Kasanko, N. McCormick, and C. Lavalle 2003. "Modelling Dynamic Spatial Processes: Simulation of Urban Future Scenarios through Cellular Automata". *Landscape and Urban Planning,* 64: 145–160.

Casson, A. 2001. "Decentralization of Policies Affecting Forests and Estate Forests and Estate Crops in Kutai Barat District, East Kalimantan". Bogor: Center for International Forestry Research (CIFOR).

de Almeyda, C. M., M. Batty, A. M. V. Monteiro, G. Câmara, B. S. Soares-Filho, G. C. Cerqueira and C. P. Pennachin 2003. "Stochastic Cellular Automata Modeling of Urban Land Use Dynamics: Empirical Development and Estimation". *Computers, Environment and Urban Systems*, 27: 481–509.

Eastman, J. R. 2006. "IDRISI Andes. Guide to GIS and Image Processing". Clark University.

Ménard, A. and D. J. Marceau 2006. "Simulating the Impact of Forest Management Scenarios in an Agricultural Landscape of Southern Quebec, Canada, Using a Geographic Cellular Automata". *Landscape and Urban Planning*, 79: 253–265.

Mertens, B. and E. F. Lambin 2000. "Land-cover-change Trajectories in Southern Cameroon". *Annals of the Association of American Geographers*, 90: 467–494.

Messina, J. P. and S. J. Walsh 2001. "2.5D Morphogenesis: Modeling Land-use Land-cover Dynamics in the Equadorian Amazon". *Plant Ecology*, 156: 75–88.

Pontius, R. G. and J. Malanson 2005. "Comparison of the Accuracy of Land Change Models: Cellular Automata Markov Versus Geomod". *International Journal of Geographic Information Science*, 19: 243–265.

Seavoy, R. E. 1973. "The Shading Cycle in Shifting Cultivation". *Annals of the Association of American Geographers*, 63: 522–528.

Singh, A. K. 2003. "Modeling Land-use Land-cover Changes Using Cellular Automata in a Geospatial Environment". Thesis, International Institute for Geo-Information Science and Earth Observation, Enschede, the Netherlands.

Torrens, P. M. and D. O'Sullivan 2001. "Editorial. Cellular Automata and Urban Simulation: Where Do We Go from Here?". *Environment and Planning*, B28, 163–168.

Veldkamp, A. and E. F. Lambin 2001. "Editorial. Predicting Land-use Change". Agriculture, Ecosystems and Environment, 85: 1–6.

Verburg, P. H., K. Kok, R. G. Pontius Jr and A. Veldkamp 2006. "Modeling Land-use and Land-cover change". In: E. F. Lambin and Geist H. J. (eds.) *Land-use and Land-cover Change: Local Processes and Global Impacts*. The IGBP Series. Berlin and Heidelberg: Springer-Verlag, pp. 117–135.

Verburg, P. H., T.C.M. de Nijs, J. R. van Eck, H. Visser and K. de Jong 2004. "A Method to Analyze Neighborhood Characteristics of Land-use Patterns". *Computers, Environment and Urban Systems*, 28: 667–690.

White, R. and G. Engelen 2000. "High-resolution Integrated Modelling of the Spatial Dynamics of Urban and Regional Systems". *Computers, Environment and Urban Systems*, 24: 383–400.

Wu, F. and C. J. Webster 1998. "Simulation of Land Development through the Integration of Cellular Automata and Multi-criteria Evaluation". *Environment and Planning B: Planning and Design,* 25: 103–126.

Yeh, A. G. and X. Li 2003. "Simulation of Development Alternatives Using Neural Networks, Cellular Automata, and GIS for Urban Planning". *Photogrammetric Engineering and Remote Sensing,* 69 (9): 1043–1052.

Zamyatin, A. and N. Markov 2005. "Approach to Land Cover Change Modelling Using Cellular Automata". In: *Proceedings of 8th AGILE Conference.* Estoril, Portugal, 587–592.

Chapter 11
Potential of the Effective Utilization of New Woody Biomass Resources in the Melak City area of West Kutai Regency in the Province of East Kalimantan

Masatoshi Sato

1. Introduction

This chapter is based on a research conducted regarding the social and technical frameworks for integrated forest resources management, such as the development of methods for developing and utilizing forest information, proposals for the local use of new woody biomass resources and local promotion measures for utilization of forest resources.

Though forest resources (tropical timber) have already been managed and utilized as important resources, the wooden agricultural waste (by-products) from rubber and oil palm plantations has also been proposed as a new biomass resource. Specifically, with regard to rubber, using the trunks after harvesting latex (after 15–20 years) is being considered. Rubber trees have been introduced in Indonesia (except in West Kutai Regency); therefore, a system for appropriate utilization of the resources needs to be created for implementation in this area. As oil palms are logged after 20–25 years, the trunks (bark, middle layers, and heartwood) can be used during these years. When palm oil is being harvested, most of the parts, including the fronds and fruit bunches (after oil extraction), are considered as usable biomass resources. However, at present, though the global use of oil palms is at the stage where they are being considered heuristically, a utilization method is yet to be developed. In the future, the use of this new woody biomass in this area is expected to be extended to entities that can enjoy the benefits, such as farmers supplying raw materials, cooperatives (among farmers and between farmers and enterprises) and corporations.

In this chapter, the potential of utilizing new woody biomass resources within this area will be discussed in light of the abovementioned situation.

2. Types and Volumes of Forest (Tropical Timber) Resources and New Woody Biomass Resources in Melak City, East Kalimantan Province, West Kutai Regency (Sato and Saito, 2008)

2.1 Forest (Tropical Timber) Resources

This chapter describes the state of timber resources in the area surrounding Melak City in West Kutai Regency, East Kalimantan Province, Indonesia, and the possibility of utilizing such resources.

As shown in Table 11–1, there are approximately 2.8 million ha of forest in the West Kutai area, which account for approximately 15% of all forests in East Kalimantan. Compared with other areas, West Kutai has a greater area of protection forests, with an extremely small area of parks and reserves and a comparatively greater area of conversion forests.

Most of the entities utilizing such forest resources are probably not individual farmers, but cooperatives or corporations.

Figure 11–1 shows forest area by land use in West Kutai. The data clearly depicts that though there was a slight decrease in conversion forest area from FY2003, it has been stable since 2004, with other forest types also remaining unchanged.

Next, as shown in Figure 11–2, which shows the annual area of reforestation and afforestation in East Kalimantan, both afforestation and forest estates showed a dramatic increase in FY2006.

Table 11–2 shows the number of enterprises involved in forest concessions and forest estates and the respective areas involved within the regencies and cities of East Kalimantan. As illustrated in the table, in West Kutai, the number and their area of enterprises involved in forest concessions is greater than those involved in forest estates.

Figure 11–3 provides a comparison of log production volumes in East Kalimantan and West Kutai. Though some fluctuation is evident in West Kutai, the log production volume is comparatively stable when compared with the figures for East Kalimantan as a whole.

Figure 11–4 shows the types and volumes of logs produced. They differ according to the production method used in East Kalimantan, which is the Indonesian selective logging and planting system (TPTI) and timber utilization permits (IPK).

Figure 11–5 shows that West Kutai has high production volumes of Meranti (*shorea spp.*), Keruing (*Dipterocarpus spp.*), Bengkirai (*Shorea spp.*), and Kapur (*Dryobalanops spp.*), and these are used for manufacturing plywood. In addition to these species, Sengon, Agathis (*Agathis spp.*), Mersawa (*A.marginat korth.*), and Ulin (*Eusideroxylon zwageri Teysm and Binn.*) are also produced. As seen in the figure, production volumes have been showing an overall tendency to decrease every year.

Table 11–1. Forest Area by Forest Land Use Consensus and Regency/Manucipality (2006)   (Unit: ha)

Regency/ Municipality	Protection Area Forest	Park and Reserve Forest	Production Forest		Conversion Forest	Total
			Limited production Forest	Definitive Production Forest		
1. Pasir	116,952	109,302	145,350	257,126	531,664	1,160,394
2. West Kutai	745,551	5,850	587,645	587,645	892,125	2,818,816
3. Kutai	213,959	11,621	507,614	781,762	1,073,009	2,587,965
4. East kutai	454,708	54,710	1,090,893	969,952	1,043,716	3,613,979
5. Berau	339,395	523,431	631,491	616,211	—	2,110,528
6. Malinau	708,647	—	1,624,356	447,910	—	2,780,913
7. Bulungan	167,748	—	493,583	461,769	542,199	1,665,299
8. Nunukan	157,855	—	299,762	113,215	470,914	1,041,746
9. Panajan P.U	21,496	—	—	42,282	483,935	547,713
10. Balikpapan	2,043	—	—	—	—	2,043
11. Samarinda	—	—	—	386	—	386
12. Tarakan	3,715	1,048	—	6,860	13,256	24,879
13. Bontang	5,230	720	—	1,141	5,248	12,339
Total	2,937,299	706,682	5,110,694	4,286,259	5,056,066	18,367,000

Source: Kalimantan Timur Dalam Angka (2007).

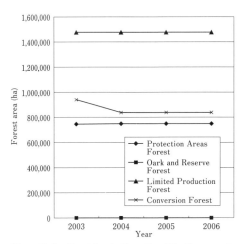

Figure 11–1. Forest Area by Forest Land Use Consensus in West Kutai
Source: Kalimantan Timur Dalam Angka (2007).

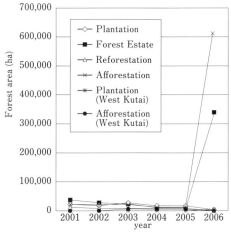

Figure 11–2. Reforestation and Rehabilitation of Forest Area (ha) in East Kalimantan and West Kutai
Source: Kalimantan Timur Dalam Angka (2006 and 2007).

Table 11–2. Number of Establishment and HPH - HTI Areas by Regency Municipality in East Kalimantan (2006)

Regency/City	Forest Concession (HPH)		Forest Estate (HTI)	
	Number of Companies	Area (ha)	Number of Companies	Area (ha)
1. Pasir	3	185,755	4	56,375
2. West Kutai	20	1,297,511	10	219,955
3. Kutai	5	321,035	67	483,531
4. East kutai	18	1,550,123	8	185,230
5. Berau	11	608,806	5	22,400
6. Malinau	12	1,039,530	—	—
7. Bulungan	7	569,485	1	5,000
8. Nunukan	3	105,080	1	25,000
9. Panajan P.U	—	—	2	83,134
10. Balikpapan	5	485,903	3	48,853
11. Samarinda	—	—	—	—
12. Tarakan	—	—	3	799,651
13. Bontang	—	—	—	—
Total	84	6,163,228	104	1,929,129

Source: Kalimantan Timur Dalam Angka (2007).

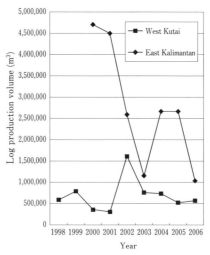

Figure 11–3. Log production volume in West Kutai and East Kalimantan
Source: Kalimantan Timur Dalam Angka (2006).

295

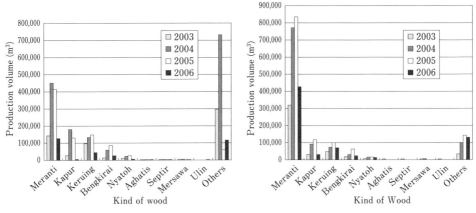

Log Production Volume (IPK:Izim Pemanfaatan Kayu, Indonesia)   Log Production Volume (TPTI:Tebang Pilih Tnm, Indonesia)

Figure 11–4.   Log Production by Kind of Wood in East Kalimantan
Source: Kalimantan Timur Dalam Angka (2006).

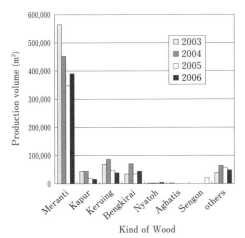

Figure 11–5.   Log Production by Kind of Wood in East Kalimantan
Source: Kalimantan Timur Dalam Angka (2006).

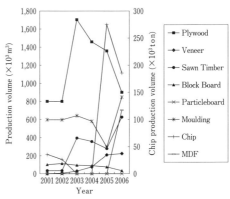

Figure 11–6.   Production of Processed Logs by Kind of Product and Regency
Source: Kalimantan Timur Dalam Angka (2007).

Table 11-3. Producution of Processed Logs by Kind of Product and Regency/City in East Kalimantan (2006)

Regency/City	Plywood (m³)	Venner (m³)	Swan timber (m³)	Block board (m³)	Particleboard (m³)	Chip (ton)	Moulding (m³)	MDF (m³)
1. Pasir	–	–	–	–	–	–	–	–
2. West Kutai	–	–	–	–	–	–	–	–
3. Kutai	–	162,461.83	158,534.99	–	–	–	–	116,520.85
4. East kutai	5,139.27	–	444,976.73	–	·	–	–	–
5. Berau	–	–	–	–	–	–	–	–
6. Malinau	–	–	–	–	–	–	–	–
7. Bulungan	–	–	–	–	–	–	–	–
8. Nunukan	–	–	–	–	–	–	–	–
9. Panajan P.U	–	–	–	–	–	–	–	–
10. Balikpapan	234,677.84	–	21,259.49	17,247.73	–	70,360.80	53,813.52	–
11. Samarinda	490,343.07	24,271.74	–	14,264.52	–	–	26,266.92	–
12. Tarakan	166,208.43	33,041	–	–	–	114,750.00	60,913.93	–
13. Bontang	–	–	–	–	–	–	–	–
Total	896,396.61	219,774.09	624,771.21	31,512.25	–	185,110.80	140,994.37	116,520.85

Source: Kalimantan Timur Dalam Angka (2007).

Table 11–3 shows production volumes according to wood-based product type in East Kalimantan. The production volumes for plywood, including veneer, are especially notable.

Figure 11–6 presents annual production volumes of forest products for East Kalimantan, which have shown a tendency to decrease after reaching a peak in 2003. In contrast, there has been a noticeable increase in the production volumes of chip, interior materials, and medium density fiberboard (MDF). Wood-based materials are not produced in West Kutai.

2.2 New Woody Biomass Resources

Non-wood forest products

The non-wood forest products in East Kalimantan and West Kutai are shown in Tables 11–4 and 11–5, respectively. Non-wood forest products include pine resin, rattan, bird's nests, agarwood, reptile skins, roofing materials, and bark. The main non-wood forest products in East Kalimantan are rattan, bird's nests, and roofing materials, whereas only rattan is produced in West Kutai, and its production volume is approximately 4–9% of that in East Kalimantan.

As it is assumed that distribution channels have already been established for these non-wood forest products, entities such as corporations or cooperatives are considered to be already benefitting from them, whereas farmers as producers receive only a portion of the profits.

Though, at present, these resources are not utilized anywhere, even outside Kalimantan, it should be possible to effectively utilize the original materials (wood) from which non-wood forest products such as pine resin, agarwood, roofing materials, and bark, are made as new woody biomass resources. Moreover, it is thought that farmers will mostly be the entities that produce and profit from the use of such resources. However, depending on the situation, there will be cases where it is possible for the farmers themselves to jointly produce products and cases where production is impossible without the cooperation of corporations. In West Kutai, in terms of non-wood forest products, only rattan is currently produced, and it is not used as a new woody biomass resource.

Plantation Crops (Sato et al., 2009)

Other than timber, the biomass resource that exists in the greatest volume is agricultural crops grown in plantations. The main plantation crops, the area under cultivation, and the production volumes in each of the regencies/cities of East Kalimantan are shown in Table 11–6. As seen in the table, the agricultural crops with the greatest area are oil palms, cocoa, rubber, coconuts, coffee, and pepper in descending order. Among these, oil palms and rubber especially stand out. In West Kutai, rubber accounts for a greater production area than oil palms.

Table 11–4. Production of Minor Forest Product in East Kalimantan

Non-wood Forest Products	Year production volume (ton)				
	2002	2003	2004	2005	2006
Pine resin	—	—	—	—	—
Rattan	1,543.70	1422.98	1671.03	1671.03	2537.63
Bird nest	—	—	—	—	2,514
Agarwood	—	—	—	—	—
Reptiles skin/Leather	—	—	—	—	—
Roofing material	—	42.72	42.72	42.72	—
Bark	—	—	—	—	—
Others	—	—	—	—	359

Source: Kalimantan Timur Dalam Angka (2006).

Table 11–5. Production of Minor Forest Product in West Kutai

Non-wood Forest Products	Year production volume (ton)	
	2005	2006
Pine resin	—	—
Rattan	153.4	93.77
Bird nest	—	—
Agarwood	—	—
Reptiles skin/Leather	—	—
Roofing material	—	—
Bark	—	—
Others	—	—

Source: Kalimantan Timur Dalam Angka (2006).

Some plantation crops are already utilized outside the area as woody biomass resources. For example, rubber trees are used for furniture, and coconut husk and coconut fibers are used for palm-shell activated carbon (deodorant), scrubbing brushes, and house mats. Rubber and oil palm production volumes are increasing in the area, and it is hoped that new methods will be found to use the large volume of resulting biomass as a resource. By doing so, it may be possible to obtain profits from resources other than crude rubber and palm oil, which will open the door to new avenues of deriving income for farmers. However, as installation of manufacturing machinery, production, and sales are required to produce products, it is difficult for individual farmers to initiate such utilization. Thus, the participation of cooperatives (farmers and farmers/corporations) or corporations will probably be essential.

Table 11–6. Planted Area (ha) of Estates by Type of Crops and Production Volume (tons) in East Kalimantan (2006)

Regency/City	Rubber	Coconut	Coffee	Pepper	Cloves	Cocoa	Oil palm	Others
1. Pasir	7,016.0	4,161.0	3,016.0	188.0	2.0	947.0	65,918.5	513.0
	6,761.0	3,570.0	1,238.5	31.0	—	59.0	596,129.0	313.0
2. West Kutai	31,076.5	1,332.0	1,287.5	87.0	—	411.0	5,371.0	2,346.0
	28,184.5	234.0	378.5	8.0	—	24.5	6,928.0	1,528.5
3. Kutai	17,533.5	13,326.5	3,870.5	10,408.0	124.5	2,463.0	37,517.5	2,042.0
	4,828.5	5,943.5	1,198.0	7,249.5	9.0	370.0	252,740.0	419.5
4. East Kutai	659.5	5,686.5	1,229.5	445.0	—	12,814.5	51,599.0	1,076.0
	25.0	2,015.0	203.5	47.5	—	3,499.0	260,162.0	566.5
5. Berau	751.5	11,695.0	2,240.0	908.5	53.0	6,190.0	11,479.0	920.0
	45.0	12,386.0	578.0	530.5	2.0	3,955.0	—	261.0
6. Malinau	—	382.5	1,335.5	110.0	53.0	3,289.5	—	123.0
	—	210.0	597.0	12.0	8.5	720.5	—	30.0
7. Bulungan	54.0	1,596.5	269.5	52.0	—	829.0	2,215.0	175.0
	—	2,669.0	32.0	10.5	—	193.0	—	765.0
8. Nunukan	—	2,685.5	3,560.0	187.0	22.5	13,038.0	34,217.5	248.5
	—	7,458.5	214.0	14.0	0.5	17,702.0	165,500.0	105.0
9. Panajan P.U	5,177.5	3,331.0	188.5	2,015.5	1.0	257.5	16,829.5	191.5
	2,896.0	2,629.0	98.5	1,935.0	—	119.0	144,130.5	127.5
10. Balikpapan	1,845.0	1,702.0	19.0	122.0	4.0	33.0	—	237.0
	823.0	4,495.5	18.0	79.5	1.0	17.5	—	227.5
11. Samarinda	838.5	1,011.6	363.0	242.5	10.5	984.2	205.0	661.5
	282.0	860.0	32.5	40.5	—	71.5	—	121.0
12. Tarakan	—	724.0	5.0	2.0	2.0	—	—	102.0
	—	1,526.0	1.5	1.0	—	—	—	—
13. Bontang	5.0	100.0	2.0	0.5	—	50.0	—	106.0
	—	701.0	22.5	—	—	43.0	—	40.5
Total	64,957.0	47,734.0	17,409.0	14,768.0	253.0	41,307.0	225,352.0	8,741.5
	43,845.0	44,697.5	4,612.5	9,959.0	21.0	26,774.0	1,425,589.5	4,505.0

Upper stage: Area, Lower: Production Volume
Source: Kalimantan Timur Dalam Angka (2007).

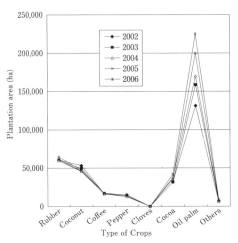 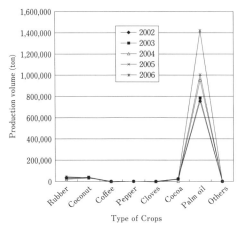

Figure 11-7. Planted Area of Estates by Type of Crops in East Kalimantan
Source: Kalimantan Timur Dalam Angka (2007).

Figure 11-8. Production Volume of Estates by Type of Crops in East Kalimantan
Source: Kalimantan Timur Dalam Angka (2007).

The annual plantation area and production volumes for the main crops under cultivation in East Kalimantan are shown in Figures 11-7 and 11-8. From these, it is clear that the plantation area and production volume of oil palms is showing an annual tendency to increase compared with rubber. On the other hand, compared with oil palms, rubber is stable in terms of both plantation area and production volume. There also seems to be little change for other crops.

3. Present State of New Woody Biomass Resources from Rubber Trees and Oil Palms in Melak City (Sato et al., 2009)

3.1 Rubber Tree Plantations

The state of rubber and oil palm plantations in West Kutai is shown in Table 11-7. As is evident, rubber tree plantations are small-scale plantations run by farmers, whereas oil palm plantations are large-scale plantations run by corporations.

The results of interviews shown in Table 11-8 reveal that most of the rubber plantations in Melak City belong to farmers (total area of 11,200 ha (including areas that cannot be harvested), with 200 ha of trees that are 20 years or older and 10,000 ha of trees younger than 20 years) and that, at present, there are 200 groups (20-30 ha per group), with an average landholding of 1-2 ha and a total harvestable area of 3,012-5,000 ha.

Table 11–7. Actual Situtation of Rubber and Oil palm Plantation in West Kutai (2006)

Products	Large-scale Plantation by Large Companies		Small-scale Plantation by Farmers	
	Area (ha)	Production Volume (ton)	Area (ha)	Production Volume (ton)
Rubber	—	—	31,076.50	28,184.50
Oil palm	5,371	6,928	—	—

Source: Kalimantan Timur Dalam Angka (2007).

Table 11–8. Actual Situation of Rubber Plantation in Melak (2006)

Total Area	11,200 ha	
Plantation period of rubber (Year of Tree)	First stage (1981–1983: More than 20 years) Area: 1,200 ha	Second stage (1992–1997: less than 20 years) Area: 10,000 ha
Space of planting	Standard 4 × 5 m (3 × 5 m, 3 × 3 m)	
Harvestable rubber plantation area	3,012–5,000 ha (Increased to 19,000 ha in 2012)	
Farmer's group number	200 groups	
Average area owned per group	20–30 ha	
Average area owned per farmer	1–2 ha	

Table 11–9. Actual Condition of Rubber Palntation in Melak (2006)

	A	B	C
Area of rubber Plantation	48 ha (2 Groups, 1 ha/Farmer)	First stage (from 1982) 36 ha, Second stage (from 1995) 474 ha, 3–9 ha/Farmer: 100 houses	69 ha (About 10 ha/Farmer)
Forest management	Weeding: once every 6 months, Fertilizer supply: from afforestation to 4–5 years, Peeling bark: From Monday to Friday, Collection of latex: Saturday/Sunday	Weeding: once every a month, Fertilizer supply: from afforestation to 5 years, Peeling bark: Monday to Friday, Collection of latex: Saturday/Sunday	—
Price of seedling	3,500 Rp/Seedling without root, 7,000 IDR/Rooted seedling		
Collection volume of latex	Ave. 280 kgs/month/ha (with fertilizer supply: 320–440 kgs/month/ha, without fertilizer supply: 160–280 kgs/month/ha)	First stage: Ave. 700 kgs/month/ha, Second stage: Ave. 500 kgs/month/ha	Ave. 320 kgs/month/ha
Price	Common products: 7,000–9,000 R/kg		Quality products: 10,000 IDR/kg

The state of rubber tree plantation cultivation in Melak City is shown in Table 11-9. This can be classified into three different groups (A, B, and C). The area owned in each case is 1 ha/farmer (A group), 3-9 ha/farmer (B group), or approximately 10 ha/farmer (C group).

Though rubber processing is conducted in South Kalimantan, a company built a plant with Filipino investment in Harjo Basuki village (monthly production capacity of 900 tons), which was due to commence operations in mid-2008. However, after 20 years, the ability of rubber trees to produce latex declines, and no method to utilize these trees after logging has been suggested in Melak City. Moreover, because logging these trees costs money, there are many cases of rubber tree plantations being left to revert to natural forest.

In contrast, in other Southeast Asian counties, rubber trees are logged after 15-20 years and used as raw materials for furniture, etc. Rubber trees can usually be utilized as normal timber and processed in local sawmills. Through negotiations with farmers, logging companies, and the government (Ministry of Agriculture and Ministry of Forestry), it is possible to formulate a system for the utilization of rubber and develop applications for this resource. Presently, there are many abandoned rubber plantations, which means that there are large volumes of waste biomass. Furthermore, as the abandoned rubber tree plantations will revert to natural forests, this poses problems from the perspective of biological diversity.

3.2 Oil palm Plantations

Oil palm plantations in the area surrounding Melak City include PT.Kedap Sayaq Dua (presently 278 ha and scheduled to expand to 6,000 ha by 2010 and 20,000 ha by 2040), which is within Melak City, and PT.Lonsum (based in Sumatra and one of the largest companies in Indonesia), which is located between Melak and Samarinda with a planted area of 5,000 ha and a total area of 43,000 ha (see Table 11-10). As there is no manufacturing plant in Melak City, crude palm oil production is conducted within 24 hours of harvest in South Kalimantan. However, PT.Lonsum was building a manufacturing plant with a processing capacity of 45 tons/hour (to be increased to 60 tons/hour in the future) in 2006. Harvesting of fruit from oil palms begins three years after planting, with the volume of oil produced reaching its maximum after 8-15 years before the trees are finally logged at 20-25 years. At present, the development of applications and technology to utilize the woody biomass resources of oil palms when they reach an age of 20-25 years is behind schedule. This is a problem in not only Indonesia but also Malaysia, which ranks second globally in palm oil production. In Malaysia, development of technology to effectively utilize waste biomass, such as the fronds (5-6 million tons/year), trunks (3-4 million tons/year: bark, middle part, and heartwood), leaves, and fruit bunches (20 million tons/year after oil extraction), from oil palms has only just begun.

Table 11–10. Actual Situation of Oil palm Plantation in Melak (2006)

Company	PT. KEDAP SAYAQ DUA	PT. PP. LONSUM
Start of Planting Time	2006	1994
Plantation Area	Total area: 25,000 ha, Present plantation area: 278 ha, by 2010: about 6,000 ha, by 2040: about 19,000 ha	Total area: 43,000 ha, Present area: 5,000 ha
Planting Space	9 × 7.8 m	9 × 9 m
Seedling Price	6,000 IDR/seeding	Home grown
Plantation Cost	30,000,000 IDR/ha to four years from planting (Seedling cultivation (13 months): 20%, Planting/Maintenance/Fertilizer supply/Spraying pesticide: 80%)	Seedling cultivation: 20%, Planting/Maintenance/Fertilizer supply/Spraying pesticide: 80%
Pruduction Volume of Fruit Bunch (FB)	2–5 years: 8 tons/ha/year, 7–15 years: 32 tons/ha/year, Production decreases from 15 years, and trees will be felled in 25 years	Collection of FB from three years of planting, ave. 16 tons/ha/year (24 tons/ha/year in Sumatra), 3 year: 6–8 tons/ha, 12 years: 16–20 tons/ha (30–40 tons/ha in Sumatra), Maximum yield in 8–15 years, Felling in 25–30 years

4. Utilization of New Woody Biomass Resources (Rubber Trees and Oil palms) in Melak City

4.1 Utilization of Rubber Trees

Rubber trees are already being used as sawtimber throughout Southeast Asia, including Indonesia (apart from Melak City). Therefore, if farmers or the relevant government department can establish rules in Melak City for the utilization of rubber trees that have been logged 15–20 years after latex has been harvested and the plantation is going to be replanted, they can be used in sawmills and furniture factories within the city. The price of rubber tree logs in Vietnam is cheaper than that of other timber, as can be seen in Table 11–11 (Sato, 2009).

Rubbers are currently used as timber for furniture, used in plywood and particleboard, and used as a substrate for growing mushrooms. It is possible to manufacture these wood-based materials in Melak City. As shown in Table 11–3 earlier, although there are sawmills, etc. in West Kutai, there are no plants for manufacturing wood-based materials. Thus, to enable such manufacturing, raw materials must be supplied to plants in Samarinda or other plants in the vicinity that can manufacture wood-based products. Apart from the need for a building for housing a plywood or particleboard manufacturing plant, the machinery shown in Tables 11–12 and 11–13 is necessary (Sato and Okuda, 2009). The machinery shown is

Table 11–11. Wood Species, Size, and Price of Log used in Wood Industry (2005)

	Diameter (cm)	Length (m)	Price (US$/m³)
Acasia	20	Less than 2 m	53–60
Rubber	20	1	36
Apitong (Keruing)	20	2	245
Kapur	30	2–3	267–333
Melaleuca	7	4	85

Table 11–12. Production Machine list for Particleboard (2005)

	Machine	Number	Price (US$)	Subtotal (US$)
01	Chipper	4	1,867	7,468
02	Chip screening machine	1	67	67
03	Dryer	2	2,000	4,000
04	Blender	2	1,000	2,000
05	Hot press	1	37,000	37,000
06	Trim saw	1 set	2,333	2,333
07	Sanding machine	1	10,000	10,000
	Total			62,868

Notes: 1) Transportation fee, human resource development fee, and Technology transfer fee: USD 667–1,000
2) Salary for person who conducted Technology transfer: USD 667–1,000
3) Production capacity: 10–12 m³/day
4) Electricity charge: 0.15 USD/particleboard

Table 11–13. Production Machine list for Plywood (2005)

	Machine	Number	Price (USD)	Subtotal (USD)
01	Rotary veneer lathe 2.6 m	1	9,000	9,000
02	Rotary veneer lathe 1.3 m	1	4,333	4,333
03	Rotary veneer lathe 0.6 m	1	2,000	2,000
04	Rotary veneer lathe 0.55 m	1	1,000	1,000
05	Hot press (Six piece press specification)	1	23,333	23,333
06	Boiler	1	4,000	4,000
07	Trim saw	1 set	1,667	1,667
08	Automatic belt sander (*)	1	3,333 (10,000*)	3,333 (10,000*)
09	Dust collector with filter expression	1	667	667
10	Blade sharpener	1	267	267
11	Glue mixer	1	333	333
12	Glue/Adhesive	1	533	533
	Total			50,466 (57,133*)

Notes: 1) Human resource development fee and Technology transfer fee: 6,667 USD
* In the case of belt sander: 10,000 USD

Table 11-14. Production Machine list for Saw Mill (2005)

	Machine	Number	Price (US$)	Subtotal (US$)
01	Carriage type twin circular saw machine	1	1,000	1,000
02	Band saw machine	1	567	567
03	Circular saw machine	1	400	400
04	Trimmer	1	100	100
05	Planer	2	867	1,734
06	Machine planning	2	667	1,334
07	Sanding machine	1	5,000	5,000
08	Others			1,333
Total				11,468

based on an example from Vietnam. Though there are already sawmills in Melak City, for further reference, see the machinery required in Table 11-14 (Sato and Okuda, 2009). Such machinery is the minimum required in Indonesia and their prices differ according to the state of affairs within the country.

In Melak City, negotiations are currently being conducted between sawmill owners and farmers regarding the price of rubber tree logs. If a consensus can be reached between both parties regarding the utilization, then as there already are sawmills and furniture factories there, it will be possible to utilize such logs on a temporary basis even without the government formulating a system for their utilization. When seeking to use rubber logs as a certain resource, it is necessary to establish the volumes that can be supplied because of the relationship between the reforestation volume and logging cycle of the rubber tree plantations. At present, individual farmers arbitrarily harvest latex and manage the rubber tree plantations themselves. Thus, it will probably be necessary for each area's producing farmers to manage the overall production volume at a certain level if it is to be used as a timber resource.

4.2 Utilization of Oil palms (Ando et al., 2010; Sato et al., 2011; Ando et al., 2010)

Oil palms are not being used as a biomass resource in Melak City. In fact, there are few examples of such use globally, and there are still many facets that need to be researched.

Considering Malaysia, which is a pioneer in the utilization of oil palms, though the trunks are already being made into plywood and packing materials for shipping, approximately two to three times the normal amount of adhesive is required compared with normal plywood and reducing this volume is a pressing challenge. After extracting the palm oil, the empty fruit bunches are used as fuel or raw material for pulp.

Therefore, in terms of waste biomass resources from oil palms, there are the fronds, which are cut off when harvesting the fruit, the trunk, which is logged after 25–30 years, gas expelled from the trunk in the ethanol manufacturing process (which is currently being investigated), and empty fruit bunches. Ways of utilizing the fronds, which are currently discarded in the plantation (see Plate 11–1), have not yet been considered. Thus, it is conceivable that every part of the oil palm could be utilized, from the fronds to the trunk (bark, middle layers and heartwood), and fruit bunches (after oil extraction).

The possibility of developing binderless boards using the abovementioned techniques is currently being explored. This particularly includes utilizing waste biomass in binderless board and plywood as one of the technologies for manufacturing functional materials using a high temperature/high pressure hot press without the use of synthetic resin adhesives derived from petrochemicals. Compared with other boards, binderless board manufactured using the bark and fronds has lower internal bonding strength and is assumed to contain factors that hinder self-bonding; thus, it has been acknowledged that such factors need to be further analyzed. Regardless of the particle size of the powder, essentially same degrees of bonding shear strength was exhibited by plywood made from using powder from the trunk of oil palms as adhesive and plywood made using urea resin adhesive or phenol resin adhesive. Although plywood made using powder from the trunk with the smallest particle size of 10 μm exhibited the strongest bonding shear strength under wet conditions, it was concluded that it was necessary to improve the water resistance of the plywood because it was not as good as that of plywood made using urea resin adhesive. Of the trunk, bark, fronds, and leaves, the most suitable raw material for making plywood was powder from the trunk, and the most suitable adhesive was oil palm veneer with a density of 0.3 to 0.5 g/cm3. Furthermore, one of the methods being considered for development using the trunk is compressed wood. With this method, the density of the surface can be increased by hot pressing the surface of the material to improve surface hardness or bending properties, making it suitable to use in many applications, such as flooring, furniture, and glued laminated timber. Incidentally, in Japan, compressed wood using low density Japanese cedar is being sold as flooring. Suggestions were made regarding the possibility of developing new materials using this concept (see Plates 11–2-11–13).

From the above results, it has been suggested that it may be possible to develop new materials in the future using oil palm trunks, such as binderless board, plywood, and composite materials.

However, a major problem that requires further technological development is the lack of clarity regarding the particular parts of oil palms that can be used as waste biomass resources and the manufacturing technology and applications for materials to be manufactured using such parts. Therefore, it is thought that to use these oil palms as a waste biomass resource in Melak City, the direction of technological development aimed at utilizing them

307

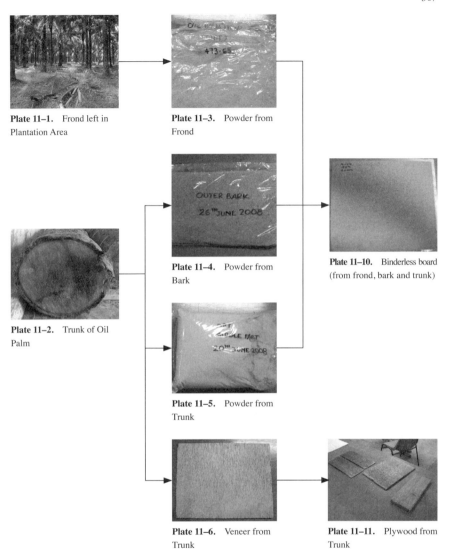

Plate 11–1. Frond left in Plantation Area

Plate 11–3. Powder from Frond

Plate 11–2. Trunk of Oil Palm

Plate 11–4. Powder from Bark

Plate 11–10. Binderless board (from frond, bark and trunk)

Plate 11–5. Powder from Trunk

Plate 11–6. Veneer from Trunk

Plate 11–11. Plywood from Trunk

Plate 11–7. Residues after Squeezing Sap

Plate 11–12. Production of Oriented Board and Boards

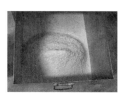

Plate 11–8. Powder of Residues after Squeezing Sap

Plate 11–13. Composite Board (after Forming) and Board

Plate 11–9. Veneer from Trunk

should be considered in light of future research results. On the other hand, with regard to the afforestation of oil palms in West Kutai, plantations are mainly under the umbrella of corporations and the use of waste biomass in terms of internal corporate policy is certainly possible because such utilization is influenced by economics (corporations). If the utilization of waste is shown to be economically viable, then there is a high possibility the utilization of waste biomass can be implemented.

5. Conclusion

This chapter has reported the possibility of effectively utilizing new woody biomass resources in Melak City in West Kutai Regency in Indonesia's East Kalimantan Province and the effectiveness of using rubber trees and oil palms as new woody biomass resources. Rubber trees especially are already being utilized elsewhere outside this area. Therefore, as there are necessary resource management structures and a market for such products, introducing such systems will make moving toward the utilization of such woody biomass resources relatively easy. As capital is required for things such as the purchase of manufacturing machinery, the involvement of cooperatives (farmers or farmers and corporations) or corporations is considered necessary because it would be difficult for individual farmers to develop the necessary production and sales systems.

On the other hand, as technological development aimed at the utilization of oil palms as a woody biomass resource is currently being implemented, it will be necessary to consider their future utilization on the basis of such results.

Acknowledgement

This research is partially supported under *Conditions for the Creation of Forest Collaborative Governance that gives Consideration to Local Characteristics* (FY2007/2010), scientific research sponsored by (Foundation A).

References

Ando, M., M. Sato, T. Sugimoto, R. Hashim 2010. "Unused Biomass in Indonesia and Malaysia". Abstract of the 60th annual meeting of the Japan Wood Research Society, R18–1415.

Ando, M., M. Sato, T. Sugimoto, R. Hashim, and O. Sulaiman 2010. "Manufacture and Properties of Plywood from Oil Palm". *Wood Industry*, 65(6): 261–265.

Sato, M., and Y. Saito 2008. "Unused Biomass in the Area of West Kutai Regency in the Province of East Kalimantan". Abstract of the 58th annual meeting of the Japan Wood Research Society, R19–1130.

Sato, M., M. Ando, S. Matsuba, F. Kawamura and R. Hashim 2009. "Unused Biomass in the

Area of West Kutai Regency in the Province of East Kalimantan (2)". Abstract of the 59th annual meeting of the Japan Wood Research Society, R15–1745.

Sato, M. 2009. "Utilization of Woody Biomass in Southeast Asia: Melaleuca in the Mekong Delta Region". *Journal of Agricultural Development Studies*, 20(1): 15–21.

Sato, M. and N. Okuda 2007. "Development of the Utilization Technology for Melaleuca Wood (3)". Abstract of the 57th annual meeting of the Japan Wood Research Society, R09–1330.

Sato, M., M. Ando, T. Sugimoto, and R. Hashim 2011. "Possibility of Utilization of Oil Palm as a Material for Wood-Based Materials". Conference proceedings of WCTE2011, ID/paper No. 202.

Final Chapter

Multifaceted Significance of Collaborative Governance and Its Future Challenges

Makoto Inoue

1. Definition and Design Guidelines of Collaborative Governance

Collaborative governance is a "mechanism for the administration of the environment and resources realized through the solidarity and collaboration of a range of diverse stakeholders centered around local residents" (Inoue, 2004). In tropical regions, there is a strong connection between the livelihoods of local people and the forest. However, because the so-called "citizens" from outside the area are only interested in certain functions of the forest, such as biodiversity, there is a gap between the locals and the outsiders. A consensus must thus be reached, and this is where the concept of collaborative governance comes into the picture.

Guidelines have been provided for creating systems to realize the concept of collaborative governance (Inoue, 2009). These guidelines were developed by examining the design principles presented by Ostrom (an American political scientist and economist) and her peers (Ostrom, 1990; McKean, 1999; Stern et al., 2002; Ostrom, 2005). The following are two of the most important design guidelines for collaborative governance (refer to the Appendix).

The first guideline is "graduated membership". This is the design guideline that stems from the concept of "open-minded localism" (Inoue, 2004). It begins with local people opening up to well-intentioned outsiders, rather than isolating themselves. However, in such cases, the outsiders may end up having more power. The concept here is to establish a graduated membership where responsibilities and obligations extend from the strongest core members (first-class members) to the weakest members (second and third class members).

The second guideline is the "commitment principle". This is a design guideline stemming from the "principle of involvement" (Inoue, 2004). It is important to respect the opinions of those who are doing their best to manage the forests (people with strong involvement and commitment) and give them importance during decision making. However, the decision making right of people who only talk about things and make no active contribution is comparatively insignificant.

Incidentally, from the viewpoint of local people, who are important actors in the utilization and management of the commons, three strategies with regard to the involvement of outsiders can be conceived (Inoue, 2010a). The first is "resistance strategy" whereby local people exclude and reject outsiders' involvement. The second is "adjustment strategy" whereby local people act in response to initiatives by outsiders. The third eclectic strategy is "collaborative governance strategy" that is the middle ground between the other two strategies. In "collaborative governance strategy", people who are deeply committed according to the "commitment principle" are highly regarded, whereas outsiders are accepted in line with "graduated membership". However, the total influence of outsiders is maintained at less than 50%. This is because once total influence exceeds 50%, it becomes an "adjustment strategy" rather than a "collaborative governance strategy".

Therefore, we can see that "collaborative governance strategy" is an "adjustment strategy" with limitations, while also partially a "resistance strategy". That is, resistance to enemies is continued, while cooperation and collaboration are extended to allies. However, in this report, activities placed under "adjustment strategy", which is separate from "collaborative governance strategy", will also be considered, which will be useful when considering how to bring "adjustment strategy" under the banner of "collaborative governance strategy".

In light of the arguments regarding sharing in other chapters, this chapter presents future challenges regarding the role of the government and the three aspects[1] of collaborative governance.

2. Role of Government in Collaborative Governance

With regard to forest utilization and administration, there is always a sense of discomfort about the concept of "governance without government" proposed by people such as Rhodes (1997). However, there is a sense of discomfort about the citizen participation theory, which tends to accord little importance to the local residents. In terms of gaining an overall picture of the main stakeholders, the local people form the base, above whom are three sectors: government, market, and civil society. With respect to debate, policies, and action, the base should be always the local resident. Conflict arises if the government (or governance) alone becomes strong or if citizens and enterprises become stronger than the local resident. Governance refers to how to set, apply, and execute rules (Kjaer, 2004). Though there may be some variation in governance according to differences in the stakeholders' involvement,

---

1) This chapter is a revision of Inoue (2012) that has been made in light of the concept of sharing and results of this project.

with regard to forest governance, in addition to the base of local people, the government also plays an important role.

The role of government in the collaborative governance of forests should also be considered. The first role of government is to firmly establish constitutional rules—one entails the three rules of commons theory[2]. In other words, the role of establishing a framework to which a diverse range of stakeholders can commit. If policies that do not acknowledge the commitment of a diverse range of stakeholders are adopted, then interested people living in cities and local residents of rural villages cannot be expected to participate as agents of collaborative governance. As clarified in the chapter written by Yokota, which is an example of the management of manmade forests on the island of Java, Indonesia, if authority is delegated in a limited fashion, residents will automatically adopt an adjustment strategy or respond in a reluctant manner, thus rendering the authority ineffective and inadequate in terms of improving their livelihoods. In Hyakumura's chapter, which considered the same type of adjustment strategy in afforestation activities in Laos, the possibility of land being controlled by certain residents who were able to participate in afforestation activities on a contract or community group basis together highlighted the fact that it was necessary for the local government to monitor the situation.

Hence, though the government's first role is to establish constitutional rules, the above examples suggest that the government's role should not be limited to this aspect. That is, within the scope of the constitutional rules set at the initiative of the government, to realize the fair utilization of resources by stakeholders other than itself and in terms of collective choice rules and operational rules, there is still room for the government to fulfill its role without infringing on the rights of the major players. The issue of how the government's role may best be concretely defined without upsetting the foundation of local autonomy will be an important theme for future consideration.

The second role of government concerns the creation of a green safety net. Assuming that residents have decision-making rights and they decided to convert a major part of the forest into a golf course or an oil palm plantation, the forest would completely disappear. Is such a thing acceptable from the perspective of biological diversity? Thus, some check-

---

2) The three rules of commons theory are as follows. (1) Operational rules: Rules that directly regulate the actions of individuals, for example, regulations regarding things that people may do, may not do, and should do. (2) Collective decision-making rules: Rules for determining how to formulate operational rules. These rules decide who will create what type of rules, and whether it is necessary to make revisions. (3) Constitutional rules: foundational rules in political systems, i.e., these are necessary procedures for creating new units for governance and formulating and changing collective decision-making rules. For example, regulations regarding who can participate in political systems, the role of such systems, the selection of people engaged in that role, and the power and authority of such people.

ing mechanisms are required for securing environmental national minimums. For the collaborative governance of forests to achieve legitimacy in the eyes of society at large, though not only the participation of a diverse range of stakeholders (ensuring democracy) but also the actual achievement of sustainable forest management (ensuring effectiveness)—and economically, if possible, its rationalized management (securing efficiency)—is sought, the involvement of government in the form of development restrictions and subsidies is essential.

In relation to this, in their chapter, Sato and Fujiwara considered the conditions for collaborative governance using research involving two case studies in Japan, in which artificial forestation is quite developed. Their results led them to the conclusion that (1) on the national level, introduction of policies to stabilize economic conditions such as the price of timber were important indicators for residents; (2) on the local level, fostering of forests that could be used in various ways in response to changes in demand that occur over time was necessary; (3) even for privately owned forest, from the perspective of environmental conservation, establishing usage restrictions was necessary rather than depending on economic incentives; and (4) it was important for local residents to continue living in rural villages in Japan where underutilization was a problem because of rural depopulation. In other words, green safety net policies must be introduced together with social safety net policies (rural depopulation countermeasures) and economic safety net policies (stabilization of economic conditions).

The third role of government involves displaying standard expertise. Kakizawa, in his chapter, pointed out the necessity of fostering foresters (forest administrators/forestry technicians) who play a central role in local forest administration in Japan. Foresters are expected to fulfill the role of coordinating the establishment of relationships among local residents, citizen groups, forest owners, business entities, and on-site technicians. This scenario will probably be played out in various places, and when that happens, consideration will have to be given to the dual role of foresters as administrators and technicians and the factors distinguishing them from general administrators and on-site forestry technicians.

3. Collaborative Governance as Resource Management Theory/Ownership Theory

Sociologist Yoshida (1981) presented the concept of social controllability as a framework for analyzing how best to control natural resources from the perspective of ownership, usufruct, and management. In line with this framework, closed commons (e.g., Iriai institution in Japan) were compared with open commons (e.g., Google's book searching service) (Inoue, 2010). Therefore, to ascertain the ownership theory attributes associated with the nature of

Final Chapter Multifaceted Significance of Collaborative Governance

different commons, it is necessary to focus on three elements of social controllability, which are management agent, usufruct agent, and belongingness (or exclusivity). What sublated the contradictions between closed commons and open commons was collaborative governance commons. Of course, this does not indicate that only the collaborative governance commons is good, but that it should be recognized as a pattern.

The management agents in the collaborative governance commons are the core members of a diverse range of interested parties, including local residents and interested people. Furthermore, as usufruct agents can be any stakeholders on a huge scale (established in the form of settlements, villages, local bodies, states or the world, etc.), there are no obvious problems. The problem probably has more to do with exclusivity in terms of the rights of management and usufruct.

Specifically, it becomes a problem of whom to exclude or whether laypeople can be excluded. The third wave debate provides a suitable reference. British sociologist Collins (2011) said that there were two aspects to decision making involving technology. He maintains that from the political aspect (the first aspect), which concludes on decision making, all citizens have the right to be involved as stakeholders. However, from the technological aspect (the second aspect), which produces the knowledge used in decision making, he maintains that only specialists and certain citizens possessing specialized knowledge should have the right to participate.

In contrast, another British sociologist Wynne (2011) criticized that the statement made by Collins excluded both ordinary people and local knowledge. This matter was debated in the context of the Chernobyl nuclear accident. As far as is known, the decision regarding whether laymen should be involved may change according to the nature of specialized knowledge. For example, it would be strange for the indigenous people of Kalimantan, who possess local knowledge, to be excluded. In that case, what about nuclear power? It is not easy to come to the conclusion that local people with no specialized knowledge should be excluded from decision making. It is necessary to debate this far more carefully.

Therefore, considering whether laypeople should be involved in forest collaborative governance, it is necessary to question the nature of the specialization required for the foresters. Whether laypeople or specialists, it is necessary for both groups to share the various risks involved in each situation. Furthermore, not only risk regarding accidents, such as at nuclear power plants, but also risks associated with everyday life, such as securing food, should be given importance.

In Ohashi's chapter, considering the customary sharing of food in indigenous societies in the Peruvian Amazon revealed that the state of resource utilization in the post-harvest stage (sharing of crops) influenced the state of resource utilization in the pre-harvest stage (sharing of land and investment of labor). In other words, when considering the ideal state for collaborative governance, it is important to examine the sharing of outputs. Furthermore, the customary sharing of food is not only done to alleviate risk but also connected to the relationships among people, enjoyment of eating together, and joy of sharing. This occurs not only among people of the same group but also with the laborers of corporations seeking to utilize the village resources. This hints at a picture of society that collaborative governance can bring to pass.

4. Collaborative Governance as Community Theory / Civil Society Theory

According to sociologist MacIver (2009), community is the focal point of social life (coexistence of social beings), and an association is an organizational structure for social life clearly established for seeking a certain common interest or various interests. Regarding the difference between community and associations, philosopher Uchiyama (2010) pointed out the importance of diverse small groups and that association could turn into community[3]. Thus, if one seeks to redefine collaborative governance, it could be said that collaborative governance is a social system where the community as a base practices the management of resources through collaboration with associations by citizens (outsiders).

In the chapter by Okuda and Inoue about endogenous development in a Japanese rural village (Kaneyama Town), it was noted that the more difficult the efforts are to lead the local society into endogenous development where the desire to protect what is important to the area is weak, the greater the tendency for local residents and organizations to accept deeper involvement of outsiders when seeking to check outside knowledge, technology and systems, etc. Therefore, collaborative governance can be an effective social system in the endogenous development of areas that transcends the management of forests.

However, outsiders are not always supporters of local society. Though even well-intentioned associations of researchers, etc., seeking to promote nature conservation proj-

---

3) According to Uchiyama (2010), community is "an entity with a shared world ... that seeks to protect without asking why. Or, it seeks to perpetuate itself. ... And, of course, has the will to persist". Furthermore, association is, "an organizational structure connected for a reason". Incidentally, "... an organization that should be connected for a reason in a village, that can change in an instant to an entity with no reason to be connected. ... The reason that happens is that if people are in a community, then they are satisfied with their existence. They accept it", noted Uchiyama, explaining the change from association to community. Further, he pointed out that community is not born from raising up the association. Furthermore, in indicating the importance of a diverse range of small groups, he noted that "Alexis de Tocqueville thought that small groups are born out of diverse entities existing and diverse 'habits of the mind and heart'... To express his view of society in today's language one could say that the more diverse the communities within a society, the healthier that society will be".

ects can have an unexpected detrimental effect, what happens if for-profit corporations become involved as players in collaborative governance? In his chapter, based on his surveys in the Solomon Islands, Tanaka pointed out that though logging companies induced some short-term happiness, logging turned into something that destroyed the local peoples' historical products—their long-term happiness—that they had accumulated over a long period of time. This case study shows that in terms of the legitimacy of involvement, which is the basis of the commitment principle, there is both short-term legitimacy that is given according to the situation at the time and long-term legitimacy that is cultivated over a long period of time. It is also suggested that local people sometimes prioritize short-term legitimacy. Furthermore, as long as that is the decision of the local people, outsiders have no right to question their decisions. However, it is appropriate to understand, that the state of the big happiness within the community itself is changing through collaborative governance (the sharing of roles, benefits, and risks). The construction of theory for when for-profit corporations are players in collaborative governance was a point that was absent from collaborative governance theory until now. It is necessary to consider what can be shared between corporations and local residents and how it is to be shared.

Okubo et al. (2011) focused on the place of out-migrants in the relationship between insiders and outsiders and outlined their importance as follows. "The first is the point that out-migrants may be involved in decision making in a way that minimizes the events of the settlement ... and are continuing with the events that have been important to the village for generations while minimizing the scope and content to a scale that they can cope with. In some cases, they may even give up on continuing such traditional events. This is a common social change shared among rural mountain villages experiencing depopulation... ." "The second is the point that even after people who live in the settlement are no longer there, consideration is being given to the possibility of outsiders playing a partial role as 'local residents'... [having] responsibility in the vertical direction to their own ancestors while, at the same time, having responsibility in the horizontal direction to their colleagues who gather together in the same place". In addition to enjoyment, if they are able to nurture these responsibilities, it may be possible for them to continue to a certain extent without actually living in the settlement.

Assigning proper focus to out-migrants seems to be a good idea. Inoue (2002) defined the local spread of human networks limited to a certain time distance as the "human network circle" and further proposed the creation of a system, based on the "human network circle" beyond the local society, for promoting the participation of out-migrant children and I-turn reserves (those who are more likely to move from urban to rural areas in the future) living especially in urban areas. The basis for this is that if the design of the "human network circle" includes out-migrants, they will be able to fulfill certain roles.

In Okubo's chapter, as a result of considering collaboration by out-migrants in rural villages experiencing depopulation, it was possible to clarify that out-migrants are able to share their roles to maintain important events while reducing the burden of each role so that the roles are easier to handle. This shows that the concept of collaborative governance is valid not only in situations of overutilization where adjustments need to be made among a diverse range of stakeholders but also in situations of underutilization where links must be formed between local residents and citizens to develop new commons.

Considering the abovementioned factors, the next challenge is as follows. In cases where the community has been weakened through depopulation, etc., it will not be possible for local residents to adopt either resistance strategies that are closed or collaborative governance strategies run by local people that are becoming more open to outsiders, such as NPOs. In such a scenario, the only option for them would be to, due to necessity, adopt adjustment strategies that rely on outsiders, such as NPOs. That is, using support for the human network circle as an intermediate measure, they will consider making the transition to public management in the future. This will design collaborative governance as a social system for the transition period to public management.

5. Collaborative Governance as Publicness Theory

Commons research is related to the theory about publicness in three aspects (Shimomura, 2011). The first is publicness in terms of resource attributes. In other words, as long as commons such as forests have positive externalities, as environmental resources, the maintenance and administration of commons will have an active relationship with outside society. The second is publicness in the political arena. According to Shimomura, "Makoto Inoue's collaborative governance theory is, in reality, publicness theory in aspects of governance. ... It is also possible to see it as publicness created by internal actors and diverse actors from outside". The third is publicness in terms of legitimacy. This refers to publicness embedded in the historical legitimacy, in which maintenance and administration of the commons has been managed by local residents over the course of many years to protect the commons.

Of these, publicness in terms of resource attributes has been the basis for public subsidies for forests with public functions emphasized by the forestry sector over many years until now. However, this did not lead to the involvement of major political power because there is a gap between the awareness of specialists and that of local residents or citizens, and publicness in terms of resource attributes was recognized only by certain people. All the more reason, therefore, for the development of information tools showing, in an easy-to-understand manner, local attributes through the geographical analysis of green tourism that will probably promote the involvement of people in collaborative governance, as shown in N. Tanaka's chapter. With its visual ease of understanding, the land cover change simulation model developed by Darmawan and Tsuyuki in their chapter examining East Kalimantan in Indonesia, will probably prompt the participation of not only the local government but also the local residents in the collaborative governance of the forests. Furthermore, the possibility of biomass utilization as a resource considered in M. Sato's chapter concerns the previously mentioned possibility of post-harvest sharing of technical information and will probably broaden the potential for the participation of others in the collaborative governance of forests.

On the other hand, there is a great possibility that publicness in terms of legitimacy will be weakened in the future because of changes in values. It is doubtful whether publicness will be acknowledged merely because someone and their ancestors lived there for a long time.

Finally, with regard to collaborative governance as it pertains to publicness in the political arena, there are major issues that remain to be addressed. The question here is whether parties involved in collaborative governance, including outsiders, should be frowned upon as being self-satisfied by members of broader society or whether they should be appreciated for making a contribution in the public arena. If local residents collaborate with outsiders because they are unable to cope with these issues by themselves, then publicness really includes all participants. However, is it going to be acknowledged as publicness on a much larger scale by people and society? If publicness in terms of resource attributes of the forest is not acknowledged as a public function, local attempts may be dismissed as merely being local people doing what they want. These hierarchical or nested problems of collaborative governance are important theoretical challenges to be considered in the future.

References

Brian, W. 2011 (Japanese edition). "Misunderstood Misunderstanding: Social Identities and Public Uptake of Science" [The title of the Japanese edition translated into Japanese by Y. Tateishi: "Gokai Sareta Gokai: Shakaiteki Aidentiti to Koshu no Kagaku Rikai"]. *Shiso*, 1046: 64–103.

Harry, C. 2011 (Japanese edition). "The Third Wave of Science Studies: Development and Politics" [The Title of the Japanese edition translated into by M. Wada: Kagakuron no Daisan no Nami: Sono Tenkai to Poritikusu]. *Shiso*, 1046: 27–63.

Inoue, M. 2002. "Sanson deno Seikatsu wo Sasaeru Jinteki Nettowaku" [People Networks That Sustain Livelihoods in Mountain Villages], *Kankyo to Kogai* [*Research on Environmental Disruption*]. 31(4): 31–38.

Inoue, M. 2004. *Komonzu no Shiso wo Motomete: Karimantan no Mori de Kangaeru* [*Looking for the Thought of Commons: From Cases in the Forest of Kalimantan*]. Tokyo: Iwanami Shoten Publisher.

Inoue, M. 2009. "Shizen Shigen 'kyouchi' no Sekkei Shishin: Rokaru kara Gurobaru he" (The Design Guidelines of Natural Resources 'Collaborative Governance': Connecting the Local and the Global). In: T. Murota (ed.) *Gurobaru Jidai no Rokaru Komonzu* [*The Local Commons in the Global Era*]. Kyoto: Minerva Shobo, pp. 3–25.

Inoue, M. 2010a. "'*Kyouchi*' Ron no Shin Tenkai: Atogaki ni Kaete" [Afterword: Developing a New Theory of '*Collaborative Governance*']. In: G. Mitsumata, Y. Suga, and M. Inoue (eds.) *Rokaru Komonzu no Kanosei: Jichi to Kankyo no Arata na Kankei* [*The Possibility of the Local Commons: Current Relations between Autonomy and Environment*]. Kyoto: Minerva Shobo, pp. 263–265.

Inoue, M. 2010b. "Han Komonzu Ron eno Apurochi" [Approach to the Theory of Pan-commons]. In: S. Yamada (ed.) *Komonzu to Bunka: Bunka ha Dare no Monoka* [*Commons and Culture: By Whom Culture Is Owned*]. Tokyo: Tokyodo Shuppan, pp. 234–262.

Inoue, M. 2012. "Kyouchiron no Shosokumen kara Shinrin Gabanansu he Idomu" [Approaching to Issues of Forest Governance from Some Aspects of the Theory of Collaborative Governance]. *Ringyo Keizai* [*Forest Economy*], 64 (10): 24–26.

Kjaer, A. M. 2004. *Governance*. Cambridge, UK: Polity Press.

McKean, M. A. 1999. "Common Property: What Is It, What Is It Good for, and What Makes It Work?" In: C. Gibson, M. A. McKean, and E. Ostrom (eds.) *Forest Resources and Institutions*. Rome: FAO, pp. 27–55.

MacIver, R. M. 2009 (Japanese edition). *Community: a Sociological Study: Being an Attempt to Set Out the Nature and Fundamental Laws of Social Life* [The title of the Japanese edition translated into by H. Naka and M. Matsumoto, *Komyuniti*]. Kyoto: Minerva Shobo.

Ostrom, E. 1990. *Governing the Commons*. Cambridge, UK: Cambridge University Press.

Ostrom, E. 2005. *Understanding Institutional Diversity*. Princeton, NJ: Princeton University Press.

Okubo, M., M. Tanaka, and M. Inoue 2011. "Matsuri wo Tooshite Mita Tashutsu-sha to Shusshin-mura tono Kakawari no Hen'you: Yamanashi-ken Hayakawa-chou Mogura Shuraku no Ba'ai" [Changes in Relationships between Out-migrants and Their Origin Village: Focusing on Traditional Festivals in Mogura Village, Hayakawa Town, Yamanashi Prefecture, Japan]. *Sonraku Shakai Kenkyuu Jaanaru* [*Journal of Rural Studies*]. 17(2): 6–17.

Rhodes, R.A.W. 1997. *Understanding Governance: Policy Networks, Governance, Reflexivity and Accountability*. Buckingham: Open University Press.

Shimomura, T. 2011. "Komonzu no Koukyosei [The Publicness of the Commons]". *Local Commons*, 15: 18–21.

Stern, P. C., T. Dietz, N. Dolšak, E. Ostrom, and S. Stonich 2002. "Knowledge and Questions after 15 years of Research". In: E. Ostrom, T. Dietz, N. Dolšak, P. C. Stern, S. Stonich, and E. U. Weber (eds.) *The Drama of the Commons*. Washington, DC: National Academy

Press, pp. 445–489.

Uchiyama, T. 2010. *Kyodotai no Kiso Riron* [*Basic Theory on the Community*]. Tokyo: Nousan Gyoson Bunka Kyokai.

Yoshida T. 1981. Shoyu Kozo no Riron (The Theory of Ownership Structure). In: S. Yasuda, T. Siobara, K. Tominaga, and T. Yoshida (eds.) *Kiso shakaigaku IV: Shakai Kozo* [*Basic Sociology IV: Social Structure*]. Tokyo: Toyo Keizai, pp. 198–244.

Appendix

Prototype Design Guidelines for "Collaborative Governance" of Natural Resources

Makoto Inoue

1. Introduction

Commonpool resources (CPRs), such as forests, wild animals, rivers, coastal zones, and oceans, are characterized by low excludability and high subtractability. Thus, it is difficult to apply management institutions in recation to private goods, which are characterized by high excludability and high subtractability, and to public goods, which are characterized by low excludability and low subtractability. This feature of CPRs has inspired social scientists to assess options to tackle the issue.

One of the influential achievements to inspire a series of debates was the eight key "design principles" (Ostrom, 1990), which were related to the long-term robustness of institutions crafted to govern common-pool resource systems (Ostrom, 2009). Agrawal (2002) combined the three landmark studies of Ostrom (1990), Wade (1988), and Baland and Platteau (1996)[1], to identify "critical enabling conditions for sustainability on the commons". The conditions, which comprise 33 factors, are organized into the four major categories of resource system characteristics, group characteristics, institutional arrangements, and external environment, which seem to be a relatively comprehensive list of factors that potentially affect CPR management (Agrawal, 2002).

Though Agrawal's list of factors is definitely useful for studying CPR management, that the list is perceptibly a mixture of factors affecting the emergence and formation of new institutions, and factors causing the robustness of existing and newly developed institutions. One is reminded of the following statement by Ostrom (2009):

> (T)hey are causal variables of a process. The design principles, on the other *hand, are an effort to understand why the results of this process are robust in* some cases and fail in others.

---

1) This appendix was presented orally at the International Association for the Study of the Commons (presented at 13th Biennial Conference of the International Association for the Study of the Commons, held in Hyderabad, India, on January 12, 2011). This paper may also be downloaded from the commons related documents database of Indiana University. (http://dlc.dlib.indiana.edu/dlc/handle/10535/7321).
Similarly, McKean (1999) identified 10 attributes of successful common property regimes, and Stern et al. (2002) clarified seven challenges of institutional design.

Nevertheless, it is not easy to distinguish between the factors for devising new institutions and those that cause robustness to existing institutions[2]. Therefore, this argument is omitted herein. Instead, focus is allotted to the importance of relations with external stakeholders. "Nested enterprises", in which the relationship of the local commons with a wider unit is encouraged, are considered one of the eight design principles (Ostrom, 1990; Ostrom, 2005). Other scholars (Berkes, 2002; Stern et al., 2002; Agrawal, 2002) also point out the significance of how institutions and the external environment are interlinked in the face of economic globalization and political democratization. However, this issue is yet to be discussed further.

By using forests as a typical CPR, this research shows the rationale for focusing on relations with external stakeholders by summarizing experiences in Japan and tropical and sub-tropical Asian countries, and propose design guidelines to devise collaborative governance with external stakeholders.

2. Significance of Collaboration with External Stakeholders

Collective forest management systems in Japan and other Asian countries have their own historical, economic, social, and political backgrounds. Though it is not so easy to bridge these systems and discuss them integrally, possibilities were sought for bridging the systems of Japan and other Asian countries by focusing on the fact that collaboration among concerned stakeholders is indispensable for sustainable forest use and management.

Communal (Iriai) forest management in Japan

Village communities (*Sonraku-kyodotai*) in Japan are defined as communal entities formed on the basis of common land and irrigation that are indispensable to sustainable agricultural production by small farmers (Mita et al., 1996). Basically, a village community has an "ethical dualism" feature (Shiobara et al., 1991), that is, closure to the outside and equality on the inside. Closure to the outside arises from the need to protect common land as the physical basis of the community. Under this principle, beneficiaries of common land are limited to villagers. Meanwhile, equality or fairness on the inside arises from the need for all members to be able to provide for themselves. Under this principle, an equal amount of labor is requested from each household for community service to maintain farm roads and waterways, each household bears equal cost for communal administration, and each household has equal access to common land and irrigation. Though access to irrigation is limited to landed farmers, and although equal cost for common purposes is often regressive and

---

[2] Gautam and Shivakoti (2005) indicated that Ostrom's design principles are useful for analyzing the institutional robustness of local forest governance systems.

unfavorable to the poor, "ethical dualism" can be considered a general feature of Japan's village communities.

Geographically Japan's rural villages comprised of a domicile (*mura*), farmland (*nora*), and woods (*yama*) on which villagers depended for their livelihood and were called *satoyama* (Mitsui, 2005). Other natural forests beyond *satoyama* areas were called the *okuyama*, which were managed by feudal domains (*han*) and the shogunate (*Bakuhu*) in the Edo period (1603–1868) and by the central government after the Meiji Restoration (1868–1877). Generally, villagers collectively managed *satoyama*, which were defined as *iriai* (communal) forests. More than half of *iriai* forests were not forested, but were actually meadows from the end of Edo to the beginning of Meiji period. It is an important fact that *iriai* forest utilization had sustained agricultural production. For example, young grass, sprouts/shoots of trees, and twigs, called *karishiki*, were scattered as green manure into the paddy fields before rice planting in spring. Grasses were utilized for compost and manure in summer. Moreover, meadows in *iriai* forests were used for roof thatching and as pastures for livestock. Trees were used as fuelwood, under a coppice system with a 20-year rotation. Edible wild plants, nuts, mushrooms, and medicinal herbs supported the livelihoods of the villagers.

Iriai rights (*iriai-ken*)[3] are defined as the rights of local people to collectively utilize and manage the *iriai* forests (Nakao, 1984). In accordance with the Civil Code of 1896, *iriai* rights are categorized into two types. First, the group of *iriai* right holders has exclusive ownership of the forestland (article 263 of the Civil Code). Second, the group has collective usufruct over *iriai* forest that stands on land owned by other individuals or entities (article 294 of the Civil Code).

There are four types of *iriai* forest-use patterns (Kawashima 1983; McKean 1992): (1) classical collective use[4], in which right-holders as individuals can enter any part of the *iriai*

---

[3] Specific features of *iriai* rights are summarized as follows (Nakao, 1984): (1) *iriai* rights shall follow the custom in each locality (*iriai* rights and forest-use patterns vary from place to place); (2) *iriai* rights shall be granted to the residents living in a certain hamlet (a household loses its *iriai* rights when it moves out of the locality); (3) *iriai* rights shall not be granted to individuals but to households; (4) *iriai* rights shall not be inherited; (5) *iriai* rights shall not be transferred to others; (6) *iriai* rights shall not be registered (land ownership of *iriai* forests can be registered legally); (7) *iriai* rights shall be effective as long as collective forest management is continued.

[4] Especially for patterns of classical collective use, specific regulations were effective in villages. In certain cases, when somebody violates the rule, that person is fined or temporarily banned from the *iriai* forest. On certain occasions, however, no sanction is enacted. Examples of the rule in clude the following: (1) Regulation of time periods: The date of when mowing starts, called *yama-no-kuchiake*, was clearly determined. For example, cutting and collecting *karishiki* was generally commenced just before rice planting. (2) Regulation in terms of use: Usually log cutting was prohibited. (3) Regulation in terms of volume: The amount of grass cut by a person is limited to the amount that could be shouldered at one time. (4) Regulation in terms of the number of people: Only one person from a household was permitted to enter the *iriai* forest at a time. (5) Regulation in terms of tools: Only sickles for mowing and hatchets for felling logs were permitted. (6) Regulation in terms of purpose: People were permitted to fell logs only for their own use.

forest to collect forest products in accordance with their own rules; (2) corporate use, in which right holders collectively harvest *iriai* forest products to generate income for common use while prohibiting access by individuals; (3) individual use, in which right-holders as individuals use segmented parts of the *iriai* forest (*wariyama*) but cannot sell their land; (4) contract use, in which all right holders retain collective ownership and can lease *iriai* forest to another parties for harvesting timber or other benefits.

It is important to understand that *iriai* rights comprise the rights of management, control, and disposal held by an *iriai* group or a corporation, and the usufruct is held by individual members of the group (Nakao, 1984). Even the *iriai* membership was decided in accordance with the custom of the village; non-farmers, collateral families, and new settlers usually do not have *iriai* rights. This means that only feudal landed farmers have *iriai* rights. Finally, even *iriai* right holders would lose their right when moving out of the village.

While the *de facto* privatization of *iriai* forests started in the Edo period, the government has been trying to modernize *iriai* forest ownership since the beginning of the Meiji period in 1868. The modernization of *iriai* forest ownership[5] refers to the government's attempts to identify the legal owners of *iriai* forests with national, municipal, and private ownership to invalidate *iriai* rights. Here, private ownership includes, for example, individuals, group of individuals, and organizations.

Regardless of the formal type of registered forestland ownership, most *de facto* and former *iriai* forests in Japan have been subject to not only the impacts of policy pressure but also economic difficulties (Inoue, 2001). The forestry sector has long since faced depression due to economic conditions such as (1) severing the relationship between forests and farmland because farmers began to buy fertilizers such as soybean cake before the WWII; (2) a sharp decrease in the demand for fuelwood due to the energy revolution or use of fossil fuels after the WWII; (3) a rapid increase in timber imports due to cheaper prices since the 1950s; and (4) low-priced domestic timber since the 1950s. Furthermore, owners of natural forests relinquish managing the forests because they cannot create new demand for fuelwood and pulp. Owners of plantation forests cannot sell their planted and tended stands of trees such as Japanese cedar because of unprofitability, and the amount of plantation forest ironically keeps increasing and seems enriched at first sight.

Since the mid-1980s, city dwellers started visiting rural areas to help manage forests as "forest volunteers" (Mitsui, 2005) for their own recreation and for social justice in terms of environmental conservation. The number of "forest volunteers" later increased, and some of them acquired technical knowledge and skills in forestry. The government cannot avoid

---

5) Yamashita et al. (2009) describes the details of the process.

incorporating forest management by civil society in national and local forestry policy. The role of the civil sector (including local people and the general public) in the sustainable management of forests is emerging and is quite important in both the private and public sector. *De facto* and former *iriai* forest owners and outsiders are seeking collaboration to manage forests even though their livelihood does not depend on the forest anymore.

Participatory forest management in tropical and sub-tropical Asia

The leading programs for participatory and decentralized forest management (Balooni and Inoue, 2007) are (1) Community Forestry in Nepal, where authority for forest management is transferred to local people or forest user groups; (2) Joint Forest Management in India (Balooni and Inoue, 2009), where local people or village forest protection and management committees (VFPMCs) collaborate with the government or Forest Department, which retains management authority; and (3) Community-based Forest Management (CBFM) in the Philippines (Pulhin et al., 2007; Balooni et al., 2008), where the government issues tenure instruments to organized local communities that provide the latter the legal basis to manage and benefit from the forest resources (Pulhin and Inoue, 2008). Other countries have also tried the following programs: Social Forestry, Individual Forestry, and Management of Customary Forests in Indonesia (Inoue, 2003a); Management of Private Woodlots, Farm Forestry, and Social Forestry in Sri Lanka (DeZoysa and Inoue, 2008); Social Forestry, Community Forestry, Woodlot Plantations, and Agroforestry in Bangladesh (Nath and Inoue, 2009); Village Forestry and Forest Management on formally allocated village territory in Laos; and Community Forestry in Cambodia.

These programs have been introduced since the 1990s in line with the promotion of decentralization in which local governments were given greater responsibilities for forest management. As observed by the Institute for Global Environmental Strategies (IGES, 2007), the impacts of decentralization have been limited by unstable and unpredictable policies, the higher-level forest administrators' desire to retain the status quo that sustains their influence, a lack of confidence amongst foresters in the ability of local communities to manage forests, and manipulation of the decentralization process by local elites for their own advantage. Decentralization can also stimulate conflicts between competing interest groups because more local stakeholders have opportunities to benefit than before, such as conflicts over boundaries between villages (Imang et al, 2009) and elite capture (Balooni et al, 2010).

Despite the above mentioned shortcomings of the current policies, decentralization has provided opportunities for governments to more effectively support participatory forest management. It may also create opportunities for new alliances to promote rural development and forest management (IGES, 2007). Furthermore, not all local people have developed appropriate local resource management systems based on traditional local knowledge; many people need support, in terms of skills for forest management, appropriate budgets, and formation and intensification of social capital by reliable outsiders such as NGOs, local governments, and scientists.

3. Three Strategies for Sustainable Forest Use and Management

These experiences indicate that collaboration among concerned stakeholders is indispensable to sustainable forest use and management in Japan and in tropical and sub-tropical Asian countries in the era of globalization in which economic and social activities are taken across the national border. In terms of the "spatial/geographical scale" of collaboration, "focal actors/stakeholders", and "attitude of local people", three strategies can be developed.

The first is "resistance strategy" or localization strategy, in which people do not want to adapt to globalization and refuse outsider involvement to preserve their autonomy. This strategy emphasizes reconstructing local systems characterized by autonomy and reciprocity. Use and management of local resources and environment might be embedded into the livelihoods of local people. The expected focal actor of local forest governance is the village community, which is characterized by exclusive membership. This strategy accords with neither "liberalism" nor "social democracy" and might be promoted under "conservative" politics.

The second is "adjustment strategy" or globalization strategy, in which people are eager to assimilate the benefits of globalization. This strategy intends to design open systems characterized by publicness. Local resources and the environment might be valued as broader social welfare, being separated from the context of the local people's livelihood. The expected focal actors of local forest governance include associations such as NGOs and NPOs that are formed in civil society, whose viewpoint conflicts inherently with that of local people. This strategy accords with neither "liberalism" nor "conservatism", and might be promoted under "social democratic" politics.

The third is "eclectic strategy" or glocalization strategy, which compromises between both strategies, in which closure and openness as well as inherent values and universal values are adjusted, and which is endowed with partial resistance strategy and limited adjustment strategy. Under this strategy, "collaborative governance" (*kyouchi* in Japanese) of natural resources might be achieved. This type of governance is organized through collaboration among various stakeholders who have various interests in local forest use and management (Inoue, 2004).

In the field, however, neither easy coordination nor satisfactory consensus for every stakeholder can be accomplished. Even though most people may state that equal participation by all stakeholders should be ensured, the voices of the people residing in forest regions, usually minorities with less political power, might not be ultimately reflected in government policies. Such typical examples can be observed in the establishment and management of national parks and other protected areas in the tropics.

Moreover, the sphere of collaborative governance is not identical to the administrative area and scale. It may be formed within a local community, beyond communities and local government, or even beyond the nation. The sphere of collaborative governance, or *kyouchi*, resembles a *mandala* of Buddhism and Hinduism[6], in which many spheres overlap with some parts.

4. Prototype Design Guidelines

To tackle the barriers obstructing the facilitation of "eclectic strategy" or glocalization strategy, prototype design guidelines were proposed for collaborative governance of forests (Inoue, 2009)[7]. These guidelines, or *kyouchi* principles, were derived and evolved from the design principles for CPRs (Ostrom, 1990; McKean, 1999; Stern et al., 2002; Ostrom, 2005), in which the importance of links with outside organizations and nested enterprises was highlighted but not developed.

A further elaboration is provided for three vital design guidelines: "graduated membership" and "commitment principle", assured by "trust building", which have the potential to make an original contribution to enriching the conditions for "group characteristics", "institutional arrangements", and "external environment", respectively, that were categorized by Agrawal (2002).

Graduated membership of executive management body

Collaborative governance, in which local people and outsiders successfully build a consensus, cannot be established if local people abide solely by their cultural traditions that are completely exclusive of outsiders. Thus, "open-minded localism" is required, wherein local people allow outsiders to use their resources and environment. This principle agrees with the principle of subsidiarity, whereby the larger-scale political and administrative unit only supplements the smaller-scale unit or basic autonomous unit.

On the basis of "open-minded localism", some of the local people act as core members (first-class members), who have the highest authority and co-operate with other graduated members (second- and third-class members), who have relatively less authority. Clear and graduated membership boundaries imply exclusion of non-members. As such, executive

---

6) *Mandala* is a figure representing the universe in which many circles of the Buddha are situated.
7) Inoue (2009) proposed nine design guidelines: 1) degree of local autonomy, 2) clearly defined resource boundary, 3) graduated membership, 4) commitment principle, 5) fair benefit distribution, 6) two-storied monitoring system, 7) two-storied sanctions, 8) nested conflict management mechanism, and 9) trust building.

bodies should deal with the exclusion issue to ensure fairness and acquire legitimacy from relevant stakeholders.

In line with the notion that participation by all local people is neither possible nor favorable (Edmonds and Wollenberg, 2001), we can propose a forest management committee, for instance, in which representatives of the local people form a core of (first-class) members; local government administration, NGOs, and academics/scientists make commitments as second-class members; and others support activities as third-class members. The second- and third-class members should be provided definite legitimacy by core members.

Commitment principle for decision-making

To avoid the deterioration of the local autonomy, it is essential for all stakeholders to consent to the "principle of involvement" (Inoue, 2003b), which recognizes the rights of stakeholders to speak in a capacity that corresponds to their degree of involvement in forest use and management. The "principle of involvement" was a concept to embrace the authority to speak out with a voice in the forum, whereas the "commitment principle" (Inoue, 2009) clearly refers to the authority to make decisions in the arena. Here, the "commitment principle" is defined as a principle for decision making in which the stakeholders' authority is recognized to an extent that corresponds to their degree of commitment to relevant activities.

Under this principle, local people who often enter and care for the forest may be expected to have greater power over the decision-making process; outsiders who say a lot without doing much may be provided less power; and the conscientious outsiders who devote their time or money to local forest management may be given more power. Thus, various stakeholders are able to agree on the legitimacy of the opinions of outsiders as well as those of local people.

Decision making is not accomplished on an equal basis or through one-person, one-vote ballots, but should be regarded as fair, equitable, and just by the stakeholders. Whether the decision is admitted as fair depends on whether the decision-making process is considered legitimate. It is vital for all members involved in the process to reach a consensus regarding what extent they should grant legitimacy to what statement made by whom. Having done so, the scale of the arena or the number of members for decision making should be limited appropriately, because all members should recognize the approximate degree of commitment toward each other. A small-scale arena is an ideal trial base for the commitment principle. Even though stakeholders are spread out over broader geographical areas, a small-scale arena can be organized under the guidelines of "graduated membership". When organizing a larger arena cannot be avoided, indicators must be identified to evaluate the degree of commitment, such as the contribution of labor and funding by individual members, and to admit the weighted right to vote, even though it is not an easy task.

Trust building with outsiders

Collaborative governance, or *kyouchi*, cannot function well unless social capital with outsiders is formed, maintained, and strengthened. Though there are a variety of definitions (Coleman, 1990; Putnam, 1993; Fukuyama, 1995), "trust" is definitely one of the important factors of social capital. Here, the distinction between "assurance" and "trust" given by social psychologists (Yamagishi, 1998) is noteworthy. "Assurance" is when a person reasons that the other has no motive to take action that would exploit theme and that there is no social uncertainty between themselves and the person, "trust" is the person's expectation of the other's intention. When "assurance" is not provided or social uncertainty is high, "trust" is vital. On the other hand, "trust" is useless when a person has no or little possibility of being cheated by others.

Local people might strengthen collaboration among villagers based on "assurance" under the "resistance strategy" or localization strategy. On the other hand, they have to provide for collaboration with heterogeneous outsiders based on "trust" under the "eclectic strategy" or glocalization strategy as well as the "adjustment strategy" or globalization strategy. Trust building with outsiders is a precondition for "graduated membership" and the "commitment principle", though it is difficult to identify the conditions for trust building.

5. Conclusions

In accordance with the need for collaboration between local people and outsiders on forest use and management in Japan and other Asian countries, three design guidelines were proposed for the collaborative governance of natural resources: "graduated membership" of executive management bodies that can make an original contribution to enrich the conditions for "group characteristics", the "commitment principle" of decision making for "institutional arrangements" and "trust building", with outsiders for the "external environment".

Players in the system of collaborative governance are assumed to be only those who are interested in specific issues such as environmental conservation, without considering the livelihood of the local people. Their interest in the issues often does not persist for long because of their fickle nature. Hence, the existence of reliable core members is indispensable for continuous activities.

If community members and their activities fail to gain the approval of the majority of society, or if they cannot acquire legitimacy in the larger society, collaborative governance will not mature into a robust system. Getting the approval of the larger society is connected to the concept of "deliberative democracy". Deliberative democracy, often called "discursive democracy", is a system of political decisions that relies on citizen deliberation to formulate sound policies (Yamaguchi, 2004). In deliberative democracy, legitimate law making can arise only through public deliberation by the people. This notion seems to have a close connection with the concept of collaborative governance, or *kyouchi*, described in this paper.

However, the "commitment principle", seems to contradict the principle of deliberative democracy (Cohen, 1997). The participants should all have an equal say in a deliberative democracy. This principle is important so that the participants can speak out freely regardless of their social status, and they are bound only by the results of deliberation. However, the commitment principle was introduced to avoid the influence of social status, because otherwise, it seems impossible for the participants to speak out freely regardless of their social status in the real world.

As the next step, a theoretical and empirical investigation to demonstrate the validity of this proposal is required.

References

Agrawal, A. 2002. "Common Resources and Institutional Sustainability". In: E. Ostrom et al. (eds.) *The Drama of the Commons*. Washington, DC: National Academy Press, pp. 41–85.

Baland, J. and J. Platteau. 1996. *Halting Degradation of natural Resources: Is There a Role for Rural Communities?* Clarendon Press.

Balooni, K. and M. Inoue. 2007. "Decentralised forest Management in South and Southeast Asia". *Journal of Forestry,* 105(8): 414–420.

Balooni, K. and M. Inoue. 2009. "Joint Forest Management in India: The Management Change Process". *IIMB Management Review,* 21(1): 1–17.

Balooni, K., J. M. Pulhin, and M. Inoue. 2008. "The Effectiveness of Decentralisation Reforms in the Philippines's Forestry Sector". *Geoforum,* 39(6): 2122–2131.

Balooni, K., J. F. Lund, C. Kumar, and M. Inoue. 2010. "Curse or Blessing? Local Elites in Joint Forest Management in India's Shiwaliks". *International Journal of the Commons,* 4(2): 707–728.

Berkes, F. 2002. "Cross-scale institutional linkages: Perspective from the bottom up". In: E. Ostrom et al. (eds.) *The Drama of the Commons*. Washington, DC: National Academy Press, pp. 293–321.

Cohen, J. 1997. Deliberation and Democratic Legitimacy. In: Bohman, J. and Rehg, W. (eds.) *Deliberative Democracy: Essays on reason and Politics*. Cambridge, MA: The MIT Press, pp. 67–91.

Coleman, J. 1990. *Foundations of Social Theory*. Cambridge, MA: Harvard University Press.

DeZoysa, M. P. and M. Inoue. 2008. "Forest Governance and Community Based Forest Management in Sri Lanka: Past, Present and Future Perspectives". *International Journal of Social Forestry,* 1 (1): pp. 27–49.

Edmonds, D. and E. Wollenberg, 2001. "A Strategic Approach to Multistakeholder Negotiations". *Development and Change*, 32: 231–253.

Fukuyama, F. 1995. *Trust: The Social Virtues and the Creation of Prosperity*. New York, NY: Free Press.

Gautam, A. P. and G. P. Shivakoti, 2005. "Conditions for Successful Local Collective Action in Forestry: Some Evidence from the Hills of Nepal". *Society and Natural Resources*, 18: 153–171.

IGES, 2007. *Decentralization and State-sponsored Community Forestry in Asia*. Hayama: Institute for Global Environmental Strategies.

Imang, N., M. Inoue, and M. A. Sardjono, 2009. "Importance of Boundaries in Customary Resource Management under Decentralized Policies: Case Study in Indigenous Kenyah Dayak, East Kalimantan, Indonesia". *Journal of Forest Economics*, 55 (3): 35–43.

Inoue, M. 2001. "Natural Resource Management by the Local People and Citizens". In: M. Inoue, and T. Miyauchi (eds.) *Sociology of the Commons*. Tokyo: Shin-yo-sha, pp. 213–235 (in Japanese).

Inoue, M. 2003a. "Participatory Forest Management Policy in South and Southeast Asia". In: Inoue, M. and H. Isozaki (eds.) *People and Forest: Policy and Local Reality in Southeast Asia, the Russian Far East, and Japan*. Dordrecht: Kluwer Academic Publishers, pp. 49–71.

Inoue, M. 2003b. "Diverse Management of Indonesian Forests: A New Governor Gives Locals a Greater Say in their Resources". *International Herald Tribune and the Asahi Shimbun*, April 4.

Inoue, M. 2004. *Komonzu no Shiso wo Motomete* [*In Search of the Principle of Commons*]. Tokyo: Iwanami Shoten, Publisher.

Inoue, M. 2009. "Shizen Shigen '*kyouchi*' no Sekkei Shishin: Rokaru kara Gurobaru he" [The Design Guidelines of Natural Resources 'Collaborative Governance': Connecting the Local and the Global]. In: T. Murota (ed.) *Gurobaru Jidai no Rokaru Komonzu* [*The Local Commons in the Global Era*]. Kyoto: Minerva Shobo, pp. 3–25.

Kawashima, T. 1983. *Kawashima Takeyoshi Chosaku-shu, Dai 8 Kan* [*The Writings of Kawashima Takeyoshi, Vol. 8*]. Tokyo: Iwanami Shoten, Publisher.

McKean, M. A. 1992. "Management of Traditional Common Lands (*Iriai-chi*) in Japan". In: Bromley, D. W. (ed.) *Making the Commons Work: Theory, Practice, and Policy*. San Francisco, CA: ICS Press, pp. 63–98.

McKean, M. A. 1999. "Common Property: What Is It, What Is It Good for, and What Makes It Work?" In: Gibson, C., M. A. McKean, and E. Ostrom (eds.) *Forest Resources and Institutions*. Rome: FAO, pp. 27–55.

Mita, M., A. Kurihara, and Y. Tanaka (eds.) 1996. *Shakaigaku Jiten (Shukusatu-ban)* [*Dictionary of Sociology (Reduced-sized Edition)*]. Tokyo: Koubundou Publishers.

Mitsui, S. 2005. "Iriai Rin'ya no Rekishi-teki Igi to Komonzu no Saisei [Historical significance of *Iriai* Forest and Restoration of the Commons]". In: Shinrin Kankyo Kenkyu-kai (eds.) *Forest Environment 2005*. Tokyo: Asahi Shimbun-sha, pp. 42–52.

Nakao, H. 1984. *Iriai Rin'ya no Horitsu Mondai [Legal Issues of Iriai Forests]*. Tokyo: Keiso Shobo Publishing.

Nath, T. K. and M. Inoue. 2009. "Forest-based Settlement Project and its Impacts on Community Livelihood in the Chittagong Hill Tracts, Bangladesh". *International Forestry Review*, 11(3): 394–407.

Ostrom, E. 1990. *Governing the Commons*. Cambridge, UK: Cambridge University Press.

Ostrom, E. 2005. *Understanding Institutional Diversity*. Princeton, NJ: Princeton University Press.

Ostrom, E. 2009. Design Principles of Robust Property Rights Institutions: What Have We Learned? In: G. K. Ingram, and Y. H. Hong (eds.) *Property Rights and Land Policies*. Cambridge, MA: Lincoln Institute of Land Policy, pp. 25–51.

Pulhin, J. M., M. Inoue, and T. Enters. 2007. "Three Decades of Community-based Forest Management in the Philippines: Emerging Lessons for Sustainable and Equitable Forest Management". *International Forestry Review*, 9(4): 865–883.

Pulhin, J. M. and M. Inoue. 2008. "Dynamics of Devolution Process in the Management of the Philippine Forests". *International Journal of Social Forestry*, 1 (1): 1–26.

Putnam, R. D. 1993. *Making Democracy Work: Civic Traditions in Modern Italy*. Princeton, NJ: Princeton University Press.

Shiobara, T., H. Matsubara, and M. Ohashi (eds.) 1991. *Shakaigaku no Kiso Chishiki [A Grounding of Sociology]*. Tokyo: Yuhikaku Publishing.

Stern, P. C., T. Dietz, N. Dolšak, N., E. Ostrom, and S. Stonich, 2002. "Knowledge and Questions after 15 years of Research". In: E. Ostrom, T. Dietz, N. Dolšak, P. C. Stern, S. Stonich, and E. U. Weber (eds.) *The Drama of the Commons*. Washington, DC: National Academy Press, pp. 445–489.

Wade, R. 1988. *Village Republics: Economics Conditions for Collective Action in South India*. San Francisco, CA: ICS Press.

Yamagishi, T. 1998. *Structure of Trust*. Tokyo: University of Tokyo Press.

Yamaguchi, Y. 2004. *Shimin Shakai-ron: Rekishi-teki Isan to Shintenkai [Civil Society: Historical Inheritance and New Development]*. Tokyo: Yuhikaku Publishing.

Yamashita, U., K. Balooni, and M. Inoue. 2009. "Effect of instituting 'Authorized Neighborhood Associations' on Communal (*Iriai*) Forest Ownership in Japan". *Society and Natural Resources*, 22(5): 464–473.

Index

action (*tavete*) 107
activity 259
adjustment strategy 71, 89, 327
afforestation 163
aging of the population 191
agricultural cooperative 178, 181
agricultural use 78
agroforestry 73
Asian Development Bank (ADB) 212
association 316

banana 229, 230, 232, 234
Basic Plan for Forests and Forestry 26
beautiful landscape 67
binderless board 306
binderless plywood 306
business-as-usual 277

cash income 91, 94
Cellular Automata (CA) 274
citizen 311
citizen participation theory 312
city resident 182
civil society theory 315
Clean Development Mechanism (CDM) 203
clear cutting 172
collaborative forest governance 71
collaborative governance 3–8, 41–42, 131, 163–164, 181, 194, 216, 218, 311, 312
Collins 315
commercial logging 100, 109, 232, 240
commitment principle 7, 311, 330
community forest 205
common property resources (CPRs) 68–69, 242
commons 3, 4, 16, 311
communal use 102
communal utilization 99, 105, 114
communal utilization rights 105
community 316
community association 146
community collaborative labor 146, 147
community forestry 326
community theory 315
community work center 43, 45, 47
community-based forest management 326
conservation 56, 109

constitutional rule 313
consumption 243, 246
contract 77, 82
corporate social responsibility (CSR) 218
corporation 300
cultural resources 4, 5
customary land 87, 112

de facto 325
decentralized forest management 326
degraded forest 205
design(ing) guideline 311, 328
destination-based industry 271
distribution 120

East Kalimantan Province 292
eclectic strategy 237
endogenous development 40–41, 47, 54–55
environmental plantation 206
existing database 259
existing tourism 250

festival 151–152
field facilitator 80, 95–96
fish 229, 231, 236
5 meshgrid x 5 meshgrid filtering method 262
floodplain 231, 242
Food and Agriculture Organization of the United Nations (FAO) 206
Forest Act 20
Forest Administration Institution 32
Forest and Forestry Basic Act 22
Forest and Forestry Revitalization Plan 19
Forest Basic Law 21
forest governance 16
forest land-use allocation map (TGHK) 275
forest management 71, 163, 314
  goals of — 93
Forest Management Plan 27
forest owners association 21, 172, 181
forest ownership 163
forest planning system 26
forest product 73, 80
forest resource management group 62–63, 82
forest resources 48–49, 58–60, 199
forest road association 177

forest volunteer   325
Forestry Basic Law   185
Forestry Management Planners   33
freshwater fisheries cooperative   179, 181

game meat   229, 231, 233, 237, 239
generosity (*hihoho*)   108
generous behavior   116
geographical analysis   267
geographical information   250
Global Research Forests Assessment   202
globalization   5, 201
governance without government   18, 312
government-run program   71
graduated membership   311, 328
great happiness   129
green safety net   313
green tourism   250
green tourism base   264

human network circle   317
human resource development   33

incentive   94
Indonesia   292
industrial plantation   206
iriai forest   48, 49, 324

Japan   135
Japanese Forester System   33
joint forest management   326
journalist   182

Kaneyama-machi   249
Kjaer   313

land-cover modeling   280
landslide   176
land-use and land-cover change (LULCC)   273
land-use plan   273
Laos   202
large-scale plantation   302
laypeople   315
lazy people (*chiquish*)   229, 232, 241
leader   94
legitimacy   99, 318
Lifestyle of Health and Sustainability (LOHAS)   253
livelihood   243, 246
local government   185, 194

local people support   80, 88
local resident   182, 193, 311
logging company   231
logging contractor   177
logging roads   172

MacIver   316
Madiun Forest District   74–75
Madiun Regency   74–75
Mahakam   275
management group   78–80, 82, 94
Markov model   277
Melak City   292
mestizos   225, 231, 237
mobile lifestyle   225
modernization process   195
Mogami Region of Yamagata Prefecture   250
Mogura   141–142
mountain village   135
Mountain Village Development law   186
multiple forest uses   194
multiple resource uses   243
Municipal Forest Plan   26
mutual assistance (*vinari tokae*)   87
mutual network   102

National Forest Plan   26
native community   225, 226
natural resources   4
neighborhood size   278
network   45, 56, 65
new tourism   252
new woody biomass resources   291
NGO   200, 216
non-forestation   172
non-reforestation   172
non-resident owner   172
non-wood forest product   297
noro   104, 105
non-timber forest product (NTFP)   200 *see also* forest product

oil palm plantation   291
open-minded localism   328
out-migrant   137, 317
outsiders   4, 223, 224, 240, 243
overexploitation   56
ownership theory   314

Index

participation of cooperative 300
Peruvian Amazon 224
PHBM 71
plantation 21, 186
   types of — 204–208
plantation crop 297
political arena 318
population concentrating 196
potential of utilizing 291
prediction scenario 277
preservation association 147
price of standing tree 191
principle of involvement 330
priority utilization right 106
privatization 166
probabilistic transition style 275
profit sharing 79, 85
profit-sharing forest system 183
profit-sharing plantation 166
profit-sharing plantation cooperative 190
profit-sharing rights 185
protection forest 28
protection of the life of the elderly and handicapped 48
public participation 16
publicness theory 318

quantitative analysis 272

rapid deforestation 278
rapid forest degradation 278
rapid forest regrowth 278
REDD-plus 188
reforestation 163
Regional Forest Plan 26
regional information 249
relationship 90, 94, 95
researcher 182
resistance strategy 327
resource attribute 318
resource management theory 314
restriction of private rights 195
road network 31
role of government 313
rubber plantation 291
rural depopulation 4, 135

satoyama 48, 67, 325

scenario analysis 277
school profit-sharing forest 188
semi-domestication 106
shared forest 182
sharing 228, 240, 241
sharing (of land by tribes) 99, 102
shifting cultivation 104
Shipibo 225, 243
silvicultural cost 165
small-scale plantation 302
social change 39–41
social controllability 314
social forestry 326
social risk 73
soil conservation 164 *see also* conservation
stakeholder 249
standard expertise 314
state-owned forest 73
state-owned profit-sharing forest 171, 182
stingy (*vusivusi*) 107
sufficiency (*isiri*) 107
suitability map 279

teak 73, 80, 95
teak plantation 93
tertiary meshgrid 262
timber production 73
tolerance (*vinamagua*) 107, 108
tourism facility 264
tourism resource 255
tourist destination 254
traditional communal forest (*iriai*) 165 *see also* iriai forest
transfer of rights 71, 92
trust 91, 92, 95
trust building 331

Ucayali River 225, 227
Uchiyama 316
underproduction 116
underuse 136
underutilization 196
unregulated clear cutting 178
utilization of forest resources 291
utilization of oil palms 303, 305
utilization of rubber trees 303

vehicle-based logging 177

village / settlement development
volunteer committee   178
volunteer group   41, 67

water and soil conservation function   172
watershed protection   164
West Kutai District   275

wodden agrcultural waste   291
wood-based product   297
Wynne   315

Yoshida   314

zoning   28